NUREG-1900, Vol. 1

Safety Evaluation Report
Related to the License Renewal of
Nine Mile Point Nuclear Station,
Units 1 and 2

I0488591

Docket Nos. 50-220 and 50-410

Constellation Energy Group, LLC

Manuscript Completed: June 2006
Date Published: September 2006

Prepared by
T. Le

Division of License Renewal
Office of Nuclear Reactor Regulation
U.S. Nuclear Regulatory Commission
Washington, DC 20555-0001

ABSTRACT

This safety evaluation report (SER) documents the technical review of the Nine Mile Point Nuclear Station Units 1 and 2 (NMPNS), license renewal application (LRA) by the staff of the U.S. Nuclear Regulatory Commission (NRC) (the staff). By letter dated May 26, 2004, Constellation Energy Group, LLC submitted the LRA for NMPNS in accordance with Title 10, Part 54, of the *Code of Federal Regulations* (10 CFR Part 54). Due to concerns with the adequacy of support for and documentation of the license renewal activities in the initial submission, the applicant submitted an amended LRA (ALRA) on July 14, 2005. Constellation Energy Group, LLC is requesting renewal of the operating licenses for NMPNS (Facility Operating License Numbers DPR-63 and NPF-69, respectively), for a period of 20 years beyond the current expiration dates of midnight August 22, 2009, for Unit 1 (NMP1) and midnight October 31, 2026, for Unit 2 (NMP2).

NMPNS is located approximately six miles northeast of Oswego, NY. The NRC issued the construction permits for NMP1 on April 12, 1965, and for NMP2 on June 24, 1974. The NRC issued the operating licenses for NMP1 on December 26, 1974 and for NMP2 on July 2, 1987. NMP1 is a boiling water reactor design with a Mark 1 containment. The nuclear steam supply system was supplied by General Electric and the balance of the plant was originally designed and constructed by Stone and Webster with the assistance of its agent, Niagra Mohawk Power Corporation. NMP1's licensed power output is 1850 megawatt thermal, with a gross electrical output of approximately 615 megawatt electric. NMP2 is a boiling water reactor design with a Mark 2 containment. The nuclear steam supply system was supplied by General Electric and the balance of the plant was originally designed and constructed by Stone and Webster. NMP2's licensed power output is 3467 megawatt thermal, with a gross electrical output of approximately 1144 megawatt electric.

This SER presents the status of the staff's review of information submitted to the staff through April 21, 2006, the cutoff date for consideration in this SER. On March 3, 2006, the staff issued a draft SER which identified two open items that had to be resolved before the staff makes a final determination on the application. The two open items have now been resolved and SER Section 1.5 summarizes these items and their resolutions. SER Section 6 provides the staff's final conclusion on the review of the NMPNS License Renewal Application dated May 26, 2004, as amended July 14, 2005, and all its subsequent supplemental letters as listed in SER Appendix B.

TABLE OF CONTENTS

Appendices

Tables

ABBREVIATIONS

115KVAC	115KV AC electrical distribution
120VAC	120V AC electrical distribution
125VDC	125V DC electrical distribution
13.8KVAC	13.8KV AC electrical distribution
24VDC	24V DC electrical distribution
4.16KVAC	4.16KV AC electrical distribution
600VAC	600V AC electrical distribution
AC	alternating current
ACI	American Concrete Institute
ACRS	Advisory Committee on Reactor Safeguards
ADAMS	Agency Document Access Management System
AERM	aging effects requiring management
AFW	auxiliary feedwater
AISC	American Institute of Steel Construction
ALARA	as low as reasonably achievable
ALRA	amended license renewal application
AMP	aging management program
AMR	aging management review
ANSI	American National Standards Institute
ARI	alternate rod insertion
ART	adjusted reference temperature
ASB	auxiliary service building
ASME	American Society of Mechanical Engineers
ASTM	American Society for Testing and Materials
ATWS	anticipated transient without scram
B24V	battery-24V-station
BSW	biological shield wall
BTP	branch technical position
BWR	boiling water reactor
BWROG	Boiling Water Reactor Owners Group
BWRVIP	boiling water reactor vessel internals project
CAP	corrective action program
CAS	compressed air system
CASS	cast austentitic stainless steel
CBF	cycle-based fatigue
CCCWS	closed-cycle cooling water system
CEG	Constellation Energy Group
CF	chemistry factor
CFR	Code of Federal Regulations
CGG	Constellation Generation Group
CI	confirmatory item
CLB	current licensing basis
CMEB	Chemical and Mechanical Engineering Branch

CNS	Constellation Nuclear Services
CR	condition report
CRB	control room building
CRD	control rod drive
CRDRL	control rod drive return line
CSH	high pressure core spray
CST	condensate storage tank
CUF	cumulative usage factor
DBA	design basis accident
DBD	design basis document
DBE	design basis event
DBTT	ductile-to-brittle transition temperature
DC	direct current
DER	deviation event report
DG	diesel generator
DGB	diesel generator building
EC	emergency condenser
ECCS	emergency core cooling systems
ECP	electrochemical cooling system
ECS	emergency cooling system
ECT	eddy current testing
EDG	emergency diesel generator
EFPY	effective full power years
EMA	equivalent margin analysis
EOL	end of life
EPRI	Electric Power Research Institute
EQ	environmental qualification
ERV	electromatic relief valve
ESF	engineered safety feature
EVT-1	enhanced VT-1 visual inspection
EYS	essential yard structures
FAC	flow-accelerated corrosion
F_{en}	environmental fatigue life correction factor
FMP	fatigue monitoring program
FP	fire protection
FPEE	fire protection engineering evaluation
FSAR	final safety analysis report
FW	feedwater
FW/HPCI	feedwater/ high pressure coolant injection
FWS	feedwater system
GALL	Generic Aging Lessons Learned Report
GDC	general design criteria or general design criterion
GE	general electric
GEIS	Generic Environmental Impact Statement
GL	generic letter

GSI	generic safety issue
GWT	ground water table
HCU	hydraulic control unit
HELB	high-energy line break
HEPA	high efficiency particulate air
HFIR	high flux isotope reactor
HPCI	high pressure coolant injection
HPCS	high pressure core spray
HVAC	heating, ventilation, and air conditioning
HWC	hydrogen water chemistry
HX	heat exchanger
I&C	instrumentation and controls
IASCC	irradiation assisted stress corrosion cracking
IBA	intermediate-break accident
IEEE	Institute of Electrical and Electronics Engineers
IGA	intergranular attack
IGSCC	intergranular stress corrosion cracking
IN	information notice
INPO	Institute of Nuclear Power Operations
IPA	integrated plant assessment
ISG	interim staff guidance
ISI	inservice inspection
ISP	integrated surveillance program
J	joule
J-R	joule-resistent
KV	kilovolt
KVA	kilovolt Amperes
LBS	leakage boundary (spatial)
LOCA	loss of coolant accident
LOOP	loss of offsite power
LPCI	low pressure coolant injection
LPCS	low pressure core spray
LR	license renewal
LRA	license renewal application
LRT	leak rate test
MCC	motor control center
MEL	master equipment list
MIC	mircrobiologically induced corrosion
MG	motor generator
MS	main steam
MSIV	main steam isolation valve
MWe	megawatt electric
MWt	megawatt thermal

NDE	non-destructive examinations
NEI	Nuclear Energy Institute
NEIL	Nuclear Electric Insurance Limited
NEPA	National Environmental Policy Act of 1969
NER	Nuclear Engineering Report
NFPA	National Fire Protection Association
NMP1	Nine Mile Point Unit 1
NMP2	Nine Mile Point Unit 2
NMPC	Niagra Mohawk Power Corporation
NMPNS	Nine Mile Point Nuclear Station
NRC	U.S. Nuclear Regulatory Commission
NSR	nonsafety-related
NSSS	nuclear steam supply system
NUMARC	Nuclear Management and Resources Council (now NEI)
NUREG	U.S. Nuclear Regulatory Commission Regulatory Guide
OCCW	open-cycle cooling water
ODSCC	outside-diameter stress-corrosion cracking
OGB	offgas building
OI	open item
ORNL	Oak Ridge National Laboratory
P&ID	piping and instrumentation diagram
PAA	program attribute assessment
PCS	primary containment structure
PEO	period of extended operation
PM	preventive maintenance
PMT	post-maintenance test
P-T	pressure-temperature
PTS	pressurized thermal shock
PUAR	plant-unique analysis report
PWR	pressurized water reactor
PWSCC	primary water stress-corrosion cracking
RAI	request for additional information
RB	reactor building
RBCLC	reactor building closed look cooling
RBEDT	reactor building equipment drain tank
RCIC	reactor core isolation cooling
RCPB	reactor coolant pressure boundary
RCS	reactor coolant system
RG	regulatory guide
RHR	residual heat removal
RI-ISI	risk-informed inservice inspection
RPS	reactor protection system
RPT	reactor recirculation pump trip
RPV	reactor pressure vessel
RSSB	radwaste solidification and storage building
RT_{NDT}	reference temperature nil ductility transition

RVID	reactor vessel integrity database
RVSP	Reactor Vessel Surveillance Program
RWB	radwaste building
RWCU	reactor water cleanup
S&W	Stone and Webster
SBA	small-break accident
SBF	stress based fatigue
SBO	station blackout
SC	structure and component
SCC	stress-corrosion cracking
SDC	shutdown cooling
SE	safety evaluation
SER	safety evaluation report
SGTB	standby gas treatment building
SGTS	standby gas treatment system
SHE	standard hydorgen reference electrode
SIA	structural integrity attached
SIL	service information letters
SOC	statements of consideration
SPH	screen and pumphouse
SR	safety-related
SRP	Standard Review Plan
SRP-LR	Standard Review Plan for Review of License Renewal Applications for Nuclear Power Plants
SSC	system, structure, and component
SSE	safe-shutdown earthquake
SWB	screenwell building
t	thickness
TAP	torus attached piping
TB	turbine building
TBCLC	turbine building closed loop cooling
TER	technical evaluation report
TLAA	time-limited aging analysis
TS	technical specification
UFSAR	updated final safety analysis report (for Nine Mile Point Unit 1)
UPS	uninterruptible power supplies
USAR	updated safety analysis report (for Nine Mile Point Unit 2)
USAS	United States of America Standards
USE	upper-shelf energy
UT	ultrasonic testing
UV	ultra violet
V	Volt
WDB	waste disposal building
WO	work order

SECTION 1

INTRODUCTION AND GENERAL DISCUSSION

1.1 Introduction

This document is a safety evaluation report (SER) on the license renewal application (LRA) for Units 1 and 2 of Nine Mile Point Nuclear Station (NMPNS), as filed by Constellation Energy Group, LLC (CEG or the applicant). By letter dated May 26, 2004, CEG submitted its application to the U.S. Nuclear Regulatory Commission (NRC) for renewal of the NMPNS operating licenses for an additional 20 years. By letter to NMPNS, dated December 7, 2004, the staff stated its concern regarding the applicant's inadequate support of license renewal activities for the initial submission. In its response, by letter dated January 3, 2005, the applicant stated that it had taken additional actions to resolve the contributing factors for past performance and agreed to provide supplemental support for the license renewal process. By letter dated March 3, 2005, NMPNS requested a 90-day grace period to recover the quality of the NMP LRA. In response to the NMPNS request, the NRC staff (the staff) informed the applicant, by letter dated March 7, 2005, that the review of the NMP LRA had been suspended and that the standard 22-month review schedule would not be met due to the suspended period. On July 14, 2005, NMPNS submitted its amended LRA (ALRA).

In the ALRA, the applicant revised the original LRA sections and tables where applicable to identify each nonsafety-related (NSR) system or NSR portion of a safety-related (SR) system that is within the scope of license renewal. In conjunction with this amended information, the applicant also identified additional NSR component types and intended function(s) and made them consistent with the standardized list of intended functions described in the staff Standard Review Plan for License Renewal (SRP-LR) and Nuclear Energy Institute (NEI) 95-10, "Industry Guidelines for Implementing the Requirements of 10 CFR Part 54 - The License Renewal Rule."

The staff prepared this report, which summarizes the results of its safety review of the LRA and ALRA for compliance with the requirements of Title 10, Part 54, of the *Code of Federal Regulations* (10 CFR Part 54), "Requirements for Renewal of Operating Licenses for Nuclear Power Plants." The NRC license renewal project manager for the NMPNS license renewal review is Ngoc B. (Tommy) Le. Mr. Le can be contacted by telephone at 301-415-1458 or by electronic mail at nbl@nrc.gov. Alternatively, written correspondence may be sent to the following address:

License Renewal and Environmental Impacts Program
U.S. Nuclear Regulatory Commission
Washington, D.C. 20555-0001
Attention: Ngoc B. (Tommy) Le, Mail Stop 0-11F1

In its May 26, 2004 submittal letter, as supplemented by its July 14, 2005, letter the applicant requested renewal of the operating licenses issued under Section 104b (Operating License No. DPR-63) and Section 103 (Operating License No. NPF-69) of the Atomic Energy Act of

1954, as amended, for Unit 1 (NMP1) and Unit 2 (NMP2), for a period of 20 years beyond the current license expiration dates of midnight August 22, 2009, for NMP1 and midnight October 31, 2026, for NMP2.

NMPNS is located approximately six miles northeast of Oswego, NY. The NRC issued the construction permits for NMP1 on April 12, 1965, and for NMP2 on June 24, 1974. The NRC issued the operating licenses for NMP1 on December 26, 1974 and for NMP2 on July 2, 1987. NMP1 is a boiling water reactor design with a Mark 1 containment. The nuclear steam supply system was supplied by General Electric and the balance of the plant was originally designed and constructed by Stone and Webster, with the assistance of its agent, Niagra Mohawk Power Corporation. NMP1's licensed power output is 1850 megawatt thermal, with a gross electrical output of approximately 615 megawatt electric. NMP2 is a boiling water reactor design with a Mark 2 containment. The nuclear steam supply system was supplied by General Electric and the balance of the plant was originally designed and constructed by Stone and Webster. NMP2's licensed power output is 3467 megawatt thermal, with a gross electrical output of approximately 1144 megawatt electric. The NMP1 updated final safety analysis report (UFSAR) and the NMP2 updated safety analysis report (USAR) contain details concerning the plant and the site.

The license renewal process consists of two concurrent reviews: (1) a technical review of safety issues and (2) an environmental review. The NRC regulations found in 10 CFR Parts 54 and 51, respectively, set forth the requirements against which license renewal applications are reviewed. The safety review for the NMPNS license renewal is based on the applicant's original LRA, ALRA, and on responses to the staff's requests for additional information. The applicant supplemented its LRA and ALRA and provided clarifications through its responses to requests for additional information in audits, meetings, and docketed correspondence. Unless otherwise noted, the staff reviewed and considered information submitted through April 21, 2006. The staff reviewed information received after this date on a case-by-case basis depending on the stage of the safety review and on the volume and complexity of the information. The public may view the LRA, ALRA, and all pertinent information and materials, including the UFSAR and USAR mentioned above, at the NRC Public Document Room, located on the first floor of One White Flint North, 11555 Rockville Pike, Rockville, MD 20852-2738 (301-415-4737 / 800-397-4209), and at the Penfield Library, Reference and Documents Department, 7060 State Route 104, State University of New York, Oswego, NY 13126. In addition, the public may find the LRA and ARLA, as well as materials related to the license renewal review, on the NRC Web Site at http://www.nrc.gov/reactors/operating/licensing/renewal/applications/nine-mile-pt.html.

This SER summarizes the results of the staff's safety review of the NMPNS LRA and ALRA, and describes the technical details considered in evaluating the safety aspects of the proposed operation for an additional 20 years beyond the term of the current operating licenses. The staff reviewed the LRA and ALRA in accordance with NRC regulations and the guidance provided in NUREG-1800, "Standard Review Plan for Review of License Renewal Applications for Nuclear Power Plants" (SRP-LR), dated July 2001.

SER Sections 2 through 4 address the staff's review and evaluation of license renewal issues that it has considered during the review of the application. Section 5 is reserved for the report of the Advisory Committee on Reactor Safeguards (ACRS). Conclusions of this report are presented in Section 6.

SER Appendix A contains a table that identifies the applicant's commitments associated with the renewal of the operating licenses. Appendix B provides a chronology of the principal correspondence, between the staff and the applicant, related to the review of the application. Appendix C is a list of the principal contributors to this SER. Appendix D is a bibliography of the references used in support of the review.

In accordance with 10 CFR Part 51, the staff prepared a draft, plant-specific supplement to NUREG-1437, "Generic Environmental Impact Statement for License Renewal of Nuclear Plants (GEIS)". This supplement discusses the environmental considerations related to renewing the licenses for NMPNS. The staff issued draft Supplement 24 to NUREG-1437 "Generic Environmental Impact Statement for License Renewal of Nuclear Plants, Regarding Nine Mile Point Nuclear Station, Units 1 and 2, Draft Report for Comment," on September 29, 2005. The final Supplement 24 to NUREG-1437 report was published on May 16, 2006.

1.2 License Renewal Background

Pursuant to the Atomic Energy Act of 1954, as amended, and NRC regulations, operating licenses for commercial power reactors are issued for 40 years. These licenses can be renewed for up to 20 additional years. The original 40-year license term was selected on the basis of economic and antitrust considerations, rather than on technical limitations; however, some individual plant and equipment designs may have been engineered on the basis of an expected 40-year service life.

In 1982, the staff anticipated interest in license renewal and held a workshop on nuclear power plant aging. This workshop led the staff to establish a comprehensive program plan for nuclear plant aging research. On the basis of the results of that research, a technical review group concluded that many aging phenomena are readily manageable and do not pose technical issues that would preclude life extension for nuclear power plants. In 1986, the staff published a request for comment on a policy statement that would address major policy, technical, and procedural issues related to license renewal for nuclear power plants.

In 1991, the staff published the license renewal rule in 10 CFR Part 54 (the Rule). The staff participated in an industry-sponsored demonstration program to apply the Rule to a pilot plant and to gain experience necessary to develop implementation guidance. To establish a scope of review for license renewal, the Rule defined age-related degradation unique to license renewal; however, during the demonstration program, the staff found that adverse aging effects occur and are managed during the period of initial license. In addition, the staff found that the scope of the review did not allow sufficient credit for existing programs, particularly the implementation of the Maintenance Rule, which could also manage plant-aging phenomena. As a result, the staff amended the license renewal rule in 1995. The amended Rule established a regulatory process that is simpler, more stable, and more predictable than the previous license renewal rule. In particular, the staff amended the Rule to focus on managing the adverse effects of aging, rather than on identifying age-related degradation unique to license renewal. The staff initiated these rule changes to ensure that important systems, structures, and components (SSCs) will continue to perform their intended functions during the period of extended operation. In addition, the revised Rule clarified and simplified the integrated plant assessment process to be consistent with the revised focus on passive, long-lived structures and components (SCs).

In parallel with these efforts, the staff pursued a separate rulemaking effort and developed an amendment to 10 CFR Part 51 to focus the scope of the review of environmental impacts on license renewal and fulfill the NRC's responsibilities under the National Environmental Policy Act of 1969.

1.2.1 Safety Review

License renewal requirements for power reactors are based on two key principles:

(1) The regulatory process is adequate to ensure that the licensing bases of all currently operating plants provide and maintain an acceptable level of safety, with the possible exception of the detrimental effects of aging on the functionality of certain SSCs, as well as a few other safety-related issues, during the period of extended operation.

(2) The plant-specific licensing basis must be maintained during the renewal term in the same manner and to the same extent as during the original licensing term.

In implementing these two principles, 10 CFR 54.4 defines the scope of license renewal as including those SSCs: (1) that are safety-related; (2) whose failure could affect safety-related functions; and (3) that are relied on to demonstrate compliance with the NRC's regulations for fire protection (FP), environmental qualification (EQ), pressurized thermal shock, anticipated transient without scram (ATWS), and station blackout (SBO).

Pursuant to 10 CFR 54.21(a), an applicant for a renewed license must review all SSCs that are within the scope of the Rule to identify SCs that are subject to an aging management review (AMR). Those SCs that are subject to an AMR perform an intended function without moving parts or without a change in configuration or properties, and are not subject to replacement based on a qualified life or specified time period. As required by 10 CFR 54.21(a), an applicant for a renewed license must demonstrate that the effects of aging will be managed in such a way that the intended function(s) of those SCs will be maintained, consistent with the current licensing basis (CLB), for the period of extended operation; however, active equipment is considered to be adequately monitored and maintained by existing programs. In other words, the detrimental effects of aging that may affect active equipment are more readily detectable and can be identified and corrected through routine surveillance, performance monitoring, and maintenance activities. The surveillance and maintenance activities programs for active equipment, as well as other aspects of maintaining the plants' design and licensing basis, are required throughout the period of extended operation.

Pursuant to 10 CFR 54.21(d), the LRA is required to include a supplement to the UFSAR and USAR. This supplement must contain a summary description of the applicant's programs and activities for managing the effects of aging and evaluation of time-limited aging analyses (TLAAs) for the period of extended operation.

License renewal also requires the identification and updating of the TLAAs. During the design phase for a plant, certain assumptions are made about the length of time that the plant can operate. These assumptions are incorporated into design calculations for several of the plant's SSCs. In accordance with 10 CFR 54.21(c)(1), the applicant must either show that these calculations will remain valid for the period of extended operation, project the analyses to the end of the period of extended operation, or demonstrate that the effects of aging on the intended function(s) will be adequately managed for the period of extended operation.

In 2001, the staff developed and issued Regulatory Guide (RG) 1.188, "Standard Format and Content for Applications to Renew Nuclear Power Plant Operating Licenses." This regulatory guide endorses Nuclear Energy Institute 95-10, "Industry Guideline for Implementing the Requirements of 10 CFR Part 54 - The License Renewal Rule," Revision 3, dated March 2001. Nuclear Energy Institute 95-10 details an acceptable method of implementing the Rule. The staff also used the SRP-LR to review the application.

In the LRA, the applicant fully utilized the process defined in NUREG-1801, "Generic Aging Lessons Learned (GALL) Report," dated July 2001. The GALL Report provides the staff with a summary of staff-approved aging management programs (AMPs) for the aging of many SCs that are subject to an AMR. If an applicant commits to implementing these staff-approved AMPs, the time, effort, and resources used to review an applicant's LRA can be greatly reduced, thereby improving the efficiency and effectiveness of the license renewal review process. The GALL Report summarizes the aging management evaluations, programs, and activities credited for managing aging for most of the SCs used throughout the industry. The report also serves as a reference for both applicants and staff reviewers to quickly identify those AMPs and activities that the staff determined can provide adequate aging management during the period of extended operation.

1.2.2 Environmental Review

Environmental protection regulations are codified in 10 CFR Part 51. In December 1996, the staff revised the environmental protection regulations to facilitate the environmental review for license renewal. The staff prepared a Generic Environmental Impact Statement (GEIS) to document its evaluation of the possible environmental impacts associated with renewing licenses for nuclear power plants. For certain types of environmental impacts, the GEIS establishes generic findings that are applicable to all nuclear power plants. These generic findings are codified in Appendix B to Subpart A of 10 CFR Part 51. Pursuant to 10 CFR 51.53(c)(3)(i), an applicant for license renewal may incorporate these generic findings in its environmental report. In accordance with 10 CFR 51.53(c)(3)(ii), an environmental report must also include analyses of those environmental impacts that must be evaluated on a plant-specific basis (i.e., Category 2 issues).

In accordance with the National Environmental Policy Act of 1969 and the requirements of 10 CFR Part 51, the staff performed a plant-specific review of the environmental impacts of license renewal, including whether new and significant information existed that the GEIS did not consider. As part of its scoping process, the staff held a public meeting on September 21, 2004, in Scriba, NY, to identify environmental issues specific to the plant. The draft, plant-specific Supplement 24 to the GEIS, dated September 29, 2005, documents the results of the environmental review and includes a preliminary recommendation with respect to the license renewal action. The staff held two other public meetings on November 17, 2005, in Scriba, NY, to discuss draft GEIS Supplement 24. After considering comments on the draft, the staff published the final, plant-specific GEIS Supplement 24 on May 16, 2006.

1.3 Principal Review Matters

The requirements for renewing operating licenses for nuclear power plants are described in 10 CFR Part 54. The staff performed its technical review of the NMPNS LRA and ALRA in accordance with NRC guidance and the requirements of 10 CFR Part 54. The standards for renewing a license are set forth in 10 CFR 54.29. This SER describes the results of the staff's safety review.

In 10 CFR 54.19(a), the NRC requires a license renewal applicant to submit general information. The applicant provided this general information in the original LRA Section 1, submitted by letter May 26, 2004, and the ALRA, submitted by letter dated July 14, 2005. The staff reviewed Section 1 of the LRA and ALRA and found that the applicant had submitted the information required by 10 CFR 54.19(a).

In 10 CFR 54.19(b), the NRC requires that each LRA include "conforming changes to the standard indemnity agreement, 10 CFR 140.92, Appendix B, to account for the expiration term of the proposed renewed license." In its LRA and ALRA, the applicant stated the following regarding this issue:

> The current indemnity agreement for NMPNS does not contain a specific expiration term for the operating licenses. Therefore, conforming changes to account for the expiration term of the proposed renewed licenses are not necessary, unless the license number is changed upon issuance of the renewed licenses.

The staff intends to maintain the original license numbers upon issuance of the renewed licenses, if approved. Therefore, conforming changes to the indemnity agreement do not need to be made and the requirements of 10 CFR 54.19(b) have been met.

In 10 CFR 54.21, the NRC requires that each LRA contain: (a) an integrated plant assessment, (b) a description of any CLB changes that occurred during the staff's review of the LRA, (c) an evaluation of TLAAs, and (d) an FSAR supplement. Sections 3, 4 and Appendix B of the LRA and ALRA address the license renewal requirements of 10 CFR 54.21(a), (b), and (c). Appendix A of the LRA and ALRA contains the license renewal requirements of 10 CFR 54.21(d).

In 10 CFR 54.21(b), the NRC requires that each year following submission of the LRA, and at least three months before the scheduled completion of the staff's review, the applicant must submit an amendment to the LRA that identifies any changes to the facility's CLB that materially affect the contents of the LRA, including the UFSAR and USAR supplements. The applicant submitted an update to the LRA, by letter dated December 20, 2005, which summarizes the changes to the CLB that have occurred during the staff's review of the original LRA. This submission satisfies the requirements of 10 CFR 54.21(b).

In 10 CFR 54.22, the NRC requires that the LRA include changes or additions to the technical specifications that are necessary to manage the effects of aging during the period of extended operation. In Appendix D of the LRA and ALRA, the applicant stated that it had not identified any technical specification changes necessary to support issuance of the renewed operating licenses for NMPNS. This adequately addresses the requirement specified in 10 CFR 54.22.

The staff evaluated the technical information required by 10 CFR 54.21 and 10 CFR 54.22 in accordance with NRC regulations and the guidance provided by the SRP-LR. SER Sections 2, 3, and 4 document the staff's evaluation of the technical information contained in the LRA and ALRA.

As required by 10 CFR 54.25, the ACRS will issue a report to document its evaluation of the staff's review of the LRA, ALRA, and associated SER. SER Section 5 will incorporate the ACRS report, once it is issued. SER Section 6 documents the staff's conclusion as required by 10 CFR 54.29.

The final, plant-specific GEIS Supplement 24 documents the staff's evaluation of the environmental information required by 10 CFR 54.23 and specifies the considerations related to renewing the licenses for NMPNS. The staff prepares this supplement separately from this SER.

1.4 Interim Staff Guidance

The license renewal program is a living program. The staff, industry, and other interested stakeholders gain experience and develop lessons learned with each renewed license. The lessons learned address the staff's performance goals of maintaining safety, improving effectiveness and efficiency, reducing regulatory burden, and increasing public confidence. Interim staff guidance (ISG) is documented for use by the staff, industry, and other interested stakeholders until it is incorporated into the license renewal guidance documents such as the SRP-LR and the GALL Report.

The following table provides the current ISG, issued by the staff, as well as the SER sections in which the staff addresses each ISG issue.

ISG Issue (Approved ISG No.)	Purpose	SER Section
GALL Report presents one acceptable way to manage aging effects (ISG-1)	This ISG clarifies that the GALL Report contains one acceptable way, but not the only way, to manage aging for license renewal.	N/A
SBO Scoping (ISG-2)	The license renewal rule 10 CFR 54.4(a)(3) includes 10 CFR 50.63(a)(1)—SBO. The SBO rule requires that a plant must withstand and recover from an SBO event. The recovery time for offsite power is much faster than that of EDGs. The offsite power system should be included within the scope of license renewal.	2.1.2.1 2.1.2.1.4 2.1.3.1.1 2.1.4.3.5

ISG Issue (Approved ISG No.)	Purpose	SER Section
Concrete AMP (ISG-3)	Lessons learned from the GALL demonstration project indicated that GALL is not clear on whether concrete requires an AMP.	3.5A.2.2 (NMP1) 3.5B.2.2 (NMP2) 3.5.C.1.2 (Common) 3.5.C.3.1 (Common)
FP System Piping (ISG-4)	This ISG clarifies the staff position for wall-thinning of the FP piping system in GALL AMPs XI.M26 and XI.M27. The staff's new position is that there is no need to disassemble FP piping, as disassembly can introduce oxygen to FP piping, which can accelerate corrosion. Instead, use a non-intrusive method, such as volumetric inspection. Testing of sprinkler heads should be performed at year 50 of sprinkler system service life, and every 10 years thereafter. This ISG eliminates the Halon/carbon dioxide system inspections for charging pressure, valve line-ups, and the automatic mode of operation test from GALL; the staff considers these test verifications to be operational activities.	3.0.3.2.14
Identification and Treatment of Electrical Fuse Holders (ISG-5)	This ISG includes electrical fuse holders AMR and AMP (i.e., same as terminal blocks and other electrical connections). The position includes only fuse holders that are not inside the enclosure of active components (e.g., inside of switchgears and inverters). Operating experience finds that metallic clamps (spring-loaded clips) have a history of age-related failures from aging stressors such as vibration, thermal cycling, mechanical stress, corrosion, and chemical contamination. The staff finds that visual inspection of fuse clips is not sufficient to detect the aging effects from fatigue, mechanical stress, and vibration.	3.0.3.3.4

ISG Issue (Approved ISG No.)	Purpose	SER Section
The ISG Process (ISG-8)	This ISG provides clarification and update to the ISG process on Improved License Renewal Guidance Documents.	N/A
Standardized Format for License Renewal Applications (ISG-10)	The purpose of this ISG is to provide a standardized license renewal application format for applicants.	N/A

1.5 Summary of Open Items and its Resolutions

An issue is considered open if the applicant has not presented sufficient basis for resolution and; therefore, has not met all applicable regulatory requirements. As a result of its review of the LRA and ALRA, including additional information submitted to the staff through April 21, 2006, the staff identified the following open items:

Open Item 3.0.3.2.17-1: Subsequent to the onsite audit and review of NMP ALRA, the staff also reviewed the applicant's Inservice Inspection Owner Activity Report, dated July 23, 2003. In this report, the applicant has stated that, for NMP1, corrosion was identified over the entire 360 degree circumference of the drywell interior surface of the liner plate at the 225 foot elevation. The applicant further stated in the report that (1) a subsequent detailed (D-VT) visual examination (VT-1) was performed and that (2) no unacceptable degradation in the visible areas of the drywell liner was found and that (3) no immediate corrective action was taken. The staff has asked the applicant to provide further discussion to address the staff concern regarding the loss of material due to corrosion for the NMP1 drywell.

On March 27, 2006, the applicant met with the staff to discuss the issue identified in OI 3.0.3.2.17-1, and by letter dated April 4, 2006, the applicant provided its response to OI 3.0.3.2.17-1. The applicant stated in the letter that during the March 2003, NMP1 refueling outage, a general visual examination of 100 percent of the accessible portion of the interior surface of the drywell shell was performed. Six localized areas, coinciding with the area coolers, were observed to have significant corrosion. In accordance with ASME Section XI Subsection IWE, a detailed visual examination (VT-1) was performed of the six localized areas. The corrosion was characterized as "major" (i.e., greater than 5 percent of the base metal was judged to be lost). The applicant further stated that the drywell is entirely within a nitrogen inerted atmosphere, and that it has performed appropriate calculation and concluded that the drywell shell will not reach its minimum design thickness until 45 years, which is 19 years beyond the end of the period of extended operation. The applicant further stated that it will continue its monitoring of the drywell shell via a proposed NMP1 Drywell Supplemental Program. This supplemental inspection program is in addition to other AMPs that have been credited for NMP1 containment aging management activities already reviewed and accepted by the staff. These AMPs include the Structural Monitoring, ASME Section XI Subsection IWE, and 10 CFR 50 Appendix J Programs. After reviewing the April 4, 2006, response to OI 3.0.3.2.17-1, the staff concludes that the applicant's response is acceptable; therefore, OI 3.0.3.2.17-1 is closed. The staff's evaluation of the applicant's response is detailed in SER Section 3.0.

Open Item 4.7B.1-1: The neutron fluence methodology for TLAA Section 4.7.1, "RPV Biological Shield (NMP2 Only)," is based on neutron fluence calculations that have been reported in SANDIA Report No. SAND 92-2420, "Accelerated 54°C Irradiated Test of Shippingport Neutron Shield Tank and HFIR Vessel Materials [January 1993]." However, the methodology for calculating the neutron fluence values reported in SANDIA Report No. SAND 92-2420 has not been approved by the staff. Therefore, the staff requested that Constellation Energy submit an updated 54 EPFY neutron fluence calculation for the biological shield wall (BSW) during the NRC's allocated review period for the amended license renewal application.

The staff also requested that the 54 EFPY neutron fluence calculation be based on a methodology that conforms to the NRC's recommendations in RG 1.190, "Calculational and Dosimetry Methods for Determining Pressure Vessel Neutron Fluence [March 2001]," and be submitted for the staff's review and approval.

The applicant responded to OI 4.7B.1-1 by letter dated March 23, 2006, and submitted a summary of a new RG 1.190 based analysis for its determination of the maximum neutron fluence at the NMP2 biological shield wall. The applicant stated that the resultant information from the calculation indicated that the maximum fluence to the shield at the end of Cycle 10 was found to be less than $2E16$ n/cm^2 and the maximum fluence at 54 EFPY was found to be $6.2E16$ n/cm^2 for fast neutrons (E>1.0 MeV). The applicant further stated that since the 54 EFPY maximum fluence value at the bioshield walls is less than the threshold fluence value ($1E17$ n/cm^2) for the susceptiblity of steel to neutron embrittlement identified in 10 CFR 50, Appendix H, the consideration of this aging effect no longer applies, and that TLAA criterion 10 CFR 54.3(a) is no longer applicable to the original analysis. Based on this information, the applicant proposed to delete its original TLAA 4.7.1 submitted in the original LRA and ALRA. The staff reviewed the applicant's response and found the response acceptable; therefore, OI 4.7B.1-1 is closed. The staff's evaluation of the applicant's response is detailed in SER Section 4.0.

1.6 Summary of Confirmatory Items

A confirmatory item is an issue that the staff has resolved, but for which the applicant has not yet formally submitted the resolution. After completing a review of the ALRA for NMP1 and NMP2, including all additional information and clarifications submitted to the staff as of April 21, 2006, the staff has identified no confirmatory items.

1.7 Summary of Proposed License Conditions

As a result of the staff's review of the LRA and ALRA, including subsequent information and clarifications provided by the applicant, the staff identified three proposed license conditions.

The first license condition requires the applicant to include the UFSAR and USAR supplements required by 10 CFR 54.21(d) in the next UFSAR and USAR updates, as required by 10 CFR 50.71(e), following the issuance of the renewed licenses.

The second license condition requires that the activities identified in the UFSAR and USAR supplements be completed in accordance with the schedule in Appendix A.

The third license condition requires the implementation of the most recent staff-approved version of the Boiling Water Reactor Vessels and Internals Project (BWRVIP) Integrated Surveillance Program (ISP) as the method to demonstrate compliance with the requirements of 10 CFR Part 50, Appendix H. Any changes to the BWRVIP ISP capsule withdrawal schedule must be submitted for NRC staff review and approval. Any changes to the BWRVIP ISP capsule withdrawal schedule which affects the time of withdrawal of any surveillance capsules must be incorporated into the licensing basis. If any surveillance capsules are removed without the intent to test them, these capsules must be stored in a manner which maintains them in a condition which would support re-insertion into the reactor pressure vessel, if necessary.

SECTION 2

STRUCTURES AND COMPONENTS SUBJECT TO AGING MANAGEMENT REVIEW

2.1 Scoping and Screening Methodology

2.1.1 Introduction

Title 10 of the *Code of Federal Regulations*, Part 54 (10 CFR Part 54), "Requirements for Renewal of Operating Licenses for Nuclear Power Plants," Section 54.21, "Contents of Application — Technical Information," requires that each application for license renewal contain an integrated plant assessment (IPA). The IPA must list and identify those structures and components (SCs) that are subject to an aging management review (AMR) from all of the systems, structures, and components (SSCs) that are within the scope of license renewal in accordance with 10 CFR 54.4.

In Section 2.1, "Scoping and Screening Methodology," of the original license renewal application (LRA) and amended license renewal application (ALRA), the applicant described the scoping and screening methodology used to identify SSCs at the Nine Mile Point Nuclear Station (NMPNS) within the scope of license renewal and the SCs that are subject to an AMR. The staff reviewed the applicant's scoping and screening methodology to determine whether it meets the scoping requirements stated in 10 CFR 54.4(a) and the screening requirements stated in 10 CFR 54.21.

In developing the scoping and screening methodology, the applicant considered the requirements of 10 CFR Part 54, the Statements of Consideration (SOC) for 10 CFR Part 54, and the guidance presented by the Nuclear Energy Institute (NEI), "Industry Guideline for Implementing the Requirements of 10 CFR Part 54 The License Renewal Rule," Revision 5, (NEI 95-10). In addition, in developing this methodology, the applicant considered the correspondence between the U.S. Nuclear Regulatory Commission (NRC) and other applicants and/or the NEI.

2.1.2 Summary of Technical Information in the Amended Application

In the original LRA and ALRA Sections 2.0 and 3.0, the applicant provided the technical information required by 10 CFR 54.21(a). In ALRA Section 2.1, "Scoping and Screening Methodology," the applicant described the process used to identify the SSCs that meet the license renewal scoping criteria under 10 CFR 54.4(a), as well as the process used to identify the SCs that are subject to an AMR, as required by 10 CFR 54.21(a)(1).

Additionally, the original LRA and ALRA Sections 2.2, "Plant Level Scoping Results," 2.3, "System Scoping and Screening Results: Mechanical Systems," 2.4, "Scoping and Screening Results: Structures and Component Supports," and 2.5, "Scoping and Screening Results: Electrical and Instrumentation and Controls Systems," amplify the process that the applicant used to identify SCs subject to an AMR. ALRA Section 3, "Aging Management Review Results," contains the following information:

Section 3.1, "Aging Management of Reactor Vessel, Internals, and Reactor Coolant System"

Section 3.2, "Aging Management of Engineered Safety Features Systems"

Section 3.3, "Aging Management of Auxiliary Systems"

Section 3.4, "Aging Management of Steam and Power Conversion Systems"

Section 3.5, "Aging Management of Structures and Component Supports"

Section 3.6, "Aging Management of Electrical and Instrumentation and Controls Components"

The original LRA and ALRA Section 4, "Time-Limited Aging Analyses," contains the applicant's identification and evaluation of time-limited aging analyses (TLAAs).

2.1.2.1 Scoping Methodology

In the original LRA and ALRA Section 2.1, the applicant described the methodology used to scope mechanical, structural, and electrical and instrumentation and controls (I&C) SSCs pursuant to the requirements of the 10 CFR 54.4(a) scoping criteria. The following sections present the applicant's scoping methodology, as described in the original LRA and ALRA.

2.1.2.1.1 Application of the Scoping Criteria in 10 CFR 54.4(a)

The applicant described the general approach to scoping safety-related (SR) and nonsafety-related (NSR) SSCs and SSCs credited with demonstrating compliance with certain regulated events in the original LRA and ALRA Section 2.1.4, "Application of License Renewal Scoping Criterion." The following sections describe the scoping approaches specific to each of the three 10 CFR 54.4(a) scoping criteria.

Application of the Scoping Criteria in 10 CFR 54.4(a)(1). In the original LRA and ALRA Section 2.1.4.1, "Safety Related Criteria Pursuant to 10 CFR 54.4(a)(1)," the applicant discussed the scoping methodology as it pertains to SR criteria in accordance with 10 CFR 54.4(a)(1). With respect to the SR criteria, the applicant stated that the SSCs within the scope of license renewal include SR SSCs that are relied upon during and following design-basis events (DBEs). The DBEs considered are consistent with the NMP Unit 1 (NMP1) and Unit 2 (NMP2) current licensing basis (CLB). As part of the process to identify the SSCs within scope of Criterion 1, NMP used a pre-established safety classification process that identifies and documents the SR functions of SSCs. The Maintenance Rule scoping documents are the primary repository of system function classifications, and the master equipment list (MEL) is the primary repository of component classifications. As a result, the Maintenance Rule scoping documents were used as the main source for identifying SR system functions that satisfy Criterion 1. Supporting information from the NMP1 updated final safety analysis report

(UFSAR) and the NMP2 updated safety analysis report (USAR), technical specifications (TSs), design documents, design drawings and MEL were reviewed to ensure all SR system functions were properly identified. Implementation of the license renewal scoping and screening procedure ensured that the UFSAR/USAR, TSs, Maintenance Rule scoping documents, design documents, design drawings and MEL were reviewed, as applicable, to ensure all system functions were identified and evaluated against this criterion.

Application of the Scoping Criteria in 10 CFR 54.4(a)(2). In the original LRA and ALRA Section 2.1.4.2, "Non-Safety Related Criteria Pursuant to 10 CFR 54.4(a)(2)," the applicant discussed the scoping methodology as it related to the NSR criteria in accordance with 10 CFR 54.4(a)(2). The applicant stated that the process used to review SSCs for 10 CFR 54.4(a)(2) applicability ensured that the SSCs within the scope of license renewal include the NSR SSCs whose failure could prevent satisfactory accomplishment of any of the Criterion 1 functions of SSCs.

The applicant reviewed UFSAR/USAR, TSs, Maintenance Rule scoping documents, design documents, design drawings and MEL to ensure all NSR SSC functional interactions were identified where an NSR SSC could fail and prevent the satisfactory accomplishment of an SR intended function. The NSR SSCs meeting Criterion 2 that are explicitly identified in the CLB, such as flood barriers, were identified.

In the original LRA, the applicant identified three additional areas to review for applicability to 10 CFR 54.4(a)(2):

(1) Supports for NSR Equipment - The applicant determined that component supports required to prevent NSR SSCs from physical interacting with SR SSCs were within the scope of license renewal. The LRA described the applicable supports as those that must remain in place such that they do not impact equipment that is required to perform an intended function in such a way as to prevent the equipment from performing its intended function. The applicant considered all NSR supports to be within the scope of license renewal if located in areas housing SR equipment.

(2) NSR SCs in Proximity to SR Equipment - The applicant reviewed NSR SCs in proximity of SR equipment in accordance with the guidance contained in NRC Interim Staff Guidance (ISG) 09, "Guidance on the Identification and Treatment of Structures, Systems, and Components which Meet 10 CFR 54.4(a)(2)." The applicant used the preventive option in order to satisfy ISG-09 and considered all NSR piping, fittings, and equipment containing water or steam to be within the scope of license renewal if the NSR SCs were located in the vicinity of SR equipment. NSR SCS were considered to be in the vicinity of SR equipment if located in the same building, corridor, and floor as SR equipment.

(3) SR/NSR Piping Interface - The applicant used plant drawings to identify classification boundaries and SR/NSR piping interfaces. The scope of the NSR piping system was extended beyond the classification change to the first seismic anchor beyond the depicted class change. The applicant determined that the piping between the depicted classification boundary and the first seismic anchor was considered to be within the scope of license renewal. In addition, the applicant considered all NSR piping, fittings, and equipment containing water or steam to be within the scope of license renewal if located in the vicinity of SR equipment. As a result, for piping containing water or steam,

the NSR portion within the scope of license renewal extended beyond the depicted class change until no longer in the vicinity of SR equipment or until the first seismic anchor was reached, whichever was furthest. The applicant defined the term "seismic anchor" as a series of supports and changes in piping geometry that combine to provide restraint to the piping in six degrees of freedom. For NMP2, the term "seismic anchor" means an actual anchor that provides restraint to the piping in six degrees of freedom.

As a result of the staff's audit of the applicant's scoping and screening methodology, the applicant revised the description of the methodology used to evaluate the 10 CFR 54.4(a)(2) criterion, and provided that revised description in the ALRA. The details of that revised methodology, and the staff's evaluation is provided in SER Section 2.1.3.1.4.

Application of the Scoping Criteria in 10 CFR 54.4(a)(3). In the original LRA and ALRA Sections 2.1.4.3, "Regulated Event Scoping Pursuant to 10 CFR 54.4(a)(3);" 2.1.4.3.1, "Fire Protection;" 2.1.4.3.2, "Environmental Qualification (EQ);" 2.1.4.3.3, "Pressurized Thermal Shock; (PTS)" 2.1.4.3.4, "Anticipated Transients Without Scram (ATWS);" and 2.1.4.3.5, "Station Blackout (SBO)," the applicant discussed the scoping methodology as it related to the regulated event criteria, in accordance with 10 CFR 54.4(a)(3). With respect to the scoping criteria related to 10 CFR 54.4(a)(3), the applicant evaluated all regulated events including fire protection, EQ, ATWS, and SBO. For each regulated event, the applicant identified and reviewed the applicable UFSAR/USAR, TSs, Maintenance Rule scoping documents, design documents, design drawings, and MEL to ensure all SSCs credited for compliance with the regulated event were identified and evaluated against these criteria. Specific scoping for each regulated event was also described in the relevant section.

In summary, the applicant included within the scope of license renewal the SSCs relied on in safety analyses or plant evaluations to perform an intended function that demonstrates compliance with NRC regulations for fire protection, EQ, ATWS, and SBO, in accordance with the criteria of 10 CFR 54.4(a)(3).

2.1.2.1.2 Documentation Sources Used for Scoping and Screening

In the original LRA and ALRA Section 2.1.1, "Introduction," the applicant stated that it had reviewed information from the following sources during the license renewal scoping and screening process:

- UFSAR/USAR
- CLB information including TSs and docketed licensing correspondence
- design-basis documents (DBDs)
- Maintenance Rule scoping documents
- controlled drawings
- MEL

The applicant stated that it used this information to identify the functions performed by plant systems and structures. It then compared these functions to the scoping criteria in 10 CFR 54(a)(1)-(3) to determine whether the associated plant system or structure performed a license renewal intended function. It also used these sources to develop the list of structures and components subject to an AMR.

2.1.2.1.3 Plant and System-Level Scoping

In the original LRA and ALRA Section 2.1.2, "Plant Level Scoping," the applicant briefly described the scoping methodology for SR and NSR systems and structures and for equipment relied upon to perform a function for any of the five regulated events described in 10 CFR 54.4(a)(3). The NMP scoping process began with the review and evaluation of plant systems and structures against the criteria outlined in 10 CFR 54.4(a)(1)-(3) to determine those systems that met the requirements for inclusion in the scope of license renewal. During the initial scoping process, all functions were defined for all systems and structures in the plant. Subsequently, those functions that are intended functions were identified, and portions of the systems and structures that perform those intended functions were identified. Systems and structures meeting the scoping criteria of 10 CFR 54.4 were, therefore, established.

2.1.2.1.4 Component-Level Scoping

After the applicant identified the intended functions of systems or structures within the scope of license renewal, it performed a review to determine which components of each in-scope system and structure supported license renewal intended functions. The applicant considered the components that supported intended functions to be within the scope of license renewal and screened them to determine whether an AMR was required.

The applicant considered three component classifications during this stage of the evaluation: (1) mechanical, (2) civil and structural, and (3) electrical. The applicant called the process of identifying the individual components of a system or structure component screening, although it also included the scoping criteria (i.e., within the scope of license renewal). The following three paragraphs discuss the scoping methodology for these component classifications.

(1) Mechanical Component Scoping - The applicant described the scoping methodology for mechanical components within SR and NSR mechanical systems in the original LRA and ALRA Section 2.1.5.1. For each mechanical system determined to be within the scope of license renewal, the applicant developed a system evaluation boundary to identify the set of structures and components necessary to perform the intended functions for the given mechanical system. These evaluation boundaries included sets of piping and instrumentation diagrams (P&IDs) for each system and the component list from the MEL database. From the system diagrams, the applicant identified components that were required to ensure the system could perform its intended functions. Then, the applicant grouped them into relevant component types associated with each function within the scope of license renewal and listed them in the scoping and screening database for further analysis.

(2) Structural Component Scoping - The original LRA and ALRA Section 2.1.5.2 discusses the scoping methodology associated with civil structures. The applicant reviewed the UFSAR/USAR, Maintenance Rule scoping results, design- and license-basis documents, regulatory requirements, the MEL, 10 CFR Part 50, Appendix B determinations, and plant drawings to determine SCs within the scope of license renewal. All SR SCs were included within the scope of license renewal, and include

items such as walls, piping and equipment supports, conduit, cable trays, electrical enclosures, instrument panels, pipe whip restraints, fire barriers, liners, sump screens, doors, blowout panels, flood barriers, missile shields, and jet impingement shields relied upon in the licensing basis. The NSR SCs listed in NEI 95-10, Appendix F; and NSR SCs that perform a function required for compliance with fire protection, ATWS, and SBO regulations were included within the scope of license renewal. The in-scope NSR SCs include missile shields that protect SR equipment; overhead handling systems that could effect SR equipment; walls, curbs, dikes, and doors that provide flood protection for SR equipment; and jet impingement shields and blowout panels that protect SR equipment from the effects of a high-energy line break (HELB). In this way, the applicant was able to compile a comprehensive list of all SCs within the scope of license renewal.

(3)   Electrical and I&C Component Scoping - The applicant described the scoping process associated with electrical and I&C systems and components in the original LRA and ALRA Sections 2.1.2 through 2.1.5. For these systems, the applicant elected to use the same methodology that it applied to mechanical and structural SSCs, typically, a bounding or spaces approach, as described in NEI 95-10. As a result, the electrical and I&C component types throughout the plant were identified with regard to specific electrical and I&C system intended functions. The applicant evaluated the electrical and I&C component types against the scoping criteria in 10 CFR 54.4(a)(1)-(3), to determine whether they perform intended functions. This was accomplished using relevant CLB documentation. During the initial scoping process, the applicant described all the electrical and I&C systems and defined their functions. Subsequently, those functions that are intended functions were identified, and portions of the electrical and I&C systems that perform those intended functions were identified.

2.1.2.2 Screening Methodology

After determining the SSCs within the scope of license renewal, the applicant implemented a process for determining which SSCs would be subject to an AMR, in accordance with the requirements of 10 CFR 54.21(a)(1). In the original LRA and ALRA Section 2.1.5, "Component Screening," the applicant discussed the screening activities as they related to the SSCs that are within the scope of license renewal. The applicant divided the screening portion of the integrated license renewal plant assessment into three engineering disciplines: mechanical, civil/structural, and electrical and I&C.

(1)   Mechanical Component Screening - The applicant stated in the original LRA and ALRA Section 2.1.5.1, that it screened each system identified to be within the scope of license renewal. This process evaluated the individual structures and components included within in-scope mechanical systems to identify specific structures and components that required an AMR. The applicant evaluated each mechanical component identified in the scoping phase. The in-scope SCs that perform an intended function without moving parts or without a change in configuration or properties (screening criterion of 10 CFR 54.21(a)(1)(i)) were identified. Active/passive screening determinations were based on the guidance in NEI 95-10, Appendix B, Revision 5. The passive, in-scope SCs that are not subject to replacement based on a qualified life or specified time period (screening criterion of 10 CFR 54.21(a)(1)(ii)) were identified as requiring an AMR. The determination of whether a passive, in-scope SC has a qualified life or specified

replacement time period was based on a review of maintenance programs and procedures.

(2) Structural Component Screening - The original LRA and ALRA Section 2.1.5.2, discusses the screening activities related to SCs within the scope of license renewal. These screening activities consisted of the identification of passive components, long-lived components, component intended functions, consumables, and component replacement based on performance or condition. The applicant relied on the guidance in NEI 95-10 to develop the plant-specific listing of passive components of interest during the review. Component supports, and fire stops and seals were considered SCs and binned in separate structural commodity groupings.

(3) Electrical/I&C Component Screening - In the original LRA and ALRA Section 2.1.5.4, the applicant described the methodology used to screen electrical and I&C components. Specifically, the applicant applied the screening methodology employed for electrical and I&C components consistent with the guidance in NEI 95-10. All passive, long-lived components, as defined by 10 CFR 54.21(a)(1)(ii), were evaluated as commodities regardless of the system or structure in which they reside in the MEL. As a result, the electrical systems results contain only active components not subject to AMR. An AMR was then conducted on a commodity basis for the entire population of passive, long-lived components. The applicant did not identify individual components that perform intended functions.

Electrical and I&C components associated with the EQ Program are replaced on a specified interval based on a qualified life. Therefore, components in the EQ Program do not meet the "long-lived" criteria of 10 CFR 54.21(a)(1)(ii). They are considered "short-lived" per the regulatory definition and are not subject to AMR. Using these screening criteria, the applicant determined that the passive electrical and I&C component commodity groups at NMPNS that require an AMR are cables and connectors (including splices, connectors, terminal blocks, and fuse holders), non-segregated/switchyard bus, containment electrical penetrations, and various switchyard components.

2.1.3 Staff Evaluation

The staff evaluated the original LRA and ALRA scoping and screening methodology in accordance with the guidance contained in U.S. Nuclear Regulatory Commission Regulatory Guide (NUREG)-1800, "Standard Review Plan for Review of License Renewal Applications for Nuclear Power Plants," (SRP-LR), Section 2.1, "Scoping and Screening Methodology." The following regulations form the basis for the acceptance criteria for the scoping and screening methodology review:

- 10 CFR 54.4(a), as it relates to the identification of plant SSCs within the scope of 10 CFR Part 54

- 10 CFR 54.4(b), as it relates to the identification of the intended functions of plant SSCs determined to be within the scope of 10 CFR Part 54

- 10 CFR 54.21(a)(1) and(2), as they relate to the methods utilized by the applicant to identify plant structures and components subject to an AMR

As part of the review of the applicant's scoping and screening methodology, the staff reviewed the activities described in the following sections of the original LRA and ALRA using the guidance contained in the SRP-LR:

- Original LRA and ALRA Section 2.1, "Scoping and Screening Methodology," to ensure that the applicant described a process for identifying SSCs that are within the scope of license renewal, in accordance with the requirements of 10 CFR 54.4(a)(1)-(3).

- Original LRA and ALRA Sections 2.2, "Plant Level Scoping Results;" 2.3, "System Scoping and Screening Results: Mechanical Systems;" 2.4, "Scoping and Screening Results: Structures and Component Supports;" and 2.5, "Scoping and Screening Results: Electrical and Instrumentation and Controls Systems," to ensure that the applicant described a process for determining structural, mechanical, and electrical components at NMPNS that are subject to an AMR in accordance with the requirements of 10 CFR 54.21(a)(1) and (2).

In addition, the staff conducted a scoping and screening methodology audit at NMPNS engineering offices in Lycoming, New York, from September 27 to October 1, 2004. The audit focused on ensuring that the applicant had developed and implemented adequate guidance to conduct the scoping and screening of SSCs in accordance with the methodologies described in the application and the requirements of 10 CFR Part 54. The staff reviewed implementation procedures and engineering reports describing the applicant's scoping and screening methodology. In addition, the staff conducted detailed discussions with the applicant on the implementation and control of the license renewal program and reviewed administrative control documentation and selected design documentation used by the applicant during the scoping and screening process. The staff reviewed the applicant's processes for quality assurance with respect to development of the original LRA and the training and qualification of the original LRA development team. The staff also reviewed a sample of system scoping and screening results reports for the feedwater/high pressure coolant injection (FW/HPCI) system and reactor building to ensure (1) that the applicant had appropriately implemented the methodology outlined in the administrative controls and (2) that the results were consistent with the CLB. The staff documented its review in an audit report dated November 9, 2004. The report identified several issues requiring additional information from the applicant prior to completion of the review. Each issue is identified and addressed in detail in this section.

2.1.3.1 Scoping Methodology

The original LRA scoping evaluations were performed by the applicant's license renewal project personnel and contractors from Constellation Nuclear Services (CNS). The staff discussed the applicant's methodology with the applicant's license renewal project management personnel and reviewed documentation pertinent to the scoping process. The staff assessed whether the scoping methodology outlined in the original LRA and CNS implementation procedures was appropriately implemented and whether the scoping results were consistent with CLB requirements. The staff also reviewed a sample of system scoping results for the following systems: FW/HPCI and reactor building (structural review).

2.1.3.1.1 Implementation Procedures and Documentation Sources Used for Scoping and Screening

The staff reviewed the applicant's scoping and screening implementation procedures to verify that the process used to identify structures and components subject to an AMR was consistent with the original LRA and SRP-LR and that the applicant appropriately implemented the procedural guidance. Additionally, the staff reviewed the scope of CLB documentation sources used to support the LRA development and the process used by the applicant to ensure that CLB commitments were appropriately considered during the scoping and screening process.

Scoping and Screening Implementation Procedures. The staff performed an on-site review of the following scoping and screening methodology implementation procedures and engineering reports: license renewal guidance (LRG)-01,"License Renewal Project Guidance," Revision 2; LRG-02, "License Renewal Scoping and Screening," Revision 4; LRG-04, "Aging Management Review for Electrical Commodities," Revision 2; LRG-08, "Work Product Review Guideline," Revision 7; LRG-09, "Site Review Guideline," Revision 5; and LRG-10, "License Renewal Application Guideline," Revision 6.

In reviewing these procedures, the staff focused on the consistency of the detailed procedural guidance with information in the original LRA and the various staff positions documented in SRP-LR and ISG documents. The staff found that the scoping and screening methodology instructions were generally consistent with the original LRA Section 2.1 and were of sufficient detail to provide the applicant with concise guidance on the scoping and screening implementation process to be followed during the LRA activities. One exception was found related to the description of the scoping and screening process used to identify electrical commodity groupings. This issue is addressed further in this SER in Section 2.1.3.1.3.

In addition to reviewing the implementing procedures, the staff reviewed supplemental design information, including the DBDs, system drawings, and selected licensing documentation the applicant relied up during the scoping and screening phases of the review. The staff found these design documentation sources to be useful for ensuring that the initial scope of SSCs identified by the applicant was consistent with the plant's CLB.

Sources of Current Licensing Basis Information. The staff reviewed the scope and depth of the applicant's CLB review to verify that the methodology was sufficiently comprehensive to identify SSCs within the scope of license renewal and SCs requiring an AMR. As defined in 10 CFR 54.3(a), the CLB is the set of staff requirements applicable to a specific plant and an applicant's written commitments for ensuring compliance with and operation within applicable NRC requirements and the plant-specific design basis that are docketed and in effect. The CLB includes certain NRC regulations; orders; license conditions; exemptions; TSs; and design-basis information documented in the most recent FSAR. The CLB also includes applicant commitments remaining in effect that were made in docketed licensing correspondence, such as applicant responses to NRC bulletins, generic letters (GLs), and enforcement actions, as well as applicant commitments documented in NRC safety evaluations or applicant event reports.

The staff determined that the original LRA and ALRA Section 2.1.1 provides a description of the CLB and related documents used during the scoping and screening process that is consistent with the guidance contained in SRP-LR and NEI 95-10. Specifically, the original LRA and ALRA

Section 2.1.1 identified the UFSAR/USAR, TSs, docketed licensing correspondence, MEL, controlled drawings, and the Maintenance Rule scoping documents. Additionally, in Section 3.2.2 of scoping implementation procedure LRG-02, the applicant provided a comprehensive listing of documents that could be used to support scoping and screening evaluations. The applicant noted that system descriptions and system intended functions were identified based on the review of applicable sections of the UFSAR/USAR, Appendix B determinations, Maintenance Rule scoping document, and design and licensing basis documents.

The NMP MEL is the applicant's primary repository for component safety classification information. During the audit, the staff reviewed the applicant's administrative controls for MEL safety classification data and concludes that the applicant had established adequate measures to control their integrity and reliability. Therefore, the staff concludes that the MEL provided a sufficiently controlled source of component data to support scoping and screening evaluations.

In LRG-02, the applicant identified topical reports as a source of information to support identification of systems and structures relied upon to demonstrate compliance with certain regulated events referenced in 10 CFR 54.4(a)(3). These reports were developed in accordance with the NMP engineering directives that describe the requirements for preparation of Nuclear Engineering Reports (NERs). These reports were developed and maintained as controlled quality documents at the NMP. The topical reports contain a listing of CLB references used for their development that is consistent with the original LRA Section 2.1.1. The staff concludes that the preparation of the topical reports in accordance with the NMPNS requirements for development of NERs provided sufficient guidance to reasonably ensure that topical reports adequately summarized CLB information for the purposes of scoping.

As part of the audit, the staff evaluated the scope and depth of the applicant's document review to provide assurance that the scoping methodology considered all SSC intended functions. In reviewing the original LRA and scoping and screening implementation procedures, the staff was unable to determine (1) the extent that the CLB was reviewed by the applicant during the development of the system description and (2) the extent that related intended function evaluations were performed during the scoping phase of the review. During discussions with the NMP license renewal project team, it was noted that an electronic document database was used to identify CLB documents pertinent to the development of system descriptions and identification of system intended functions. However, the staff remained unable to determine the extent to which that electronic database was used for those purposes at the time of the audit.

In RAI 2.1-3, dated November 22, 2004, the staff requested that the applicant provide a detailed description of the methodology used to develop system descriptions and identify the system intended functions. The staff also requested that the applicant describe the controls and processes, including proceduralized controls, used to ensure that the electronic CLB document database was complete and accurate.

In its response, by letter dated December 22, 2004, the applicant stated, in part, that the system descriptions and system intended functions were developed in accordance with LRG-02, which identified the primary sources for description and intended function information. As part of the review process, the applicant described the use of several levels of review and approval including an independent license project engineer review, discipline lead review, supervisor review, system engineer review, and, finally, project manager review and approval. This review process was implemented to ensure a high confidence that system descriptions were accurate and all functions have been properly identified. The specific documents used for the generation of the system descriptions and intended functions were also referenced in the individual system and structure scoping and screening report for ease of verification.

With respect to the electronic document database, the applicant clarified in its December 22, 2004, letter that the electronic file contained correspondence between the staff and NMP up to February 2003. The latter correspondence was not entered into the system but evaluated as part of the review process in hard copy format. The applicant also clarified that the electronic file contained documents that were part of the CLB and were used to support the development of position papers and reports for use during the license renewal evaluation. These records were researched specifically to ensure that all functions were properly identified for the fire protection, ATWS, and SBO regulated events. Specific documents reviewed included NMP responses to the issuance of new regulations (i.e., ATWS and SBO), NRC safety evaluations, NMP responses to the safety evaluations, as applicable, and NMP responses to GLs. The electronic files were also researched when specific questions arose during scoping and aging management program reviews.

On the basis of the supplemental information provided by the applicant in response to the staff's request for information, and the clarification as to what extent that information was reviewed and applied to the license renewal evaluation, the staff found that the applicant has adequately addressed the staff's request for additional information. Therefore, the staff's concern described in RAI 2.1-3 is resolved.

Conclusion. On the basis of a review of information provided in the original LRA and ALRA Section 2.1, a review of the applicant's detailed scoping and screening implementation procedures, and the results from the scoping and screening audit including the applicant's responses to the staff's RAI, the staff concludes that the applicant's scoping and screening methodology considered a scope of CLB information consistent with the guidance contained in SRP-LR and NEI 95-10, and is, therefore, acceptable.

2.1.3.1.2 Quality Assurance Controls Applied to LRA Development

The staff reviewed the quality assurance controls used by the applicant to provide reasonable confidence that the original LRA scoping and screening methodologies were adequately implemented. Although the applicant did not develop the original LRA under a 10 CFR Part 50, Appendix B, quality assurance program, the staff determined that the applicant utilized the following quality assurance processes during the original LRA development:

- Implementation of the scoping and screening methodology was governed by written procedures and guidelines.

- Although much of the original LRA development was performed by contractors, the applicant developed procedures to govern the conduct of owner acceptance reviews of contractor work products. For example, License Renewal Project Guidance LRG-08 "Work Product Review Guideline," Revision 7; and LRG-09 "Site Review Guideline," Revision 5, describe the process used by the applicant and CEG to review license renewal project documents developed by the CEG staff. Documents subject to this acceptance review included scoping and screening review reports, AMR reports, TLAAs, and aging management program (AMP) attribute and alternatives reports.

- The original LRA was reviewed and approved by the Nuclear Safety Review Board and the Station Operation Review prior to submittal to the staff. Additionally, the applicant developed procedural guidance for a final review of the original LRA prior to submittal to the staff.

- The applicant planned to retain certain license renewal documents, such as AMRs, individual system scoping reports, TLAAs, and topical reports, as quality records or controlled documents.

- The applicant performed an industry peer review and several quality assurance assessments of license renewal activities.

Conclusion. On the basis of review of pertinent original LRA development guidance, discussion with the applicant's license renewal staff, and review of quality audit reports, the staff concludes that these quality assurance activities provided additional assurance that original LRA development activities were performed consistently with the original LRA descriptions, and that this consistency is maintained in the ALRA.

2.1.3.1.3 Training

The staff reviewed the applicant's training process to ensure the guidelines and methodology for the scoping and screening activities would be performed in a consistent and appropriate manner. The screening and scoping of SSCs for license renewal was accomplished by CEG personnel. The CEG LRA team included personnel who had gained previous license renewal experience working on the Calvert Cliffs 1 and 2 LRA. The CEG LRA team was supplemented with additional CEG personnel that were provided with LRA-specific training. The purpose of the training was to provide a framework for ensuring that the personnel assigned to the technical portion of the original LRA acquired a fundamental level of knowledge of the license renewal process and regulatory requirements.

The training program for these personnel consisted of "check-outs" administered by individuals with LRA experience, required reading of selected documents, and lectures by personnel experienced in various LRA topics. A "check-out" is defined as a short interview between a qualification trainee and a subject matter expert to determine whether the trainee has an adequate understanding of a particular subject. With the exception of CEG personnel with prior license renewal experience, each CEG person assigned to license renewal maintained a training qualification record as part of the application development process. The results of the scoping and screening activities accomplished by CEG personnel were reviewed by CEG personnel. Personnel with prior experience on LRA preparation provided lectures on such topics as, scoping, boundaries, screening, AMRs, and TLAA. A check list was developed and used by CEG personnel to complete their reviews. The check list provided general guidance on what was required to be reviewed. Reviewers were required to use the check list, and the check lists were maintained as a permanent record.

The staff reviewed completed qualification and training records of several of the applicant's license renewal personnel and also reviewed completed check lists. The staff did not identify any adverse findings. Additionally, based on discussions with the applicant's license renewal personnel during the audit, the staff verified that the applicant's license renewal personnel were knowledgeable on the license renewal process requirements and the specific technical issues within their areas of responsibility.

Conclusion. On the basis of discussions with the applicant's license renewal project team responsible for the scoping and screening process, and a review of selected design documentation in support of the process, the staff concludes that the applicant's personnel understood the requirements of the original LRA and adequately implemented the scoping and screening methodology established in the original LRA.

2.1.3.1.4 Application of the Scoping Criteria in 10 CFR 54.4(a)

Application of the Scoping Criteria in 10 CFR 54.4(a)(1). In part, 10 CFR 54(a)(1) requires that the applicant consider all SR SSCs that are relied upon to remain functional during and following DBEs to ensure the following functions:

- To maintain the integrity of the reactor coolant pressure boundary.

- To shut down the reactor and maintain it in a safe-shutdown condition.

- To prevent or mitigate the consequences of accidents which could result in potential offsite exposures comparable to those referred to in 10 CFR 50.34(a)(1), 10 CFR 50.67(b)(2), or 10 CFR 100.11 to be within the scope of license renewal.

With regard to identification of DBEs, SRP-LR Section 2.1.3, "Review Procedures," states:

> The set of design basis events as defined in the rule is not limited to Chapter 15 (or equivalent) of the UFSAR. Examples of design basis events that may not be described in this chapter include external events, such as floods, storms, earthquakes, tornadoes, or hurricanes, and internal events, such as a high-energy-line break. Information regarding design basis events as defined in 10 CFR 50.49(b)(1) may be found in any chapter of the facility UFSAR, the Commission's regulations, NRC orders, exemptions, or license conditions within the CLB. These sources should also be reviewed to identify systems, structures and components that are relied upon to remain functional during and following design basis events (as defined in 10 CFR 50.49(b)(1)) to ensure the functions described in 10 CFR 54.4(a)(1).

The staff's review of original LRA Section 2.1 identified areas in which additional information was necessary to complete the review of the applicant's scoping and screening results. The applicant responded to the staff's RAIs as discussed below.

In RAI 2.1-1, dated November 22, 2004, the staff stated that during the scoping and screening methodology audit, the staff questioned how non-accident DBEs, particularly DBEs that may not be described in the UFSAR/USAR, were considered during scoping. The staff noted that limiting the review of DBEs to those described in the UFSAR/USAR accident analysis could result in omission of SR functions described in the CLB and requested the applicant provide a list of all DBEs that were evaluated as part of the license renewal review. However, during the audit, the staff was unable to identify such as listing. Therefore, the staff requested in RAI 2.1-1, that the applicant provide a list of DBEs evaluated as part of the license renewal scoping process, and describe the methodology used to ensure that all DBEs (including conditions of normal operation, anticipated operational occurrences, design-basis accidents, external events, and natural phenomena) were addressed during license renewal scoping

In its response, by letter dated December 22, 2004, the applicant stated, in part, that the methodology used to ensure that all DBEs, including operational occurrences, abnormal operating transients, anticipated and abnormal operational occurrences, design-basis accidents, and the general design criteria, were addressed during license renewal scoping was to utilize the NMP controlled documents and databases that identified those SSCs and functions classified as SR. These documents and databases consist of the NMP1 UFSAR and NMP2 USAR, safety-class boundary drawings, Appendix B determinations, Maintenance Rule scoping documents, and MEL, as well as additional CLB information identified in the individual scoping and screening reports developed for each NMP1 and NMP2 system and structure. The applicant also provided a detailed listing of the various DBEs for each unit and a description of the design and configuration control processes used to ensure that all SSCs required to perform an SR function are properly evaluated and identified.

The staff reviewed the additional information provided by the applicant, and discussed the response at a meeting on February 2, 2005, to verify that HELBs were specifically considered within the DBE evaluations. On the basis of providing (1) a detailed listing of the DBEs for each unit including HELBs; (2) a description of the design and configuration control processes used to identify the SSCs credited for DBE mitigation; and (3) a description of the processes and sources of DBE information used to perform the scoping evaluation consistent with the requirements of 10 CFR 54.4(a)(1), the staff found that the applicant has adequately addressed the staff's RAI. Therefore, the staff's concern described in RAI 2.1-1 is resolved.

The applicant's approach to satisfying the scoping requirements of 10 CFR 50.54(a)(1) was to identify and describe all plant systems and structures and evaluate those against the SR criteria. As part of this process, the applicant reviewed various licensing basis documents to identify SR intended functions associated with the NMP units. To accomplish this, the applicant performed scoping of SR SSCs in accordance with LRG-02 Sections 3.3 and 3.7.2. LRG-09 Section 3.2 was used to direct the review of scoping activities by the NMP staff. The applicant classified SSCs as either SR or NSR, using the information provided in the Maintenance Rule scoping document and the component-specific safety classification field in the MEL.

LRG-02 Section 3.8, "Component List," requires that the MEL be used to populate the license renewal database with components of systems or structures within the scope of license renewal. The MEL safety classification field was reviewed to ensure that any system or structure that has a component identified as SR was considered for inclusion in the scope of the license renewal project. Additionally the MEL safety-classification and associated MEL drawings provided a starting point for identifying specific mechanical and structural components required to meet the 10 CFR 54.4(a)(1) criteria. The staff reviewed the safety classification criteria used to determine the NMP safety classification to verify consistency with 10 CFR 54.4(a)(1) criteria. The staff determined that the nuclear SR definition used by the applicant in its safety classification program did not include all the exposure limitations referenced in 10 CFR 54(a)(1)(iii). Specifically, NMP plant procedure NIP-DES-02, "Safety Classification of Items and Activities," did not include a reference to the offsite exposure limitations contained in 10 CFR 50.67(b)(2) for use of an alternate source term. However, during discussions with the applicant it was determined that NMP had not requested a licensing basis change to use the alternate source term criteria; therefore, the requirements of 10 CFR 50.67(b)(2) do not currently impact the license renewal program.

As part of the audit discussions related to the determination of SR SSCs, the staff questioned whether some components classified as SR in the facility database might not perform any of the SR intended functions of 10 CFR 54.4(a)(1) due to plant-unique considerations or preferences. The applicant stated that these components may have been considered outside the scope of 10 CFR 54.4(a)(1). During the audit, the applicant described the process used to evaluate components classified as SR that did not perform an SR intended function. As part of the process, the applicant stated that the safety-classification of many SR components was re-evaluated in order to reconcile differences between scoping determinations and facility database information or the Maintenance Rule scoping results.

In RAI 2.1-2, dated November 22, 2004, the staff requested that the applicant provide a description of the process used during license renewal scoping activities to disposition components classified as SR that do not perform an SR intended function. In particular, the staff requested that the applicant provide a description of any components or structures classified as SR in the facility safety-classification database that were not included within the scope of license renewal under the 10 CFR 54.4(a)(1) criteria. Additionally, the staff asked the applicant to describe the process used to reconcile the facility database safety classification information with scoping intended function determinations.

In its response, by letter dated January 31, 2005, the applicant stated, in part, that during the scoping and screening process for the NMPNS original LRA, the applicant identified a small percentage of components as SR, but not required to meet any intended function for compliance with 10 CFR 54.4(a)(1). These discrepancies were entered into the NMP corrective action program (CAP) for resolution. The applicant provided a discussion of the specific cases where such re-classifications were identified. These included several components in NSR systems that were classified as SR in the plant component database, called Master Equipment List for NMP1 and NMP 2 (MEL1, MEL2) when the license renewal project was started, but have since been reclassified as NSR as a result of detailed review of the CLB as part of the renewal process. These components were entered into the CAP and reclassified in accordance with the design change process. Secondly, the applicant explained that there were instances of components identified as SR during the scoping and screening process that have been removed from the plant via the modification process. These components have been moved to the plant historical database. Thirdly, the applicant identified several components in MEL1 classified as SR that have been abandoned in place. These components are not within the scope of license renewal as they perform no system function and, therefore, do not perform any license renewal intended function.

The applicant also noted that all components reclassified as NSR that contain liquid and are in the vicinity of SR equipment are still considered within the scope of license renewal for criterion 10 CFR 54.4(a)(2).

Additionally, the applicant provided a description of the process used to evaluate these components to ensure proper classification and disposition within the license renewal evaluation. Generically, the process began when a license renewal team member identified an apparent discrepancy. A license renewal team member would then review the situation with a system and/or design engineer to obtain more information. If it still appeared that there was a component identified as SR that did not support an SR system function, it would be elevated to license renewal project supervision. If it could not be resolved at that point, or if the plant database required a revision, the issue was entered into the CAP for resolution. For any resolution that required a change to a design document or the plant database, the design and/or configuration change process was used. Both of these processes required a review and approval of the change by an individual other than the preparer. The resolutions of these discrepancies were then fed back to the license renewal team member for proper incorporation into the scoping and screening process.

On the basis of the supplemental information provided by the applicant in response, including identification of the types of components that were re-classified, and a description of the process for evaluation and disposition of such components, the staff found that the applicant adequately addressed the RAI. Therefore, the staff's concern described in RAI 2.1-2 is resolved.

To provide additional assurance that the applicant adequately implemented its SR scoping methodology, the staff reviewed a sample of the license renewal scoping results for the FW/HPCI system and the reactor building (structural review), and discussed the methodology and results with the applicant's personnel who were responsible for these evaluations. The staff verified that the applicant had identified and used pertinent engineering and licensing information to identify the SSCs required to be within the scope of license renewal.

Conclusion. On the basis of this sample review, discussions with the applicant, review of the applicant's scoping process, and RAI responses, the staff determined that, the applicant's methodology for identifying systems and structures meets the scoping criteria of 10 CFR 54.4(a)(1) and is therefore adequate.

Application of the Scoping Criteria in 10 CFR 54.4(a)(2). In part, 10 CFR 54(a)(2) requires that the applicant consider all NSR SSCs whose failure could prevent satisfactory accomplishment of any of the functions identified in 10 CFR 54(a)(1)(i), (ii), or (iii) to be within the scope of license renewal.

By letters dated December 3, 2001, and March 15, 2002, the NRC issued its position to the NEI to provide staff expectations for determining what SSCs meet 10 CFR 54.4(a)(2). The December 3, 2001, letter provided specific examples of operating experience which identified pipe failure events (summarized in Information Notice (IN) 2001-09, "Main Feedwater System Degradation in Safety-Related ASME Code Class 2 Piping Inside the Containment of a Pressurized Water Reactor") and the approaches the staff considers acceptable to determine which piping systems should be included within the scope of license renewal based on 10 CFR 54.4(a)(2). The March 15, 2002, letter further described the staff's expectations for the evaluation of non-piping SSCs to determine which additional NSR SSCs are within the scope of license renewal. The position stated that applicants should not consider hypothetical failures, but rather should base their evaluation on the plant's CLB, engineering judgment and analyses, and relevant operating experience. The paper further described operating experience as all documented plant-specific and industry-wide experience that can be used to determine the plausibility of a failure. Documentation would include NRC generic communications and event reports, plant-specific condition reports (CRs), industry reports such as Significant Operating Experience Reports (SOERs), and engineering evaluations.

The applicant implemented the scoping and screening process in accordance with LRG-02, "License Renewal Scoping and Screening." Paragraph 3.4.1 of LRG-02 states that NSR SSCs whose failure could affect the satisfactory accomplishment of any SR functions were considered within the scope of license renewal. The procedure further specified the various NSR SSCs that were considered within the scope of license renewal, such as NSR features which protect SR SSCs from missiles; certain overhead handling systems; walls, curbs and dikes which provide flood barriers to SR SSCs; NSR whip restraints, jet impingement shields, and blowout panels which provide SR SSCs from the effects of a HELB; NSR piping attached to SR piping up to and including the first equivalent anchor; NSR piping in the vicinity of SR equipment; and supports. The applicant used the UFSAR/USAR, the plant component database, P&IDs, DBD source documents, Maintenance Rule documents, safety class boundary drawings, the CLB, and plant and industry operating experience to identify NSR SSCs for inclusion within the scope of license renewal.

As part of its evaluation of the 10 CFR 54.4(a)(2) criterion, the applicant prepared a topical report titled, "Scoping and Screening Aging Management Review NSR Piping (NSR Piping Report)," to document the review and evaluation performed to identify those SSCs which met 10 CFR 54.4(a)(2). To facilitate that evaluation, the applicant divided the potential NSR/SR interactions into four separate categories: NSR SSC Within the Vicinity of SR SSCs, NSR SSCs Attached to SR SSCs, NSR SSCs Providing Functional Support to SR SSCs, and Fail-Safe Components. Each category is discussed in detail below.

(1) NSR SSC Within the Vicinity of SR SSCs - The applicant's NSR Piping Report contains the rationale for inclusion of NSR piping attached to SR piping and NSR piping located within the vicinity of SR piping. The piping effects considered by the applicant included spray, flooding, pressure and temperature rise, pipe whip, and jet impingement. The applicant had utilized the preventative option as defined in NEI 95-10 and identified each structure or area containing SR SSCs and NSR SSCs. The applicant then identified all NSR piping systems located within areas containing SR SSCs which contained fluids. The applicant then removed the NSR SSCs which contained air, gas, or oil from within the scope of license renewal. The NSR SSCs that contained water or steam and were within the vicinity of SR SSCs were determined to be within the scope of license renewal.

Discussion with the applicant indicated that a conservative definition of "in the vicinity" had been defined during the scoping process as "within the building, corridor, or floor." In practice, the applicant had applied the scoping criteria to all NSR SSCs located within the same building as SR SSCs (buildings identified as SR). In addition, the applicant had provided an analysis, "Technical Basis for Materials-Environment Group Inputs to the ConRAD Database" (the data base for equipment and components for NMPNS license renewal project), which provided the basis for the exclusion of NSR oil-filled pipe within the vicinity of SR SSCs, from the scope of license renewal. The plant analysis indicated that there were no aging affects associated with oil-filled NSR piping systems based on both plant and industry-wide experience.

(2) NSR SSCs Attached to SR SSCs - The original LRA states that for NSR SSCs attached to SR SSCs, the scope of the NSR piping system was extended beyond the classification change to the first seismic anchor beyond the depicted class change. The applicant determined that the piping between the depicted classification boundary and the first seismic anchor was considered to be within the scope of license renewal. As a

result, for piping containing water or steam, the NSR portion within the scope of license renewal extended beyond the depicted class change until no longer in the vicinity of SR equipment or until the first seismic anchor was reached, whichever is furthest. Paragraph 3.4.2.7 of LRG-02 states that for NSR SCs directly attached to SR SSCs, the NSR piping and supports, up to and including the first equivalent anchor beyond the NSR/SR interface, were within the scope of license renewal.

During the audit, the applicant indicated that this approach had been implemented by considering all NSR piping components within a building containing SR SSCs as within scope of license renewal. However, the staff noted that the statement in the original LRA, "extended beyond the depicted class change until no longer in the vicinity of SR equipment or until the first seismic anchor is reached, whichever is furthest," had not been effectively implemented. The staff found that the applicant had not evaluated beyond the vicinity of the SR SSCs (outside the building) to verify the seismic anchor (or equipment acting as the seismic anchor) and had not verified that the appropriate anchor/equipment had been included within the scope of license renewal.

In RAI 2.1-4(a), dated November 22, 2004, the staff stated that during the audit, it was noted that in some cases where NSR plant equipment provided a termination point for NSR piping attached to SR piping, the NSR piping was placed within the scope of license renewal, but the plant equipment (such as a heat exchanger) was not considered to be within the scope of license renewal. For cases where an entire pipe run, including both SR and NSR piping, are analyzed as part of the CLB to establish that it could withstand DBE loads, the SRP-LR provides explicit scoping criteria. Specifically, SRP-LR Section 2.1.3.1.2 indicates that the scoping methodology include (1) the NSR piping up to its anchors, and (2) the associated piping anchors as being within the scope of license renewal under 10 CFR 54.4(a)(2). Because in some instances plant equipment was used as a termination point for the NSR piping within the scope of license renewal, this plant equipment appears to be equivalent to an associated piping anchor as described in SRP-LR.

Therefore, the staff requested that the applicant provide additional information regarding the SR/NSR interface evaluation as follows: the definition of equivalent anchor that was used for the purposes of the 10 CFR 54.4(a)(2) evaluation; the method used to identify the first seismic anchor for NSR pipe attached to SR pipe, within the scope of license renewal; confirmation that the NSR piping, associated plant equipment, and their supports, up to and including the first seismic anchor, were within the scope of license renewal and subject to aging management review; and how plant equipment identified as the termination point for NSR piping was evaluated during the scoping process.

In its letter dated December 22, 2004, as supplemented by letter dated July 14, 2005, the applicant described the revised scoping methodology and complete re-scoping effort that was applied to the 10 CFR 54.4(a)(2) criterion as a result of the questions resulting from the staff's methodology audit. As part of those responses, the applicant stated, in part, that the re-scoping was performed consistent with the guidance in NEI 95-10, Revision 5, except for those portions of the guidance with which the staff had taken exceptions. For those cases, the applicant's scoping methodology followed the staff's position rather than the NEI guideline. The applicant provided a revised Section 2.1.4.2 in its ALRA, which describes in detail the re-scoping effort associated with this criterion. As a result of the re-scoping effort, the applicant included all NSR SSCs that are within the boundaries of the equivalent anchor locations, including the equivalent anchors themselves. As part of the ALRA, the applicant defined the equivalent anchor for each unit consistent with the CLB for the plants; described the processes used to identify each equivalent anchor location (including review of plant drawings and performance of plant walk-downs), and ensured that all NSR SSCs within the boundaries up to and including the equivalent anchor were identified and included within the scope of renewal. The staff verified that the ALRA description was consistent with the prior response to the RAI and the results of the staff's audit of the scoping and screening methodology. On the basis of the supplemental information provided by the applicant in response to the staff's RAI, and the incorporation of that information into the ALRA, the staff's concern described in RAI 2.1-4(a) is resolved.

(3) NSR SSCs Providing Functional Support to SR SSCs - The staff determined that LRG-02, "License Renewal Scoping and Screening," paragraph 3.4.3.2, stated that malfunctions of NSR equipment that result in a challenge to SR equipment (where the SR function is maintained) is not within the scope of license renewal.

In RAI 2.1-4(b), dated November 22, 2004, the staff requested that the applicant provide the basis for this position and all applications of this position during the scoping process.

In its responses, by letter dated December 22, 2004, as supplemented by letter dated July 14, 2005, the applicant stated that its scoping methodology was revised and did not use the "NSR Safety Systems and Components (SSCs) which Functionally Interact with SR SSCs" criterion from NEI 95-10, Revision 4, to exclude from scope any NSR SSCs that could inhibit an SR SSC from performing its intended functions. As a result of this effort, NMP now includes all NSR SSCs that are within the boundaries of the equivalent anchors (including the equivalent anchor) within the scope of license renewal and subject to an AMR. The original LRA Section 2.1.4.2 was revised, as reflected in the ALRA, to describe the methodology used in the NMP NSR re-scoping effort. The staff verified that the ALRA description was consistent with the prior response to the RAI, and the results of the staff's audit of the scoping and screening methodology. On the basis of the supplemental information provided by the applicant in response to the staff's request for information, and the incorporation of that information into the ALRA, the staff's concern described in RAI 2.1-4(b) is resolved.

(4) Fail-Safe Components - In RAI 2.1-4(c), dated November 22, 2004, the staff stated that LRG-02, "License Renewal Scoping and Screening," paragraph 4.1.2, stated that fail-safe components are components whose failure (through interaction with the failed NSR SSC) cannot prevent the accomplishment of an SR function since the NSR SSC causes the SR SSC to attain a fail-safe state. Therefore, the staff requested that the applicant provide the basis for this position and all applications of this position during the scoping process.

In its responses by letter dated December 22, 2004, as supplemented by letter dated July 14, 2005, the applicant stated that based on the staff's audit, its scoping methodology was revised to ensure that all NSR SSCs within the vicinity of SR SSCs were included within the scope of renewal regardless of whether the SR SSC was active or passive. As a result, the applicant reviewed its scoping results and verified that it did not exclude any NSR SSCs with potential for interaction with SR SSCs based on the fail-safe logic. The staff verified that the ALRA description was consistent with the prior response to RAI 2.1-4(c), and the results of the staff's audit of the scoping and screening methodology. On the basis of the supplemental information provided by the applicant in response to the staff's request for information, and the incorporation of that information into the ALRA submittal, the staff found that the applicant has adequately addressed the staff's request for additional information. Therefore, the staff's concern described in RAI 2.1-4(c) is resolved.

To provide additional assurance that the applicant adequately implemented its NSR scoping methodology, the staff reviewed a sample of the license renewal scoping results for the FW/HPCI system. The staff verified that the applicant had identified and used pertinent engineering and licensing information to identify the SSCs required to be within the scope of license renewal in accordance with the 10 CFR 54.4(a)(2) criteria.

On the basis of the sample review, discussions with the applicant, and review of the applicant's scoping process, the staff determined that the applicant's methodology for identifying systems and structures meeting the scoping criteria of 10 CFR 54.4(a)(2) was adequate.

Application of the Scoping Criteria in 10 CFR 54.4(a)(3). In part, 10 CFR 54(a)(3) requires that the applicant consider all SSCs relied on in safety analyses or plant evaluations to perform a function that demonstrates compliance with the regulations for fire protection, EQ, PTS, ATWS, and SBO to be within the scope of license renewal.

The applicant documented its methodology for performing the scoping of SSCs in accordance with 10 CFR 54.4(a)(3) in implementation procedures LRG-01 and LRG02 and the NERs developed by the applicant for certain regulated events for the applicable NMP unit.

The applicant performed the initial scoping for regulated events by evaluating CLB information relevant to each regulated event to identify whether the structure or system met the scoping criteria of 10 CFR 54.4(a)(3). For ATWS and SBO, the applicant developed an NER describing the relevant 10 CFR Part 54 requirements, a functional description of the implementation of that requirement at the NMPNS, specific information regarding systems and components credited for the event, the process to identify the scoping boundaries associated with the systems credited, the intended functions applicable to the requirement, information on how to record the results of the evaluation in the license renewal database and appropriate MEL, a list of CLB information sources used for the analysis, and a list of systems and components determined to be within scope for the given regulated event.

By letter dated April 1, 2002, the staff provided guidance on the scoping of equipment relied on to meet the requirements of the SBO rule, 10 CFR 50.63. In this letter, the staff noted that, consistent with the requirements specified in 10 CFR 54.4(a)(3) and 10 CFR 50.63(a)(1), the plant system portion of the offsite power system that is used to connect the plant to the offsite power source should be included within the scope of the SBO rule. In the original LRA and ALRA Section 2.1.4.3.5, the applicant stated that based on the guidance in the April 1, 2002, letter for SBO recovery, an additional evaluation was performed at NMP to determine, and bring within the scope of license renewal, components credited for recovery of the offsite power system. For each of the systems credited for SBO recovery, a scoping/screening report was developed. Additionally, an AMR was performed for all long-lived, passive structures and components within these systems. The scoping effort identified structures and components of the offsite power system for each plant required to restore power from the onsite switchyard down to the SR busses in the plant. The applicant also stated that the plant offsite power system and these structures and components were classified as satisfying the criteria in

10 CFR 54.4(a)(3) and were included within the scope of license renewal. The staff determined that the applicant's approach to scoping SSCs relied on to demonstrate compliance with the SBO rule was consistent with the staff's April 1, 2002, interim guidance.

For EQ, the master list of EQ components is detailed in each unit's MEL. Systems that contain components identified in the EQ MEL, as defined by 10 CFR 50.49, are within the scope of license renewal.

For fire protection, NMP1 UFSAR Sections X.10A, "Fire Hazards Analysis;" X.10B, "Safe Shutdown Analysis;" and X.K, "Fire Protection Program;" and NMP2 USAR Section 9.5.1, "Fire Protection Systems," describe the station fire protection and post-fire safe shutdown equipment. Fire protection, detection, mitigation, confinement, and safe shutdown equipment used at the station were reviewed during the scoping process.

Evaluations were performed on equipment needed to meet the fire protection requirements of Appendix A to Branch Technical Position APCSB 9.5-1, "Guidelines for Fire Protection for Nuclear Power Plants," as well as those needed to meet 10 CFR 50, Appendix R and 10 CFR 50.48. These evaluations were used as fire protection scoping basis documents. Structures and systems that contain components relied on to protect SR structures and components and equipment required to mitigate offsite release from a fire or explosion are within the scope of license renewal.

SRP-LR, Section 2.1.3.1.3, "Regulated Events," states that all SSCs that are relied upon in the plant's CLB (as defined in 10 CFR 54.3), plant-specific experience, industry-wide experience (as appropriate), and safety analyses or plant evaluations to perform a function that demonstrates compliance with NRC regulations identified under 10 CFR 54.4(a)(3), are required to be included within the scope of the 10 CFR Part 54. As part of the original LRA review, the staff evaluated the scope and depth of the applicant's document review to provide assurance that the scoping methodology considered all SSC intended functions.

During the scoping and screening methodology audit, the applicant identified several technical position papers as a documentation source for license renewal scoping under 10 CFR 54.4(a)(3). In reviewing the original LRA, scoping and screening implementation procedures, and evaluation of the feedwater system during the audit, the staff was informed by the applicant that two technical position papers (ATWS and SBO) had not been adequately reviewed and incorporated into the original LRA during its verification activities. This discrepancy was identified by the applicant during the audit and documented in CR-NM-2004-4466, dated September 30, 2004. CR-NM-2004-4466 states that the original LRA Section 2.3.4.B.3 is incomplete because it does not reference an SBO event in the description of why components in the NMP2 feedwater system are within the scope of license renewal. CR-NM-2004-4466 also states that an extent of condition review is necessary to determine whether there are similar instances affecting other system descriptions in the original LRA.

RAI 2.1-5, dated November 22, 2004, requested that the applicant describe the methodology used to develop technical position papers and specifically describe the actions taken to ensure that both NMP1 and NMP2 license renewal scoping and screening reports adequately address the new ATWS and SBO DBDs, as well as any potentially affected the original LRA sections.

In its response, by letter dated January 31, 2005, the applicant stated, in part, that the technical position papers used at NMP are controlled in accordance with engineering administrative procedure NEP-DES-02, "Engineering Evaluations." The general methodology employed involves the preparation of the engineering evaluation, a technical review or design verification, and approval by the responsible supervisor. The preparer is directed to "Perform the evaluation and document in sufficient detail to allow a technically qualified reviewer/design verifier to understand the purpose, inputs, evaluation criteria, assumptions, method, references, and conclusions of the evaluation, and to conclude adequacy without recourse to the originator." Design verification is required when the evaluation involves SR systems, structures, or components. The evaluation is documented as an NER. This administrative procedure also applies to the review and acceptance of vendor-supplied documents. These documents also require a review and approval by NMP prior to use. However, as noted above, these two technical position papers (ATWS and SBO) were not reviewed and approved for use. It is this error that led to the apparent discrepancy with the feedwater system described above.

In response to the finding, the applicant performed additional reviews to ensure that all the required NMP2 SCs were properly identified within the scope of license renewal and to compare the systems listed in the original LRA to those identified in the approved engineering reports. As a result of this comparison, the applicant determined that the original LRA Section 2.3.4.B.3, "NMP2 Feedwater System," did not need to be identified as within scope for the SBO regulated event since the feedwater components credited for SBO (reactor coolant/containment isolation valves) were already properly included in the original LRA Section 2.3.2.B.5, "NMP2 Primary Containment Isolation System." This system includes the reactor coolant/containment isolation valves for all systems and is properly credited for being within scope for the SBO regulated event. Therefore, the applicant determined that the apparent discrepancy identified during the audit was determined to be incorrectly characterized.

However, during the review to ensure all systems were properly identified for the ATWS and SBO regulated events, the applicant discovered that the NMP2 common electrical system should have been identified as within scope for the SBO regulated event. Therefore, ALRA Section 2.5.B.4 was revised to also include the SBO regulated event as a criterion for this system. On the basis of the supplemental description of the development and approval of technical position papers, and the review of the extent of condition of the apparent discrepancy, including the identification of the NMP2 common electrical system as within scope for the SBO regulated event, and the incorporation of that information into ALRA Section 2.5.B.4, the staff's concern described in RAI 2.1-5 is resolved.

The staff reviewed a sample of the license renewal database 10 CFR 54.4(a)(3) scoping results and discussed the methodology and results with the applicant's license renewal project personnel. From the discussion, the staff concludes that the applicant had identified and used pertinent engineering and licensing information to compile the SSCs required to be within scope in accordance with the 10 CFR 54.4(a)(3) criteria.

On the basis of the above review and discussions with the applicant, the staff determined that the applicant's methodology for identifying systems and structures meeting the scoping criteria of 10 CFR 54.4(a)(3) was adequate.

2.1.3.1.5 Plant-Level Scoping of Systems and Structures.

The applicant documented its methodology for performing the scoping of SSCs in accordance with 10 CFR 54.4(a) in implementation procedures LRG-01 and LRG-02. The applicant's approach to system and structure scoping was consistent with the methodology described in the original LRA and ALRA Section 2.1.2. Specifically, LRG-02 specified that the personnel performing license renewal scoping use CLB documents and describe the system or structure including a list all functions that the system or structure is required to accomplish. Sources of information regarding the CLB for systems included the USAR, DBDs, MEL database, Maintenance Rule scoping reports, control drawings, and docketed correspondence. The applicant then compared identified system or structures function lists to the scoping criteria to determine whether the functions met the scoping criteria of 10 CFR 54.4(a). The applicant documented the results of the plant-level scoping process in accordance with Section 3.2.3 to LRG-02. The database information included a description of the structure or system, a listing of functions performed by the system or structure, information pertaining to system realignment (as applicable), identification of intended functions, the 10 CFR 54.4(a) scoping criteria met by the system or structure, references, and the basis for the classification of the system or structure intended functions. During the scoping methodology audit, the staff reviewed a sampling of scoping reports and concludes that the applicant's scoping results in the LR database and scoping results reports contained an appropriate level of detail to document the scoping process.

Conclusion. On the basis of a review of the original LRA and ALRA, the scoping and screening implementation procedures, and a sampling review of system and structure scoping results during the methodology audit, the staff concludes that the applicant's scoping methodology for systems and structures was adequate. In particular, the staff determined that the applicant's methodology reasonably identified systems and structures within the scope of license renewal and their associated intended functions.

2.1.3.1.6 Component-Level Scoping

After the applicant had identified systems and structures within the scope of license renewal and their associated intended functions, a review was performed to identify the components of each system and structure within the scope of license renewal that supported an intended function. As described in the original LRA and ALRA Section 2.1.5, a component is considered to be within the scope of license renewal if it fulfills a system intended function.

Mechanical Component Scoping. The original LRA and ALRA Section 2.1.5.1, "Mechanical Systems," and LRG-02 Section 3.6, "Component Scoping and Screening," provided the applicant's proceduralized guidance for scoping mechanical system components. To identify system components required to perform a system intended function, the applicant initially generated a listing of mechanical system components based on information derived from controlled system diagrams and the MEL. Procedure LRG-02 discusses in detail how to (1) determine system boundaries, (2) indicate components within a specific flow path that are required for performance of intended functions, and (3) determine and identify system and interdisciplinary interfaces (e.g., mechanical/structural, mechanical/electrical, structural/electrical). The staff reviewed the results of the boundary evaluation and discussed the process further with the applicant. The staff verified that mechanical system evaluation boundaries were established for each system within the scope of license renewal. These boundaries were determined by mapping the pressure boundary associated with system-level license renewal intended functions onto the controlled system drawings. The applicant included the mechanical component types in the scoping and screening database and the applicant performed further review was performed to ensure all component types were identified. If a component type was not already in the MEL, the component type was created for use in the license database. A preparer and an independent reviewer performed a comprehensive evaluation of the boundary drawings to ensure the completeness and accuracy of the review results.

The staff conducted detailed discussions with the applicant's license renewal project management personnel and reviewed documentation pertinent to the scoping process. The staff assessed whether the applicant had appropriately applied the scoping methodology outlined in the original LRA and implementation procedures and whether the scoping results were consistent with CLB requirements. The staff determined that the applicant's proceduralized methodology was consistent with the description provided in the original LRA and ALRA Section 2.1.5.1 and the guidance contained in SRP-LR, Section 2.1, and was adequately implemented.

The staff reviewed the process of scoping for the FW/HPCI system. The staff verified that the applicant had identified and highlighted system P&IDs to develop the system boundaries in accordance with the procedural guidance. The applicant was knowledgeable about the process and conventions for establishing boundaries as defined in the license renewal implementation procedures. Additionally, the staff verified that the applicant had independently verified the results in accordance with the governing procedures. Specifically, other LR personnel knowledgeable about the system had independently reviewed the marked-up drawings to ensure accurate identification of system intended functions. The applicant performed additional cross-discipline verification and independent reviews of the resultant highlighted drawings before final approval of the scoping effort.

On the basis of the above staff review regarding the applicant's detailed scoping implementation procedures and a sampling review of mechanical components scoping results for the FW/HPCI system, the staff concludes that the applicant's methodology for identifying mechanical components within the scope of license renewal met the requirements of 10 CFR 54.4(a).

Structural Component Scoping. The applicant performed its structural scoping in accordance with the methodology defined in CNS procedure LRG-02, "LR Scoping and Screening." The procedure describes the source design documentation to be used for the evaluation of structures and is used to evaluate plant structures and to determine their functions. UFSAR/USAR, Maintenance Rule scoping results, design and license basis documents, regulatory requirements, the MEL, 10 CFR 50 Appendix B determinations, and plant drawings were reviewed. From this review, a scoping report for each plant structure was developed. The scoping report describes the functions for each structure and indicates the applicable 10 CFR 50.54(a)(1)-(3) criteria. Sections 2.4.A, 2.4.B, and 2.4.C of the original LRA and ALRA provide a complete plant-specific list of structures within the scope of license renewal.

The staff conducted detailed discussions with the applicant's license renewal project management personnel and reviewed documentation pertinent to the scoping process. The staff assessed whether the applicant had appropriately applied the scoping methodology

2-24

outlined in the original LRA and implementation procedure and whether the scoping results were consistent with CLB requirements. Component supports, and fire stops and seals were binned in separate structural commodity groupings. The staff reviewed scoping reports for the NMP1 reactor building and the NMP1 materials handling/heavy loads. In general, the staff determined that the applicant's overall approach to license renewal structural scoping was adequate.

The staff reviewed the scoping procedure, discussed the structural scoping methodology with the applicant's cognizant engineers, and reviewed several plant structural scoping reports to verify proper implementation of the scoping process for SCs. The staff determined that the applicant's proceduralized scoping methodology was consistent with the description provided in Section 2.1.4 of the original LRA and ALRA and the guidance contained in SRP-LR Section 2.1. Based on these audit activities, the staff did not identify any discrepancies between the methodology documented and the implementation results.

On the basis of a review of information contained in the original LRA and ALRA, the applicant's scoping implementation procedure, and a sampling review of SC scoping reports, the staff concludes that the applicant's methodology for identifying SCs within the scope of license renewal met the requirements of 10 CFR 54.4(a).

Electrical and I&C Component Scoping. SRP-LR Section 2.5.3.1, "Components Within the Scope of License Renewal," states that an applicant may use the plant spaces approach in scoping electrical and I&C components. In the plant spaces approach, an applicant may indicate that all electrical and I&C components located within a particular area are either within or not within the scope of license renewal. The applicant did not choose the typical electrical and I&C scoping approach, using instead an approach similar to that used for mechanical systems and structures.

The staff reviewed NMP procedures LRG-01, "License Renewal Project Guidance," Revision 2, and LRG-02, "License Renewal Scoping and Screening," and determined that adequate guidance was provided to the engineers performing the electrical and I&C license renewal scoping process. SSCs were evaluated to determine whether they were within the scope of license renewal using NMP licensing and design-basis information and regulatory requirements. System descriptions were developed and intended functions were identified and documented in ConRAD. Additionally, SSCs were evaluated to determine whether they provided a license renewal intended function. ConRAD was updated to reflect these conclusions. In unique cases, such as regulated events, NERs were developed to identify components required to support these events. Because these documents were developed late in the project, management deferred entering these components into ConRAD until the annual update. The majority of these components have been entered into the NMP MEL. The components were evaluated by the electrical license renewal engineers, and passive, long-lived components were evaluated in an AMR. This essentially completed the scoping process per LRG-02.

As part of the review, the staff noted that electrical commodities (i.e., cables, connectors, non-segregated bus, electrical penetrations, etc.) were identified and addressed separately from the electrical system scoping evaluations. LRG-01, Section 3.2.2.2.1, stated that commodities are groupings of components that perform the same intended functions and may be associated with many plant systems and structures. Standard groupings of electrical commodities have been well established by prior license renewal applicants and emboddied in the industry guidance on the preparation of the LRAs. A separate guideline, LRG-04, "Aging Management Review of Electrical Commodities," was developed by the applicant to govern the evaluation of electrical commodities with respect to aging effects and management of those effects.

The applicant conducted a search of cable design and procurement specification documents, contracts, plant modification packages, controlled drawings, the plant equipment database, and the electrical cable database (TRAC 2000) to February 21, 2006, identify all components required to perform license renewal intended functions. The staff discussed the electrical scoping methodology with the applicant's cognizant engineers, and reviewed several plant electrical packages to verify proper implementation of the scoping process for electrical components. The staff.also compared a sample of electrical components identified in the documentation to the electrical commodity list in the license renewal database to ensure consistency.

In RAI 2.1-6, dated November 22, 2004, the staff stated that during the audit it noted that the applicant's engineering staff had an adequate understanding of the process used to scope electrical and I&C components. However, the staff did identify an issue regarding the level of detail in the associated procedures describing the scoping process. Specifically, the staff was unable to determine the specific activities performed by the applicant's staff to identify the applicable intended functions, plant electrical equipment required to perform those functions, and subsequent development of the electrical commodity list from which the aging management reviews were conducted. Therefore, the staff requested that the applicant provide a detailed description of the methodology used for the scoping and screening of electrical and I&C components.

In its response, by letter dated December 22, 2004, the applicant addressed the staff's request and provided (1) a detailed description of the process used to identify the intended functions, (2) equipment necessary to perform those functions, and (3) development of the electrical commodities evaluated as part of the AMR. Specifically, the applicant clarified that the methodology used to determine whether an electrical or I&C component supported an intended function is described in project procedure LRG-02. Section 3.7 of LRG-02 requires that the electrical or functional boundary be described for the intended functions of electrical systems. This activity identifies a group of components that support a specific intended function. For example, all electrical and I&C components that are identified on the EQ list for a system are the group of components that support the EQ intended function. Additionally, the applicant described the process used to identify intended functions. Specifically, electrical systems were identified based upon those defined in the MEL, UFSAR/USAR, and Maintenance Rule scoping reports. The boundaries of each electrical system are based upon the components assigned to the system as well as any descriptions in the UFSAR/USAR and/or other DBD. The MEL was used as the design document/database that assigned components to a particular system. The electrical systems and components defined in MEL were imported into the NMP license renewal database, ConRAD. The information contained in ConRAD for each electrical system included a system description, list of system functions, identification of which functions met any of the license renewal scoping criteria, a list of NMP documents from which this information was derived, and any corresponding comments.

The staff found the above response acceptable because the applicant provided additional details regarding the process for identifying intended functions and those components necessary to perform those functions. Therefore, the staff's concern described in RAI 2.1-6 is resolved.

On the basis of a review of information contained in the original LRA and ALRA, the applicant's detailed scoping implementation procedures, and a sampling review of electrical commodity scoping results, the staff concludes that the applicant's methodology for identifying electrical commodities within the scope of license renewal met the requirements of 10 CFR 54.4(a).

2.1.3.2 Screening Methodology

The staff reviewed the methodology used by the applicant to determine whether mechanical, structural, and electrical and I&C components within the scope of license renewal would be subject to further aging management review. The applicant provided the staff with a detailed discussion of the processes used for each discipline and provided administrative documentation that described the screening methodology. The staff also reviewed the screening results reports for the FW/HPCI system and reactor building. The staff noted that the applicant's screening process was performed in accordance with its written requirements and was consistent with the guidance provided in the staff's SRP-LR and NEI 95-10, Revision 5. The staff determined that the screening methodology was consistent with the requirements of 10 CFR Part 54, and that the screening methodology will identify SCs that meet the screening criteria of 10 CFR 54.21(a)(1).

The staff reviewed the screening methodology used by the applicant to determine whether mechanical, structural, and electrical components within the scope of license renewal would be subject to further aging management evaluation. The applicant described its screening process in the original LRA and ALRA Section 2.1.5. In general, the applicant's screening approach consisted of evaluations to determine which in-scope structures and components were passive and long-lived. Passive, long-lived structures and components were then subject to further AMR.

The staff evaluated the applicant's screening methodology against the criteria contained in 10 CFR 54.21(a)(1) and (2), using the review guidance contained in SRP-LR Section 2.1.3.2, "Screening." According to 10 CFR 54.21(a)(1), the applicant's IPA must identify and list those SCs subject to an AMR. Further, 10 CFR 54.21(a)(1) requires that SCs subject to an AMR shall encompass those structures and components that (1) perform an intended function, as described in 10 CFR 54.4, without moving parts or a change in configuration or properties, and (2) are not subject to replacement based on a qualified life or specified time period. Per 10 CFR 54.21(a)(2), the applicant must describe and justify the methods used to meet the requirements of 10 CFR 54.21(a)(1). In the original LRA and ALRA, the applicant described screening methodologies that were unique to the mechanical, structural, and electrical disciplines. The following sections describe the staff evaluation of the applicant's screening approach for each of these disciplines.

2.1.3.2.1 Mechanical Component Screening

The staff reviewed the methodology used by the applicant to determine whether mechanical components within the scope of license renewal would be subject to further AMR. For mechanical components, the applicant applied a screening process to each mechanical system determined to be within the scope of license renewal in order to determine the types of mechanical component commodities within the systems and the various materials and environments to be considered in the AMR. The applicant then established evaluation boundaries for the various plant mechanical systems, in order to further identify individual mechanical components for review.

The listing of mechanical components was facilitated by combining these items into commodity groups from a review of each boundary drawing. The applicant placed these commodity groups into the license renewal database and evaluated them in accordance with the screening criteria described in LRG-02. The applicant provided the staff with a detailed discussion of the process and provided screening report information from the license renewal database that described the screening methodology, as well as a sample of the screening results reports for a selected group of SR and NSR systems. The staff determined that the screening methodology was consistent with the requirements of 10 CFR Part 54 and that implementation of the methodology will identify SCs that meet the screening criteria of 10 CFR 54.21(a)(1).

During the audit, the staff reviewed the methodology used by the applicant to identify and list the mechanical components and commodities subject to an AMR, as well as the applicant's technical justification for this methodology. The staff discussed the methodology and results with the applicant's cognizant engineers and senior staff. The staff also examined the applicant's results from the implementation of this methodology by reviewing the FW/HPCI system identified as within the scope of license renewal. The review included the evaluation boundaries and resultant in-scope components, the corresponding component-level intended functions, and the resulting list of mechanical components and commodity groups subject to an AMR.

The staff reviewed several summary screening reports that list a breakdown of the mechanical components within the scope of license renewal. Each report lists several categories, including component type, whether an AMR was required, material, and an extensive comment section. The staff also reviewed a sample of the mechanical drawing packages assembled by the applicant and discussed the process and results with the cognizant engineers who performed the review. The staff did not identify any discrepancies between the methodology documented and the implementation results.

Conclusion. On the basis of a review of the original LRA and ALRA, the scoping and screening implementation procedures, and a sampling review of system and screening results, the staff determined that the applicant's mechanical component screening methodology was consistent with the guidance contained in the SRP-LR and was capable of identifying those passive, long-lived components within the scope of license renewal that are subject to an AMR.

2.1.3.2.2 Structural Component Screening

The staff reviewed the methodology used by the applicant to determine whether structures within the scope of license renewal would be subject to further AMR. For structures, the applicant determined the types of structural elements utilized and the various materials and environments to be considered in the AMR. Generally, the boundary for a structure is the entire building including base slabs, foundations, walls, beams, slabs, and steel superstructure. A listing of all the SCs that exist in each plant structure was developed identifying the various types of structural elements, materials, and environments. The applicant created a database to compile the results. The database identifies each individual SCs and indicates whether the SC is subject to AMR. Each SC is identified as a component (e.g., door, gate, anchor support, strut, fastener, or siding) or as a material (e.g., concrete, polymer, or steel). From this review a screening report for each plant structure was developed.

The listing of structural elements was facilitated by placing component supports, and fire stops and seals in separate commodity groups. The applicant provided the staff with a detailed discussion describing the screening methodology, as well as the screening reports for a selected group of structures. The staff determined that the screening methodology was consistent with the requirements of 10 CFR Part 54 and that implementation of the methodology will identify SCs that meet the screening criteria of 10 CFR 54.21(a)(1).

During the audit of the applicant's license renewal screening process, the staff reviewed the methodology used by the applicant to identify and list the SCs and structural commodities subject to an AMR, as well as the applicant's technical justification for this methodology. The staff discussed the methodology and results with the applicant's cognizant engineers and senior staff. The staff also examined the applicant's results from the implementation of this methodology by reviewing a sample of NMP1 plant structures identified as being within the scope of license renewal. The review included the evaluation of in-scope components, the corresponding component-level intended functions, and the resulting list of SCs and structural commodity groups subject to an AMR.

The staff reviewed several screening reports that list a breakdown of the SCs within the scope of license renewal. The reports reviewed by the staff included those for the NMP1 reactor building and NMP1 materials handling/heavy loads. The staff also discussed the process and results with the applicant. The staff did not identify any discrepancies between the methodology documented and the implementation results.

Conclusion. On the basis of a review of the original LRA and ALRA, the scoping and screening implementation procedures, and a sampling review of structural screening results, the staff determined that the applicant's SC screening methodology was consistent with the guidance contained in the SRP-LR and was capable of identifying those passive, long-lived components within the scope of license renewal that are subject to an AMR.

2.1.3.2.3 Electrical and I&C Component Screening

The staff reviewed the methodology used by the applicant to determine whether electrical components within the scope of license renewal would be subject to further AMR. For electrical components, the applicant applied a screening process by identifying electrical commodities within electrical systems. The LRA engineers identified all electrical and I&C component types in use at NMPNS based on the listing provided by Appendix B to NEI 95-10, NUREG-1801, "Generic Aging Lessons Learned (GALL) Report," Electric Power Research Institute (EPRI) Electrical Handbook, and from a review of plant documents, controlled drawings, the plant equipment database, and cable database. All passive, long-lived electrical components were evaluated as commodities regardless of the system or structure in which they reside in the MEL. As a result, the electrical systems only contain active components that are not subject to AMR. An AMR was then conducted on a commodity basis for the entire population of passive, long-lived components. Identification of individual components that perform intended functions was not performed. The passive electrical and I&C component commodity groups at NMPNS was based on a review of the UFSAR/USAR, the MEL, DBDs, previous LRAs, and NEI 95-10.

The applicant's list of electrical and I&C commodity groups included cables and connectors (including splices, connectors, terminal blocks, and fuse holders); non-segregated/switch yard bus; containment electrical penetrations; and switchyard components.

The interface of electrical and I&C components with other types of components, and the assessments of these interfacing components, are provided in the appropriate mechanical or civil structural sections. For example, the assessment of electrical racks, panels, frames, cabinets, cable trays, conduits, and their supports is provided in the civil/structural assessment section of the original LRA.

Components with unique identification numbers in the MEL that are identified as part of a system, but are defined as part of a commodity, are not addressed as part of the system. They do not appear on the list of components for that system in the system scoping and screening report. Commodities are treated generically, and a list of unique identification numbers from the MEL that make up a commodity is not provided unless noted otherwise. In this way, components are moved from their actual systems to commodity groups.

The staff discussed the methodology and results with the applicant's cognizant engineers and senior staff. The staff also examined the applicant's results from the implementation of this methodology by reviewing several electrical/I&C commodity reports and samples from the license renewal database. The review verified that the applicant's staff had consistently applied the screening criteria to identify those electrical/I&C commodity groups subject to an AMR. The staff determined that the NMPNS electrical screening process was consistent with criteria in 10 CFR 54.21(a)(1)(ii) and excluded those components or commodity groups that are subject to equipment qualification requirements. The staff did not identify any discrepancies between the methodology documented and the implementation results.

The staff also reviewed the applicant's approach to scoping and screening of electrical fuse holders. In license renewal ISG-5, "Identification and Treatment of Electrical Fuse Holders for License Renewal," dated March 10, 2003, the staff stated that, consistent with the requirements specified in 10 CFR 54.4(a), fuse holders (including fuse clips and fuse blocks) are considered to be passive electrical components. Fuse holders would be scoped, screened, and included in the AMR in the same manner as terminal blocks and other types of electrical connections that are currently being treated in the process. This staff position applies only to fuse holders that are not part of a larger assembly, but support SR and NSR functions in which the failure of a fuse precludes a safety function from being accomplished (10 CFR Part 54.4(a)(1) and(2)). As described in the original LRA and ALRA Section 2.1.6.5, "Identification and Treatment of Electrical Fuse Holders for License Renewal," fuse holders (including fuse clips and fuse blocks) are passive, long-lived electrical components that are within the scope of license renewal and subject to an AMR as part of the cables and connections commodity. Additionally, NMPNS credits the Fuse Holder Inspection Program for identifying potential age-related degradation for fuse holders. The staff determined that this was consistent with the ISG.

Conclusion. On the basis of a review of the original LRA and ALRA, the scoping and screening implementation procedures, and a sampling review of electrical system screening results, the staff determined that the applicant's electrical and I&C screening methodology was consistent with the guidance contained in SRP-LR and was capable of identifying passive, long-lived components within the scope of license renewal that are subject to an AMR.

2.1.3.2.4 Consumables

Paragraph 3.1.2.4 of procedure LRG-01, "License Renewal Project Guidance," Revision 2, discusses consumables. Paragraph 3.1.2.4.2 states that structural sealants should be identified as subcomponents, and if they are determined to perform an intended function in support of a larger structure, they must be within the scope of license renewal and subject to AMR. The staff reviewed the screening report for the NMP1 reactor building and noted that structural sealants (e.g., neoprene, calking, and urethane) were identified as a component within the scope of license renewal and subject to an AMR.

2.1.3.2.5 Plant Insulation

The staff's review of the original LRA Section 2.1 identified an area in which additional information was necessary regarding plant insulation to complete the review of the applicant's scoping and screening results. The applicant responded to the staff's RAI as discussed below.

In RAI 2.1-7, dated November 22, 2004, the staff stated that during the audit the applicant was unable to adequately describe the evaluation performed to determine whether any insulation installed in the plant was required to support any system intended functions identified during the scoping process. Therefore, the staff requested that the applicant describe any intended functions performed by insulation or the basis for determining that insulation (e.g. piping insulation) did not meet the scoping criteria described in 10 CFR 54.4(a)(1), (a)(2) or (a)(3).

In its response, by letter dated January 31, 2004, the applicant stated, in part, that an evaluation of thermal insulation used at NMP1 and NMP2 was performed to determine whether plant insulation was credited for performing any license renewal functions per 10 CFR 54.4(a)(1), (2), or (3). The applicant also provided a discussion of each scoping criteria and an evaluation of plant insulation with respect to each. Based on this review, the only intended function to meet the license renewal scoping criteria was fire wrap, used for fire protection, which meets 10 CFR 54.4(a)(3) and is included within the scope of license renewal. Specifically, these structural steel fire protection coatings are within the scope of license renewal and subject to an AMR. They are included as component type, "Fire Wrap in Air," in ALRA Table 2.4.C.2-1. The AMR of the fire wrap is addressed in ALRA Section 3.5.2.C.2 and Table 3.5.2.C-2.

Conclusion. On the basis of the supplemental information provided by the applicant which describes the analysis of plant insulation in response to RAI 2.1-7, and the incorporation of that information into the ALRA submittal, the staff found that the applicant has adequately addressed the staff's concern.

2.1.4 Evaluation Findings

The staff's review of the information presented in the original LRA and ALRA Section 2.1, the supporting information in the scoping and screening implementation procedures, calculations and reports, and the information presented during the scoping and screening audit formed the basis of the staff's safety determination. The staff verified that the applicant's scoping and screening methodology was consistent with the requirements of 10 CFR Part 54. On the basis of this review, the staff concludes that there is reasonable assurance that the applicant's methodology for identifying the SSCs within the scope of license renewal and the structures and components requiring an AMR is consistent with the requirements of 10 CFR 54.4 and 10 CFR 54.21(a)(1).

2.2 Plant-Level Scoping Results

2.2.1 Introduction

In ALRA Section 2.1, the applicant described the methodology for identifying the NMPNS SSCs within the scope of license renewal. In ALRA Section 2.2, the applicant used the scoping methodology to determine which of the SSCs are required to be included within the scope of license renewal. The staff reviewed the plant-level scoping results to determine whether the applicant had properly identified all plant-level systems and structures relied upon to mitigate DBEs, as required by 10 CFR 54.4(a)(1), or whose failure could prevent satisfactory accomplishment of any of the SR functions, as required by 10 CFR 54.4(a)(2), as well as the systems and structures relied on in safety analyses or plant evaluations to perform a function required by one of the regulations referenced in 10 CFR 54.4(a)(3).

2.2.2 Summary of Technical Information in the Amended Application

In ALRA Tables 2.2.-1 and 2.2-2, the applicant provided a list of the plant systems, structures, and commodities for NMP1 and NMP2, identifying those systems, structures, and commodities that are within the scope of license renewal. Based on the DBEs considered in the plant's CLB, other CLB information relating to NSR systems and structures, and certain regulated events, the applicant identified those plant-level systems and structures that are within the scope of license renewal, as defined by 10 CFR 54.4.

In the ALRA Section 2 tables that identify the component types requiring an AMR for the various systems, the applicant, on several occasions, listed "NSR Piping, Fittings, and Equipment" as a component type. This component type was introduced to incorporate the results from 10 CFR 54.4(a)(2) scoping, and it was described in the system description sections as "NSR Piping Fittings and Equipment Containing Liquid" in the buildings that were identified in each ALRA section. The SSCs making up this component type thus varied from system to system.

In the ALRA, the applicant revised the methodology used to determine the NSR SSCs that are within the scope of license renewal in accordance with the requirements of 10 CFR 54.4(a)(2). The applicant revised the LRA sections and tables where applicable to identify each NSR system or NSR portion of an SR system that is within the scope of license renewal. In conjunction with this change, the applicant also identified the specific NSR component types and intended function(s) and made them consistent with the standardized list of intended functions in SRP-LR and NEI 95-10. The component type, "NSR Piping, Fittings, and Equipment," and its associated intended function of, "Prevent Failure from Affecting SR Equipment," is no longer used in the NMP original LRA, and this change is reflected in the applicable ALRA sections.

In the ALRA, the applicant also revised LRA Section 2.1.4.2, "Non-Safety Related Criteria Pursuant to 10 CFR 54.4(a)(2)," to provide a detailed description of the NSR scoping criteria. As a result of the staff screening and methodology audit, the applicant implemented a revised spatial methodology in addressing systems meeting 10 CFR 54.4(a)(2), 4 mechanical systems for NMP1 and 10 mechanical systems for NMP2, that were previously identified in the original LRA Tables 2.2-1 and 2.2-2 as not being within scope were brought within the scope of license renewal. In addition, three mechanical systems for NMP2 that were previously identified in the original LRA Table 2.2-2 as within scope were deleted from the scope of license renewal.

2.2.3 Staff Evaluation

In ALRA Section 2.1, the applicant described its methodology for identifying the systems, structures, and commodities that are within the scope of license renewal and subject to an AMR. The staff reviewed the scoping and screening methodology and provided its evaluation in SER Section 2.1. To verify that the applicant properly implemented its methodology, the staff focused its review on the implementation results, as shown in ALRA Tables 2.2-1, "NMP1 Plant Level Scoping Results," and 2.2-2, "NMP2 Plant Level Scoping Results," to confirm that there were no omissions of plant-level systems and structures within the scope of license renewal.

The staff determined whether the applicant properly identified the systems and structures within the scope of license renewal in accordance with 10 CFR 54.4. The staff reviewed selected systems and structures that the applicant did not identify as falling within the scope of license renewal to verify whether the systems and structures have any intended functions that would require their inclusion within the scope of license renewal. The staff's review of the applicant's implementation was conducted in accordance with the guidance described in SRP-LR Section 2.2, "Plant-Level Scoping Results."

The staff sampled the contents of the UFSAR/USAR based on the systems, structures, and commodities listed in ALRA Tables 2.2-1 and 2.2-2 to determine whether there were systems or structures that may have intended functions within the scope of license renewal, as defined by 10 CFR 54.4, but were omitted from within the scope of license renewal.

In reviewing ALRA Section 2.2, the staff identified areas in which additional information was necessary to complete the evaluation of the applicant's plant-level scoping results. Therefore, the staff issued RAIs concerning each specific issue to determine whether the applicant properly applied the scoping criteria of 10 CFR 54.4 and the screening criteria of 10 CFR 54.21(a)(1). The following paragraphs describe the staff's RAIs and the applicant's related responses.

In RAI 2.2-1, dated November 19, 2004, the staff stated that during the original LRA review the staff identified license renewal drawings in multiple original LRA sections that all, or in part, appeared to conflict with the original LRA. The staff discussed the apparent discrepancies with the applicant to determine whether they were intentional or editorial in nature. The applicant identified a large number of the discrepancies as editorial and agreed that corrections to the original LRA or LR drawings would be required to correct the discrepancies.

In order to complete its review, the staff requested that the applicant correct the LR drawings for the following original LRA sections in which the apparent discrepancies were identified:

2.3.3.A.4	2.3.3.A.8	2.3.3.A.16	2.3.3.A.17	2.3.3.A.20	2.3.3.A.21
2.3.3.A.23	2.3.3.B.1	2.3.3.B.13	2.3.3.B.14	2.3.3.B.15	2.3.3.B.21
2.3.3.B.25	2.3.3.B.27	2.3.3.B.29	2.3.3.B.30	2.3.3.B.31	2.3.4.A.5
2.3.4.B.2					

The staff also requested that the applicant identify those LR drawings that have been corrected and the corrections made to the drawings.

In its response, by letter dated December 22, 2004, the applicant stated that for each of the original LRA sections identified in RAI 2.2-1 answers have been provided to each staff's specific question from those original LRA sections that address drawing issues. The responses to those specific RAIs identify where there are drawing anomalies and whether a change to the original LRA was required. It is the applicant's understanding that the original LRA, the docketed LRA supplemental letters, and the docketed responses to staff RAIs serve as the bases for the results of the staff's review. The drawings that were submitted concurrent with, but separate from, the original LRA were provided as information-only aids to assist the NRC reviewers with their evaluations. The applicant did not intend them to be part of the formal application. Therefore, the applicant does not intend to revise these drawings and resubmit them as part of the original LRA review process.

The applicant further stated that it does plan, upon completion of the original LRA review and approval process, to update the LR drawings, the scoping and screening reports, the AMR reports, and the program basis documents, to be consistent with the content of the final staff safety evaluation. With the exception of the program basis documents, which will be controlled documents, the remaining documents, including the drawings, will not be controlled but will be archival documents maintained within the NMPNS documentation and drawing system for historical reference purposes.

Based on its review, the staff found the applicant's response to RAI 2.2-1 acceptable because the applicant has adequately addressed discrepancies associated with the identified LRA sections. The information to resolve these discrepancies were included in the response to applicable RAIs. Therefore, the staff's concern described in RAI 2.2-1 is resolved.

In RAI 2.2-2, dated November 19, 2004, the staff stated that during the original LRA review, the staff identified in multiple LRA sections, apparent omissions of component types that were described in the original LRA, from the LRA component type tables. The staff discussed the apparent omissions of component types from the LRA component type tables with the applicant to determine whether they were intentional or editorial in nature. The staff noted that during the original LRA review the applicant agreed to describe where the following component types were represented in the component type tables if they were intentionally omitted, and to include those component types in component type tables that had unintentionally omitted components. Therefore, the staff requested in RAI 2.2-2 that the applicant explain how it represented the following component types in the original LRA: flanges, bolting, orifices, tubing, vacuum breakers, elbows, unions, tees, couplings, thermowells, compressors, reducers, caps, floor drains, flexible hoses, expansion joints, vents, diffusers, manholes, and piping.

In its response, by letter dated December 22, 2004, the applicant provided the following summary of how each of the components identified in RAI 2.2-2, when subjected to AMR, were represented in the AMR results sections of the original LRA:

- Flanges, tubing, elbows, unions, tees, couplings, reducers, caps, floor drains, vents, and piping were all included with the component type "Piping and Fittings."

- Bolting, where not specifically identified as its own component type within a system, was included with the component for which it was a subcomponent. For example, it would be included with component types "Piping and Fittings," "Pumps," "Valves," etc., as applicable. In general, bolting was identified as its own component type within a system when it was identified as a separate component type in the GALL Report for that respective system. When not identified as its own component type, bolting was managed for aging based on its material in the applicable air environment.

- Orifices were identified as their own component type, as either "Orifices" or "Flow Elements." However, not all "Flow Elements" were orifices. The terms "Orifices" or "Flow Elements" were used depending on how the components were called out in the plant Mechanical Equipment List databases. The term "Flow Elements" was also used for other types of flow measurement devices such as venturis.

- Vacuum breakers were included with the component type "Valves."

- Thermowells were included with "Piping and Fittings" when they were fabricated of the same material as the piping in which they were a subcomponent. If they were fabricated of a different material than the piping in which they were a subcomponent, they were identified separately as a "Temperature Element."

- Compressors were identified as either "Pumps" or as a "Chiller" subcomponent.

- Expansion joints were included with the component type of "Bellows" or "Piping and Fittings."

- Flexible hoses were included with the component type of "Flexible Hoses," "Flex Hoses," or "Piping and Fittings."

- Diffusers were included with the component type of "Piping and Fittings" or "Structural Steel."

- Manways in large components such as tanks or heat exchangers are included with the component type in which they are a subcomponent, since consistent with that component, they also serve as a pressure boundary and are typically fabricated of the same material.

Based on its review, the staff found the applicant's response to RAI 2.2-2 acceptable because the applicant adequately explained how the component types in question are represented in the AMR results sections of the original LRA. Therefore, the staff's concern described in RAI 2.2-2 is resolved.

The staff reviewed the changes described in the ALRA and evaluated them against the information in the original LRA, the RAIs stemming from the original LRA review, and their own prior evaluation conclusions.

2.2.4 Conclusion

The staff reviewed ALRA Section 2.2, the applicant's responses to RAIs 2.2-1 and 2.2-2, and the supporting information in the UFSAR and USAR to determine whether any systems and structures within the scope of license renewal had not been identified by the applicant. The staff's review did not identify any omissions. On the basis of this review, the staff concludes that the applicant properly identified the systems and structures that are within the scope of license renewal in accordance with 10 CFR 54.4.

2.3 Scoping and Screening Results: Mechanical Systems

This section documents the staff's review of the applicant's scoping and screening results for mechanical systems. Specifically, this section discusses the following mechanical systems for NMP1 and NMP2:

- reactor vessel, internals, and reactor coolant systems
- engineered safety features systems
- auxiliary systems
- steam and power conversion systems

In accordance with the requirements of 10 CFR 54.21(a)(1), the applicant identified and listed passive, long-lived system, structure, and components (SSCs) that are within the scope of license renewal and subject to an AMR. To verify that the applicant properly implemented its methodology, the staff focused its review on the implementation results. This approach allowed the staff to confirm that there were no omissions of mechanical system components that meet the scoping criteria and are subject to an AMR.

Staff Evaluation Methodology. The staff's evaluation of the information provided in the ALRA was performed in the same manner for all mechanical systems. The objective of the review was to determine if the components and supporting structures for a specific mechanical system, that appeared to meet the scoping criteria specified in 10 CFR Part 54, were identified by the applicant as within the scope of license renewal, in accordance with 10 CFR 54.4. Similarly, the staff evaluated the applicant's screening results to verify that all long-lived, passive components were subject to an AMR in accordance with 10 CFR 54.21(a)(1).

Scoping. To perform its evaluation, the staff reviewed the applicable ALRA section and associated component drawings, focusing its review on components that had not been identified as within the scope of license renewal. The staff reviewed relevant licensing basis documents, including the NMP1 UFSAR and NMP2 USAR, for each mechanical system to determine if the applicant had omitted components with intended functions delineated under 10 CFR 54.4(a) from the scope of license renewal. The staff also reviewed the licensing basis documents to determine if all intended functions delineated under 10 CFR 54.4(a) were specified in the ALRA. If omissions were identified, the staff requested additional information to resolve the discrepancies.

Screening. Once the staff completed its review of the scoping results, the staff evaluated the applicant's screening results. For those systems and components with intended functions, the staff sought to determine: (1) if the functions are performed with moving parts or a change in configuration or properties, or (2) if they are subject to replacement based on a qualified life or specified time period, as described in 10 CFR 54.21(a)(1). For those that did not meet either of these criteria, the staff sought to confirm that these mechanical systems and components were subject to an AMR as required by 10 CFR 54.21(a)(1). If discrepancies were identified, the staff requested additional information to resolve them.

2.3A NMP1 Scoping and Screening Results: Mechanical Systems

2.3A.1 Reactor Vessel, Internals, and Reactor Coolant Systems

In ALRA Section 2.3.1.A, the applicant identified the structures and components of the NMP1 reactor vessel, internals, and reactor coolant systems that are subject to an AMR for license renewal.

The applicant described the supporting structures and components of the reactor vessel, internals, and reactor coolant systems in the following sections of the ALRA:

- 2.3.1.A.1   NMP1 reactor pressure vessel
- 2.3.1.A.2   NMP1 reactor pressure vessel internals
- 2.3.1.A.3   NMP1 reactor pressure vessel instrumentation system
- 2.3.1.A.4   NMP1 reactor recirculation system
- 2.3.1.A.5   NMP1 control rod drive system
- 2.3.1.A.6   NMP1 reactor coolant pressure boundary components in other systems

The staff's review findings regarding ALRA Sections 2.3.1.A.1 through 2.3.1.A.6 are presented in SER Sections 2.3A.1.1 through 2.3A.1.6, respectively.

2.3A.1.1 NMP1 Reactor Pressure Vessel

2.3A.1.1.1 Summary of Technical Information in the Amended Application

In ALRA Section 2.3.1.A.1, the applicant described the reactor pressure vessel (RPV). The NMP1 RPV contains and supports the reactor core, reactor internals, and the reactor coolant/moderator. The RPV forms part of the reactor coolant pressure boundary (RCPB) and serves as a barrier against leakage of radioactive materials to the drywell.

The RPV is a vertical, cylindrical pressure vessel with hemispherical heads. The cylindrical shell and hemispherical heads are fabricated from low alloy carbon steel that is clad on the interior with weld overlay. The top head is secured to the vessel with studs and nuts and includes two concentric sealings in the vessel head flange area to prevent reactor coolant leakage. The RPV is supported by a steel skirt welded to the bottom head. The base of the skirt is continuously supported by a ring girder and sole plate fastened to a concrete foundation, which carries the load to the reactor building foundation slab.

The RPV contains SR components that are relied upon to remain functional during and following DBEs. The failure of NSR SSCs in the RPV could potentially prevent the satisfactory accomplishment of an SR function.

The RPV's intended functions within the scope of license renewal include the following:

*   pressure boundary - maintains the integrity of the RCPB

*   containment - provides a fission product containment barrier

*   physical support - provides vertical and horizontal support for the core and other reactor vessel internals

*   core cooling - together with the reactor vessel internals, provides a means to distribute coolant to the fuel assemblies located in the core and provides a floodable volume to at least two-thirds core height following DBEs

*   maintains mechanical and structural integrity of NSR components to prevent spatial interactions with SR components

In ALRA Table 2.3.1.A.1-1, the applicant identified the following RPV component types that are within the scope of license renewal and subject to an AMR:

*   bottom head
*   nozzles
*   nozzle safe ends
*   penetrations: core differential pressure, CRD stub tube, flux monitor, instrumentation, vessel drain
*   support skirt and attachment welds
*   thermal sleeves
*   top head
*   top head (closure studs and nuts)
*   top head (flanges)
*   top head (leak detection line)
*   top head (nozzles)
*   valves
*   vessel shell (flange)
*   vessel shells: beltline, lower shell, upper nozzle shell, upper RPV shell
*   vessel shell welds (including attachment welds)

2.3A.1.1.2 Staff Evaluation

The staff reviewed ALRA Section 2.3.1.A.1 and the UFSAR using the evaluation methodology described in SER Section 2.3. The staff conducted its review in accordance with the guidance described in SRP-LR Section 2.3, "Scoping and Screening Results: Mechanical Systems."

In conducting its review, the staff evaluated the system functions described in the ALRA and UFSAR in accordance with the requirements of 10 CFR 54.4(a) to verify that the applicant had not omitted from the scope of license renewal any components with intended functions delineated under 10 CFR 54.4(a). The staff then reviewed those components that the applicant had identified as being within the scope of license renewal to verify that the applicant had not omitted any passive and long-lived components that should be subject to an AMR in accordance with the requirements of 10 CFR 54.21(a)(1).

During the staff review it was noted that ALRA Table 2.3.1.A.1-1 does not list a thermal shield that provides shielding from gamma and neutron radiation for such SR SSCs as the reactor vessel and the internals. Such shielding, which can reduce irradiation-induced embrittlement of the vessel and/or the internals, is not a design feature of the NMP1 RPV nor of any boiling water reactors (BWRs) manufactured by General Electric (GE); therefore, it is not appropriate to consider such a component in the NMP1 scoping evaluation.

The staff's review of original LRA Section 2.3.1.A.1 identified areas in which additional information was necessary to complete the evaluation of the applicant's scoping and screening results. The applicant responded to the staff's RAIs as discussed below.

In RAI-2, dated November 17, 2004, the staff requested the applicant to provide additional information pertaining to the reactor recirculation nozzles, and thermal sleeves for core spray, feedwater, and CRD return line. In its response, by letter dated December 17, 2004, the applicant stated that there are no low pressure coolant injection (LPCI) couplings installed in NMNP1. This response resolved the staff's concern described in RAI-2.

In RAI 2.3-1, dated October 11, 2005, the staff requested that the applicant indicate whether the liquid poison pressure nozzle is considered an RPV penetration requiring an AMR.

In its response, by letter dated October 28, 2005, the applicant indicated that the liquid poison pressure nozzle is part of the core differential pressure penetration. NMP1 utilizes a "pipe within a pipe" design similar to many other older BWR designs. The liquid poison pressure nozzle is considered a RPV penetration and is part of the license renewal scope, falling under the "Penetrations: Core Differential Pressure" subset in ALRA Table 2.3.1.A.1-1. Therefore, the staff's concern described in RAI 2.3-1 is resolved.

2.3A.1.1.3 Conclusion

The staff reviewed the ALRA and the RAI responses to determine whether any SSCs that should be within the scope of license renewal had not been identified by the applicant. No omissions were identified. In addition, the staff performed a review to determine whether any components that should be subject to an AMR had not been identified by the applicant. No omissions were identified. On the basis of its review, the staff concludes that the applicant had adequately identified the RPV components that are within the scope of license renewal, as required by 10 CFR 54.4(a), and the RPV components that are subject to an AMR, as required by 10 CFR 54.21(a)(1).

2.3A.1.2 NMP1 Reactor Pressure Vessel Internals

2.3A.1.2.1 Summary of Technical Information in the Amended Application

In ALRA Section 2.3.1.A.2, the applicant described the RPV internals. NMP1 RPV internals provide support for the core and other internal components, maintain fuel configuration (coolable geometry) during normal operation and accident conditions, and provide reactor coolant flow distribution through the core.

The RPV internals consist of the components internal to the RPV. The main components are the reactor core (fuel, channels, control rods, and instrumentation), core shroud (including the shroud support), core shroud stabilizers (shroud repair brackets and tie-rod assemblies), core support, top grid, control rod guide tubes, feedwater sparger, core spray spargers, liquid poison sparger and steam separator and dryer. All of the RPV internals, except the shroud support and springs in the fuel assemblies, are fabricated from stainless steel. The shroud support plates, spaces, tie rods, head bolts, and associated welds are Inconel. The shroud support essentially sustains all of the vertical weight of the core structure (except the fuel assembly weights transmitted to the guide tube) and the steam separator assembly. Each guide tube, with its fuel support casting, bears the weight of four fuel assemblies and rests on a control rod drive (CRD) housing welded to the stub tube mounted on the vessel bottom head.

The RPV internals contain SR components that are relied upon to remain functional during and following DBEs. The failure of NSR SSCs in the RPV internals could potentially prevent the satisfactory accomplishment of an SR function.

RPV internals components subject to AMR are located inside the RPV and extend from the bottom head to the top guide (excluding the fuel assemblies and control rods). Additionally, the steam dryer assembly is subject to an AMR.

The intended functions within the scope of license renewal include the following:

- provides spray shield or curbs for directing flow
- provides structural support to NSR components whose failure could prevent accomplishment of SR function(s)
- provides pressure retaining boundary
- provides structural and/or functional support to SR equipment

In ALRA Table 2.3.1.A.2-1, the applicant identified the following RPV internals component types that are within the scope of license renewal and subject to an AMR:

- CRD assemblies (includes drive mechanism and housing)
- control rod guide tubes
- core plates and bolts
- core shroud
- core shroud head bolts and collars
- core shroud support structures: clamps, core plate spacers, support plates, support rings, support welds, tie rod assemblies
- core spray lines and spargers
- incore instrumentation dry tubes and guide tubes
- liquid poison spray line and sparger
- orificed fuel supports
- steam dryer assembly
- top guide

2.3A.1.2.2 Staff Evaluation

The staff reviewed ALRA Section 2.3.1.A.2 and the UFSAR using the evaluation methodology described in SER Section 2.3. The staff conducted its review in accordance with the guidance described in SRP-LR Section 2.3.

In conducting its review, the staff evaluated the system functions described in the ALRA and UFSAR in accordance with the requirements of 10 CFR 54.4(a) to verify that the applicant had not omitted from the scope of license renewal any components with intended functions delineated under 10 CFR 54.4(a). The staff then reviewed those components that the applicant had identified as being within the scope of license renewal to verify that the applicant had not omitted any passive and long-lived components that should be subject to an AMR in accordance with the requirements of 10 CFR 54.21(a)(1).

The staff's review of original LRA Section 2.3.1.A.2 identified areas in which additional information was necessary to complete the evaluation of the applicant's scoping and screening results. The applicant responded to the staff's RAIs as discussed below.

In RAI-4, dated November 17, 2004, the staff noted that the steam separator assembly consists of a base into which are welded an array of standpipes, with a steam separator located at the top of each standpipe. The staff requested that the applicant provide justification why these standpipes and steam separators are not included within the scope of license renewal.

In its response, by letter dated December 17, 2004, the applicant stated that the steam separators and their standpipes are not included within the scope of license renewal, since they are not SR components that perform a license renewal intended function, and referred to an evaluation contained in Boiling Water Reactor Vessel Internals Project (BWRVIP)-06-A. Also, the staff was concerned about the possibility that failure of these components could prevent the accomplishment of SR functions of nearby components (e.g., the creation of loose parts that might hit and damage SR components). The staff noted that this consideration was also addressed in BWRVIP-06-A, and the evaluation was accepted by the staff in letters dated September 15, 1998, and September 16, 2003. Therefore, the staff's concern described in RAI-4 is resolved.

In RAI-5 dated November 17, 2004, the staff requested that the applicant indicate where the feedwater sparger is identified as a vessel internal component requiring an AMR. In its response, by letter dated December 17, 2004, the applicant indicated that this, too, was not included within the scope of license renewal. The applicant stated that, per BWRVIP-06-A, "The sole purpose of the feedwater spargers is to control thermal mixing and extend the life of the vessel and internals. The failure of feedwater spargers or associated brackets would not prevent injection of coolant makeup and are not required to safety shut down the reactor." On this basis, the staff accepted the exclusion of the feedwater sparger from within the scope of license renewal. Therefore, the staff's concern described in RAI-5 is resolved.

In RAI-7 dated November 17, 2004, the staff requested that the applicant indicate whether the core shroud stabilizers should be identified as reactor vessel internal components requiring AMR. In its response, by letter dated December 17, 2004, the applicant stated that the core shroud stabilizer components are part of the "Core Shroud Support Structures" (Tie Rod Assemblies) listed in original LRA Table 2.3.1.A.2-1. The AMR for these components is contained in original LRA Table 3.1.2.A-2. The staff reviewed the response and found that the aging management of core shroud stabilizer is properly addressed. Therefore, the staff's concern described in RAI-7 is resolved.

In RAI-8, dated November 17, 2004, the staff requested that the applicant indicate whether the core shroud vertical weld repair should be identified as a reactor vessel internal component requiring AMR. In its response, by letter dated December 17, 2004, the applicant stated that the core shroud vertical weld repair components are part of the "Core Shroud Support Structures" (Clamps) listed in original LRA Table 2.3.1.A.2-1. The AMR for these components is described in original LRA Table 3.1.2.A-2. The staff reviewed the response and found that the aging management of the core shroud vertical weld repair components are properly addressed; therefore, the staff's concern described in RAI-8 is resolved.

In the RAI response, the applicant also verified that the liquid poison spray line and sparger are included in the scope of license renewal, and revised the original LRA Table 2.3.1.A.2-1 accordingly. The applicant also indicated that the core shroud stabilizer components (i.e., the tie rod assemblies) and the core shroud vertical weld repair components are part of the core shroud support structures.

2.3A.1.2.3 Conclusion

The staff reviewed the ALRA and RAI responses to determine whether any SSCs that should be within the scope of license renewal had not been identified by the applicant. No omissions were identified. In addition, the staff performed a review to determine whether any components that should be subject to an AMR had not been identified by the applicant. No omissions were identified. On the basis of its review, the staff concludes that the applicant had adequately identified the RPV internals components that are within the scope of license renewal, as required by 10 CFR 54.4(a), and the RPV internals components that are subject to an AMR, as required by 10 CFR 54.21(a)(1).

2.3A.1.3 NMP1 Reactor Pressure Vessel Instrumentation System

2.3A.1.3.1 Summary of Technical Information in the Amended Application

In ALRA Section 2.3.1.A.3, the applicant described the RPV instrumentation system. The NMP1 RPV instrumentation system monitors and transmits values for key reactor vessel operating parameters during normal and emergency operations. This information is indicated on meters, chart recorders and hydraulic indicator units located in the control room, remote shutdown panels and instrument rooms. The parameters monitored are reactor vessel temperature, water level and pressure, core differential pressure, core spray sparger differential pressure, and reactor safety valve position. This system also provides control signals to various systems, such as the reactor protection, automatic depressurization, ATWS, feedwater/high pressure coolant injection (FW/HPCI), and shutdown cooling systems.

The RPV instrumentation system consists of piping, valves, and excess flow check valves that provide a fluid path from the RPV to various instrumentation.

The RPV instrumentation system contains SR components that are relied upon to remain functional during and following DBEs. In addition, the RPV instrumentation system performs functions that support fire protection, EQ, ATWS, and SBO.

The intended function, within the scope of license renewal, is to provide pressure retaining boundary.

In ALRA Table 2.3.1.A.3-1, the applicant identified the following RPV instrumentation system component types that are within the scope of license renewal and subject to an AMR:

- closure bolting
- condensing pots
- piping and fittings
- temperature equalizing columns
- valves

2.3A.1.3.2 Staff Evaluation

The staff reviewed ALRA Section 2.3.1.A.3 and the UFSAR using the evaluation methodology described in SER Section 2.3. The staff conducted its review in accordance with the guidance described in SRP-LR Section 2.3.

In conducting its review, the staff evaluated the system functions described in the ALRA and UFSAR in accordance with the requirements of 10 CFR 54.4(a) to verify that the applicant had not omitted from the scope of license renewal any components with intended functions delineated under 10 CFR 54.4(a). The staff then reviewed those components that the applicant had identified as being within the scope of license renewal to verify that the applicant had not omitted any passive and long-lived components that should be subject to an AMR in accordance with the requirements of 10 CFR 54.21(a)(1).

2.3A.1.3.3 Conclusion

The staff reviewed the ALRA to determine whether any SSCs that should be within the scope of license renewal had not been identified by the applicant. No omissions were identified. In addition, the staff performed a review to determine whether any components that should be subject to an AMR had not been identified by the applicant. No omissions were identified. On the basis of its review, the staff concludes that there is reasonable assurance that the applicant had adequately identified the RPV instrumentation system components that are within the scope of license renewal, as required by 10 CFR 54.4(a), and the RPV instrumentation system components that are subject to an AMR, as required by 10 CFR 54.21(a)(1).

2.3A.1.4 NMP1 Reactor Recirculation System

2.3A.1.4.1 Summary of Technical Information in the Amended Application

In ALRA Section 2.3.1.A.4, the applicant described the reactor recirculation system. The NMP1 reactor recirculation system controls reactor power level by varying the reactor coolant flow. The reactor recirculation system is part of the RCPB and consists of five external loops. Each loop draws suction from the downcomer annulus region of the RPV and discharges reactor coolant to the RPV lower plenum. Each loop consists of a variable speed pump, blocking valves, bypass line, and associated instrumentation. The reactor recirculation pumps are controlled by separate variable frequency motor generator sets, each having associated controls and instrumentation. Other systems that connect directly to the reactor recirculation system piping are the emergency cooling system, shutdown cooling system, reactor water cleanup system and the sampling system.

The reactor recirculation system contains SR components that are relied upon to remain functional during and following DBEs. The failure of NSR SSCs in the reactor recirculation system could potentially prevent the satisfactory accomplishment of an SR function. In addition, the reactor recirculation system performs functions that support fire protection, ATWS, and SBO.

The intended functions within the scope of license renewal include the following:

- maintains mechanical and structural integrity of NSR components to prevent spatial interactions with SR components

- provides pressure retaining boundary

- provides structural support to attached SR piping and components

In ALRA Table 2.3.1.A.4-1, the applicant identified the following reactor recirculation system component types that are within the scope of license renewal and subject to an AMR:

- closure bolting
- flow elements
- piping and fittings
- pumps
- pump seal flanges
- valves

2.3A.1.4.2 Staff Evaluation

The staff reviewed ALRA Section 2.3.1.A.4 and the UFSAR using the evaluation methodology described in SER Section 2.3. The staff conducted its review in accordance with the guidance described in SRP-LR Section 2.3.

In conducting its review, the staff evaluated the system functions described in the ALRA and UFSAR in accordance with the requirements of 10 CFR 54.4(a) to verify that the applicant had not omitted from the scope of license renewal any components with intended functions delineated under 10 CFR 54.4(a). The staff then reviewed those components that the applicant had identified as being within the scope of license renewal to verify that the applicant had not omitted any passive and long-lived components that should be subject to an AMR in accordance with the requirements of 10 CFR 54.21(a)(1).

2.3A.1.4.3 Conclusion

The staff reviewed the ALRA to determine whether any SSCs that should be within the scope of license renewal had not been identified by the applicant. No omissions were identified. In addition, the staff performed a review to determine whether any components that should be subject to an AMR had not been identified by the applicant. No omissions were identified. On the basis of its review, the staff concludes that there is reasonable assurance that the applicant had adequately identified the reactor recirculation system components that are within the scope of license renewal, as required by 10 CFR 54.4(a), and the reactor recirculation system components that are subject to an AMR, as required by 10 CFR 54.21(a)(1).

2.3A.1.5 NMP1 Control Rod Drive System

2.3A.1.5.1 Summary of Technical Information in the Amended Application

In ALRA Section 2.3.1.A.5, the applicant described the CRD system. The NMP1 CRD system changes core reactivity level by positioning the control rods within the reactor core in response to manual control signals, and scrams the reactor in response to manual or automatic signals. The system also provides high-pressure makeup to the RPV to compensate for leakage rates of up to 25 gpm, or for break flows caused by certain small line breaks. The CRD system also provides water to the reactor vessel level instrumentation reference leg backfill system and to the keep-full system for the emergency cooling system.

The CRD system consists of two redundant pumps, filters, strainers, control valves, hydraulic control units, CRD mechanisms, scram discharge volume, isolation valves and associated piping, valves, controls and instrumentation. The normal water supply for the pumps is the condensate system with backup supplies available from the condensate storage tanks and the demineralized water storage tank. The discharge of each pump provides water directly to the reactor level instrumentation reference leg backfill system, emergency cooling system keep-full system and the CRD water filters. The CRD system also supplies cooling water to the CRD mechanisms and charging water to the hydraulic control units. Drive water is provided to the directional control valves, and the remaining water is provided directly to the RPV. Following a reactor scram, the water discharged from the CRD mechanisms is collected in the scram discharge volume.

The CRD system contains SR components that are relied upon to remain functional during and following DBEs. The failure of NSR SSCs in the CRD system could potentially prevent the satisfactory accomplishment of an SR function. In addition, the CRD system performs functions that support fire protection, EQ, and ATWS.

The intended functions within the scope of license renewal include the following:

- provides filtration
- maintains mechanical and structural integrity of NSR components to prevent spatial interactions with SR components
- provides pressure retaining boundary

- maintains mechanical and structural integrity of NSR components that provide structural support to attached SR components

In ALRA Table 2.3.1.A.5-1, the applicant identified the following CRD system component types that are within the scope of license renewal and subject to an AMR:

- accumulators
- closure bolting
- filters
- heat exchangers
- piping and fittings
- pumps
- tank
- valves

2.3A.1.5.2 Staff Evaluation

The staff reviewed ALRA Section 2.3.1.A.5 and the UFSAR using the evaluation methodology described in SER Section 2.3. The staff conducted its review in accordance with the guidance described in SRP-LR Section 2.3.

In conducting its review, the staff evaluated the system functions described in the ALRA and UFSAR in accordance with the requirements of 10 CFR 54.4(a) to verify that the applicant had not omitted from the scope of license renewal any components with intended functions delineated under 10 CFR 54.4(a). The staff then reviewed those components that the applicant had identified as being within the scope of license renewal to verify that the applicant had not omitted any passive and long-lived components that should be subject to an AMR in accordance with the requirements of 10 CFR 54.21(a)(1).

The staff's review of original LRA Section 2.3.1.A.5 identified an area in which additional information was necessary to complete the evaluation of the applicant's scoping and screening results. The applicant responded to the staff's RAI as discussed below.

In RAI-16 dated November 17, 2004, the staff requested that the applicant indicate where CRD hydraulic control units, flow elements and indicators, pumps, and rupture discs should be identified as control rod drive system components requiring AMR. In its response, by letter dated December 17, 2004, the applicant stated that each of the components listed in this RAI is within the scope of license renewal except flow indicators, which are considered active components. The applicant further stated that original LRA Section 2.3.1.A.5 and Table 2.3.1.A.5-1 address the CRD system for scoping and screening and for AMR, original LRA Section 3.1.2.A.5 and Table 3.1.2.A.5 contains the hydraulic control units which are under "Accumulators" component type. Furthermore, the applicant stated that flow elements and pumps are included with the "NSR Piping, Fittings and Equipment" component type, and rupture disks are included with the "Valves" component type. The staff reviewed the applicant's response and found that the components for CRD are properly addressed. Therefore, the staff's concern described in RAI-16 is resolved

2.3A.1.5.3 Conclusion

The staff reviewed the ALRA and RAI response to determine whether any SSCs that should be within the scope of license renewal had not been identified by the applicant. No omissions were identified. In addition, the staff performed a review to determine whether any components that should be subject to an AMR had not been identified by the applicant. No omissions were identified. On the basis of its review, the staff concludes that there is reasonable assurance that the applicant had adequately identified the CRD system components that are within the scope of license renewal, as required by 10 CFR 54.4(a), and the CRD system components that are subject to an AMR, as required by 10 CFR 54.21(a)(1).

2.3A.1.6 NMP1 Reactor Coolant Pressure Boundary Components in Other Systems

2.3A.1.6.1 Summary of Technical Information in the Amended Application

In ALRA Section 2.3.1.A.6, the applicant stated that the components requiring AMR that have RCPB functions have been maintained in the plant system to which they are normally assigned, rather than grouped with other RCPB components in the reactor vessel internals and reactor coolant system. ALRA Table 2.3.1.A.6-1 presents a list of plant systems having RCPB components evaluated in the GALL Report as part of the reactor vessel, internals and reactor coolant system.

For each of these systems, applicable system descriptions, USAR references, license renewal boundary diagram references, system intended functions, and complete listings of component groups requiring an AMR are presented in the application section indicated in ALRA Table 2.3.1.A.6-1. AMR results for RCPB components are presented in their sections as follows:

- NMP1 core spray system (ALRA Section 2.3.2.A.3)
- NMP1 emergency cooling system (ALRA Section 2.3.2.A.4)
- NMP1 feedwater/high pressure coolant injection system (ALRA Section 2.3.4.A.3)
- NMP1 liquid poison system (ALRA Section 2.3.3.A.11)
- NMP1 main steam system (ALRA Section 2.3.4.A.5)
- NMP1 reactor water cleanup system (ALRA Section 2.3.3.A.19)
- NMP1 sampling system (ALRA Section 2.3.3.A.20)
- NMP1 shutdown cooling system (ALRA Section 2.3.3.A.22)

2.3A.1.6.2 Staff Evaluation

The staff reviewed ALRA Section 2.3.1.A.6 to determine whether there is reasonable assurance that the RCPB components in other systems components within the scope of license renewal and subject to an AMR have been identified in accordance with 10 CFR 54.4 and 54.21(a)(1). The staff conducted its review in accordance with the guidance described in SRP-LR Section 2.3 and is described below.

In conducting its review the staff selected system functions described in the UFSAR set forth in 10 CFR 54.4 to verify that components having intended functions were not omitted from the scope of the rule. The staff also focused on components not identified as subject to an AMR to determine if any components were omitted. As part of the evaluation, the staff determined whether the applicant had properly identified the SSCs within the scope of license renewal and subject to an AMR, pursuant to 10 CFR 54.4(a) and 10 CFR 54.21(a)(1). The staff reviewed the relevant portions of the UFSAR for the RCPB components in other systems and associated components and compared the information in the UFSAR with the information in the original LRA to identify those portions that the original LRA did not identify as within the scope of license renewal and subject to an AMR. The staff then reviewed the SCs that were identified as not being within the scope of license renewal to verify that (1) these SCs have none of the intended functions delineated under 10 CFR 54.4(a), and (2) for those SCs that have an applicable intended function(s), verify that they either perform this function(s) with moving parts or a change in configuration or properties, or that they are subject to replacement based on a qualified life or specified time period, as described in 10 CFR 54.21(a)(1).

The staff also reviewed the UFSAR for any functions delineated under 10 CFR 54.4(a) not identified as intended functions in the original LRA, to verify that the SSCs with such functions will be adequately managed so that the functions will be maintained consistent with the CLB for the extended period of operation.

2.3A.1.6.3 Conclusion

The staff reviewed the ALRA to determine whether any SSCs that should be within the scope of license renewal had not been identified by the applicant. No omissions were identified. In addition, the staff performed a review to determine whether any components that should be subject to an AMR had not been identified by the applicant. No omissions were identified. On the basis of its review, the staff concludes that there is reasonable assurance that the applicant had adequately identified the RCPB components in other systems components that are within the scope of license renewal, as required by 10 CFR 54.4(a), and the RCPB components in other systems components that are subject to an AMR, as required by 10 CFR 54.21(a)(1).

2.3A.2 Engineered Safety Features Systems

In ALRA Section 2.3.2.A, the applicant identified the structures and components of the NMP1 engineered safety features (ESF) systems that are subject to an AMR for license renewal.

The applicant described the supporting structures and components of the ESF systems in the following sections of the ALRA:

- 2.3.2.A.1   NMP1 automatic depressurization system
- 2.3.2.A.2   NMP1 containment spray system
- 2.3.2.A.3   NMP1 core spray system
- 2.3.2.A.4   NMP1 emergency cooling system

The staff's review findings regarding ALRA Sections 2.3.2.A.1 through 2.3.2.A.4 are presented in SER Sections 2.3A.2.1 through 2.3A.2.4, respectively.

2.3A.2.1 NMP1 Automatic Depressurization System

2.3A.2.1.1 Summary of Technical Information in the Amended Application

In ALRA Section 2.3.2.A.1, the applicant described the automatic depressurization system. The automatic depressurization system reduces RPV pressure for small line breaks when there is no feedwater flow. When RPV pressure is reduced to the low pressure permissive setpoint of the core spray system, sufficient inventory makeup is available to maintain adequate core cooling.

The automatic depressurization system consists of six solenoid-operated relief valves that discharge to the torus. Three relief valves are located on each main steam line. The discharge piping also contains vacuum breakers.

The automatic depressurization system contains SR components that are relied upon to remain functional during and following DBEs. In addition, the automatic depressurization system performs functions that support fire protection and EQ.

The component types subject to an AMR that perform the system intended functions for the automatic depressurization system are part of, and evaluated in, the main steam system. No additional components within the automatic depressurization system are subject to an AMR.

2.3A.2.1.2 Staff Evaluation

The staff reviewed ALRA Section 2.3.2.A.1 and the UFSAR using the evaluation methodology described in SER Section 2.3. The staff conducted its review in accordance with the guidance described in SRP-LR Section 2.3.

In conducting its review, the staff evaluated the system functions described in the ALRA and UFSAR in accordance with the requirements of 10 CFR 54.4(a) to verify that the applicant had not omitted from the scope of license renewal any components with intended functions delineated under 10 CFR 54.4(a). The staff then reviewed those components that the applicant had identified as being within the scope of license renewal to verify that the applicant had not omitted any passive and long-lived components that should be subject to an AMR in accordance with the requirements of 10 CFR 54.21(a)(1).

2.3A.2.1.3 Conclusion

The staff reviewed the ALRA to determine whether any SSCs that should be within the scope of license renewal had not been identified by the applicant. No omissions were identified. In addition, the staff performed a review to determine whether any components that should be subject to an AMR had not been identified by the applicant. No omissions were identified. On the basis of its review, the staff concludes that there is reasonable assurance that the applicant had adequately identified the automatic depressurization system components that are within the scope of license renewal, as required by 10 CFR 54.4(a), and the automatic depressurization system components that are subject to an AMR, as required by 10 CFR 54.21(a)(1).

2.3A.2.2 NMP1 Containment Spray System

2.3A.2.2.1 Summary of Technical Information in the Amended Application

The purpose of the containment spray system is to prevent containment pressure and temperature from exceeding its design values following loss of coolant accidents. The containment spray system consists of two redundant loops that take suction from the torus and discharge to one of two drywell spargers and a torus sparger. Each loop consists of two redundant trains. Each train consists of a suction header, pump, heat exchanger, common test return line and associated piping and valves. The heat exchangers are cooled by a dedicated containment spray raw water pump that takes suction from the circulating water intake tunnel and discharges to the discharge tunnel. Each raw water train consists of a pump, strainer and associated piping and valves. The containment spray system instrumentation and controls are included within this system.

The containment spray system contains SR components relied upon to remain functional during and following DBEs. The failure of NSR SSCs in the containment spray system could potentially prevent the satisfactory accomplishment of an SR function. In addition, the containment spray system performs functions that support fire protection and EQ.

The intended functions within the scope of license renewal include the following:

- provides heat transfer

- provides pressure retaining boundary

- converts liquid into spray

- maintains mechanical and structural integrity of NSR components that provide structural support to attached SR components

- provides flow restriction

In ALRA Table 2.3.2.A.2-1, the applicant identified the following containment spray system component types that are within the scope of license renewal and subject to an AMR:

- bolting
- filters/strainers
- flow elements
- flow orifices
- heat exchangers
- nozzles
- piping and fittings
- pumps
- valves

2.3A.2.2.2 Staff Evaluation

The staff reviewed ALRA Section 2.3.2.A.2 and UFSAR Section VII.B using the evaluation methodology described in SER Section 2.3. The staff conducted its review in accordance with the guidance described in SRP-LR Section 2.3.

In conducting its review, the staff evaluated the system functions described in the ALRA and UFSAR in accordance with the requirements of 10 CFR 54.4(a) to verify that the applicant had not omitted from the scope of license renewal any components with intended functions delineated under 10 CFR 54.4(a). The staff then reviewed those components that the applicant had identified as being within the scope of license renewal to verify that the applicant had not omitted any passive and long-lived components that should be subject to an AMR in accordance with the requirements of 10 CFR 54.21(a)(1).

2.3A.2.2.3 Conclusion

The staff reviewed the ALRA to determine whether any SSCs that should be within the scope of license renewal had not been identified by the applicant. No omissions were identified. In addition, the staff performed a review to determine whether any components that should be subject to an AMR had not been identified by the applicant. No omissions were identified. On the basis of its review, the staff concludes that there is reasonable assurance that the applicant had adequately identified the containment spray system components that are within the scope of license renewal, as required by 10 CFR 54.4(a), and the containment spray system components that are subject to an AMR, as required by 10 CFR 54.21(a)(1).

2.3A.2.3 NMP1 Core Spray System

2.3A.2.3.1 Summary of Technical Information in the Amended Application

In ALRA Section 2.3.2.A.3, the applicant described the core spray system. The purpose of the core spray system is to prevent fuel damage following any postulated LOCA. For small line breaks, the automatic depressurization system is used in conjunction with the core spray system to prevent fuel damage.

The core spray system contains SR components that are relied upon to remain functional during and following DBEs. The failure of NSR SSCs in the core spray system could potentially prevent the satisfactory accomplishment of an SR function. In addition, the core spray system performs functions that support fire protection and EQ.

The intended functions within the scope of license renewal include the following:

- provides heat transfer

- maintains mechanical and structural integrity of NSR components to prevent spatial interactions with SR components

- provides pressure retaining boundary

- maintains mechanical and structural integrity of NSR components that provide structural support to attached SR components
- provides flow restriction

In ALRA Table 2.3.2.A.3-1, the applicant identified the following core spray system component types that are within the scope of license renewal and subject to an AMR:

- accumulators
- bolting
- filters/strainers
- flow elements
- flow orifices
- heat exchangers
- level gauges
- piping and fittings
- pumps
- valves

2.3A.2.3.2 Staff Evaluation

The staff reviewed ALRA Section 2.3.2.A.3 and the UFSAR using the evaluation methodology described in SER Section 2.3. The staff conducted its review in accordance with the guidance described in SRP-LR Section 2.3.

In conducting its review the staff evaluated the system functions described in the ALRA and UFSAR in accordance with the requirements of 10 CFR 54.4(a) to verify that the applicant had not omitted from the scope of license renewal any components with intended functions delineated under 10 CFR 54.4(a). The staff then reviewed those components that the applicant had identified as being within the scope of license renewal to verify that the applicant had not omitted any passive and long-lived components that should be subject to an AMR in accordance with the requirements of 10 CFR 54.21(a)(1).

2.3A.2.3.3 Conclusion

The staff reviewed the ALRA to determine whether any SSCs that should be within the scope of license renewal had not been identified by the applicant. No omissions were identified. In addition, the staff performed a review to determine whether any components that should be subject to an AMR had not been identified by the applicant. No omissions were identified. On the basis of its review, the staff concludes that there is reasonable assurance that the applicant had adequately identified the core spray system components that are within the scope of license renewal, as required by 10 CFR 54.4(a), and the core spray system components that are subject to an AMR, as required by 10 CFR 54.21(a)(1).

2.3A.2.4 NMP1 Emergency Cooling System

2.3A.2.4.1 Summary of Technical Information in the Amended Application

In ALRA Section 2.3.2.A.4, the applicant described the emergency cooling system (ECS). The purpose of the ECS is to remove decay heat from the RPV fuel in the event that RPV feedwater capability is lost and the main condenser is not available. This system serves as an alternate heat sink when the RPV is isolated from its normal heat sink (i.e., the main condenser). The emergency cooling system consists of two redundant loops connected to the RPV on the steam supply side and to the reactor recirculation system on the condensate return side. Steam side vents are connected to each loop that removes non-condensable gases to the main steam lines or torus (for accident conditions). Drain lines are also provided on each loop's steam lines.

The ECS contains SR components that are relied upon to remain functional during and following DBEs. The failure of NSR SSCs in the ECS could potentially prevent the satisfactory accomplishment of an SR function. In addition, the ECS performs functions that support fire protection, EQ, and SBO.

The intended functions within the scope of license renewal include the following:

- provides heat transfer
- maintains mechanical and structural integrity of NSR components to prevent spatial interactions with SR components
- provides removal and/or holdup of fission products
- provides pressure retaining boundary
- maintains mechanical and structural integrity of NSR components that provide structural support to attached SR components

In ALRA Table 2.3.2.A.4-1, the applicant identified the following ECS component types that are within the scope of license renewal and subject to an AMR:

- bolting
- heat exchangers
- level gauges
- piping and fittings
- tanks
- valves

2.3A.2.4.2 Staff Evaluation

The staff reviewed ALRA Section 2.3.2.A.4 and the UFSAR using the evaluation methodology described in SER Section 2.3. The staff conducted its review in accordance with the guidance described in SRP-LR Section 2.3.

In conducting its review, the staff evaluated the system functions described in the ALRA and UFSAR in accordance with the requirements of 10 CFR 54.4(a) to verify that the applicant had not omitted from the scope of license renewal any components with intended functions delineated under 10 CFR 54.4(a). The staff then reviewed those components that the applicant had identified as being within the scope of license renewal to verify that the applicant had not omitted any passive and long-lived components that should be subject to an AMR in accordance with the requirements of 10 CFR 54.21(a)(1).

2.3A.2.4.3 Conclusion

The staff reviewed the ALRA to determine whether any SSCs that should be within the scope of license renewal had not been identified by the applicant. No omissions were identified. In addition, the staff performed a review to determine whether any components that should be subject to an AMR had not been identified by the applicant. No omissions were identified. On the basis of its review, the staff concludes that there is reasonable assurance that the applicant had adequately identified the ECS components that are within the scope of license renewal, as required by 10 CFR 54.4(a), and the ECS components that are subject to an AMR, as required by 10 CFR 54.21(a)(1).

2.3A.3 Auxiliary Systems

In ALRA Section 2.3.3.A, the applicant identified the structures and components of the NMP1 auxiliary systems that are subject to an AMR for license renewal.

The applicant described the supporting structures and components of the auxiliary systems in the following sections of the ALRA:

- 2.3.3.A.1   NMP1 administration building heating, ventilation, and air conditioning (HVAC) system
- 2.3.3.A.2   NMP1 circulating water system
- 2.3.3.A.3   NMP1 city water system
- 2.3.3.A.4   NMP1 compressed air systems
- 2.3.3.A.5   NMP1 containment systems
- 2.3.3.A.6   NMP1 control room HVAC system
- 2.3.3.A.7   NMP1 diesel generator building ventilation system
- 2.3.3.A.8   NMP1 emergency diesel generator system
- 2.3.3.A.9   NMP1 fire detection and protection system
- 2.3.3.A.10   NMP1 hydrogen water chemistry system
- 2.3.3.A.11   NMP1 liquid poison system
- 2.3.3.A.12   NMP1 miscellaneous non-contaminated vents and drains system
- 2.3.3.A.13   NMP1 neutron monitoring system
- 2.3.3.A.14   NMP1 process radiation monitoring system
- 2.3.3.A.15   NMP1 radioactive waste disposal building HVAC system

- 2.3.3.A.16   NMP1 radioactive waste system
- 2.3.3.A.17   NMP1 reactor building closed loop cooling water system
- 2.3.3.A.18   NMP1 reactor building HVAC system
- 2.3.3.A.19   NMP1 reactor water cleanup system
- 2.3.3.A.20   NMP1 sampling system
- 2.3.3.A.21   NMP1 service water system
- 2.3.3.A.22   NMP1 shutdown cooling system
- 2.3.3.A.23   NMP1 spent fuel pool filtering and cooling system
- 2.3.3.A.24   NMP1 technical support center HVAC system
- 2.3.3.A.25   NMP1 turbine building closed loop cooling water system
- 2.3.3.A.26   NMP1 turbine building HVAC system
- 2.3.3.A.27   NMP1 electric steam boiler system
- 2.3.3.A.28   NMP1 makeup demineralizer system

The staff's review findings regarding ALRA Sections 2.3.3.A.1 through 2.3.3.A.28 are presented in SER Sections 2.3A.3.1 through 2.3A.3.28, respectively.

2.3A.3.1 NMP1 Administration Building Heating, Ventilation, and Air Conditioning (HVAC) System

2.3A.3.1.1 Summary of Technical Information in the Amended Application

In ALRA Section 2.3.3.A.1, the applicant described the administration building HVAC system. The administration building HVAC system is designed to provide equipment ventilation and personnel comfort. The administration building HVAC system supplies air to the administration building and its extension. This system consists of a rooftop air conditioning unit, supply fans, exhaust fans, and associated ductwork. Individual heating and air conditioning units are provided throughout the original administration building and the administration building extension for personnel comfort. The administration building HVAC system louvered penthouse damper assembly also provides outside air to the control room HVAC system.

The administration building HVAC system contains SR components that are relied upon to remain functional during and following DBEs.

The only components requiring an AMR for the administration building HVAC system are the louvered penthouse damper assembly and cooling coil tubes that are shared with the control room HVAC system and are evaluated in that system. The remaining in-scope components for the administration building HVAC system are active components. Therefore, there are no components requiring an AMR for the administration building HVAC system.

2.3A.3.1.2 Staff Evaluation

The staff reviewed ALRA Section 2.3.3.A.1 and UFSAR Section III.E.1.2.2 using the evaluation methodology described in SER Section 2.3. The staff conducted its review in accordance with the guidance described in SRP-LR Section 2.3.

In conducting its review, the staff evaluated the system functions described in the ALRA and UFSAR in accordance with the requirements of 10 CFR 54.4(a) to verify that the applicant had not omitted from the scope of license renewal any components with intended functions delineated under 10 CFR 54.4(a). The staff then reviewed those components that the applicant had identified as being within the scope of license renewal to verify that the applicant had not omitted any passive and long-lived components that should be subject to an AMR in accordance with the requirements of 10 CFR 54.21(a)(1).

2.3A.3.1.3 Conclusion

The staff reviewed the ALRA to determine whether any SSCs that should be within the scope of license renewal had not been identified by the applicant. No omissions were identified. In addition, the staff performed a review to determine whether any components that should be subject to an AMR had not been identified by the applicant. No omissions were identified. On the basis of its review, the staff concludes that there is reasonable assurance that the applicant had adequately identified the administration building HVAC system components that are within the scope of license renewal, as required by 10 CFR 54.4(a), and the administration building HVAC system components that are subject to an AMR, as required by 10 CFR 54.21(a)(1).

2.3A.3.2 NMP1 Circulating Water System

2.3A.3.2.1 Summary of Technical Information in the Amended Application

In ALRA Section 2.3.3.A.2, the applicant described the circulating water system. The NMP1 circulating water system provides cooling water from Lake Ontario to the main condenser. Lake water is drawn from the intake tunnel through two parallel gates, three trains of mechanical rakes and traveling screens, to the suction of two redundant circulating water pumps. Each pump discharges in a separate line to one side of the condenser divided water box. Fish screens and sluice valves are installed in each line to prevent debris backwashing into the inlet tunnel. After leaving the condenser, the circulating water is discharged back into the lake. The circulating water system consists of the following subsystems: main condenser circulating water, screen washing, hydraulic fluid to tempering gate, and main condenser circulating water box vents.

The circulating water system contains SR components that are relied upon to remain functional during and following DBEs. The failure of NSR SSCs in the circulating water system could potentially prevent the satisfactory accomplishment of an SR function. In addition, the circulating water system performs functions that support fire protection.

The intended functions within the scope of license renewal include the following:

- provides filtration
- maintains mechanical and structural integrity of NSR components to prevent spatial interactions with SR components
- provides pressure retaining boundary

In ALRA Table 2.3.3.A.2-1, the applicant identified the following circulating water system component types that are within the scope of license renewal and subject to an AMR:

- actuator
- bolting
- circulating water gates
- expansion joints
- filter
- piping and fittings
- pumps
- tank
- traveling screens and rakes
- valves

2.3A.3.2.2 Staff Evaluation

The staff reviewed ALRA Section 2.3.3.A.2 and UFSAR Section XI.B.4 using the evaluation methodology described in SER Section 2.3. The staff conducted its review in accordance with the guidance described in SRP-LR Section 2.3.

In conducting its review, the staff evaluated the system functions described in the ALRA and UFSAR in accordance with the requirements of 10 CFR 54.4(a) to verify that the applicant had not omitted from the scope of license renewal any components with intended functions delineated under 10 CFR 54.4(a). The staff then reviewed those components that the applicant had identified as being within the scope of license renewal to verify that the applicant had not omitted any passive and long-lived components that should be subject to an AMR in accordance with the requirements of 10 CFR 54.21(a)(1).

The staff's review of original LRA Section 2.3.3.A.2 identified areas in which additional information was necessary to complete the review of the applicant's scoping and screening results. The applicant responded to the staff's RAIs as discussed below.

In RAI 2.3.3.A.2-1, dated November 19, 2004, the staff stated that because of the unique interface between the circulating water system, the emergency service water pumps, and the intake structure, the staff needed more information to complete its review to understand the configuration of the components requiring an AMR. This information was not clearly depicted in license renewal (LR) drawings LR-18022-C, sheet 1 and LR-26941-C. Therefore, the staff requested that the applicant supply the following UFSAR figures: circulating water system; circulating water channels under the screen and pump house - normal operation; circulating water channels under the screen and pump house - special operations; and intake and discharge tunnels plan and profile.

In its response, by letter dated December 22, 2004, the applicant provided copies of UFSAR Figures III-19, III-20, III-21, and XI-4 for the staff to complete its review. The staff found the applicant's response to RAI 2.3.3.A.2-1 acceptable because the UFSAR figures have been reviewed. Therefore, the staff's concern described in RAI 2.3.3.A.2-1 is resolved.

In RAI 2.3.3.A.2-2, dated November 19, 2004, the staff requested that the applicant provide information on the intended function of "NSR Functional Support" listed in the original LRA Table 2.3.3.A.2-1. The applicant response, by letter dated December 22, 2004, has been subsequently incorporated in the ALRA as discussed below.

In its ALRA, dated July 14, 2005, the applicant stated that this intended function is no longer used, instead, identified specific NSR intended functions and made them consistent with the standardized list of intended functions in the SRP-LR and NEI 95-10. Based on the information submitted in the ALRA, the staff's concern described in RAI 2.3.3.A.2-2 is resolved.

2.3A.3.2.3 Conclusion

The staff reviewed the ALRA and RAI responses to determine whether any SSCs that should be within the scope of license renewal had not been identified by the applicant. No omissions were identified. In addition, the staff performed a review to determine whether any components that should be subject to an AMR had not been identified by the applicant. No omissions were identified. On the basis of its review, the staff concludes that there is reasonable assurance that the applicant had adequately identified the circulating water system components that are within the scope of license renewal, as required by 10 CFR 54.4(a), and the circulating water system components that are subject to an AMR, as required by 10 CFR 54.21(a)(1).

2.3A.3.3 NMP1 City Water System

2.3A.3.3.1 Summary of Technical Information in the Amended Application

In ALRA Section 2.3.3.A.3, the applicant described the city water system. The city water system provides hot and cold domestic water to various areas within the station. Cold water is distributed to the lab, decontamination room, laundry, administration building, emergency showers and two electric hotwater heaters. Hot water is supplied to the lab and administration building. The system is supplied by the offsite water system. The city water system contains one SR breaker since a hot water circulating pump is powered from a SR powerboard.

The city water system contains SR components that are relied upon to remain functional during and following DBEs. The failure of NSR SSCs in the city water system could potentially prevent the satisfactory accomplishment of an SR function.

The intended function, within the scope of license renewal, is to maintain mechanical and structural integrity to prevent spatial interactions.

In ALRA Table 2.3.3.A.3-1, the applicant identified the following city water system component types that are within the scope of license renewal and subject to an AMR:

- bolting
- flow orifice
- piping and fittings
- pumps
- tanks
- valves

2.3A.3.3.2 Staff Evaluation

The staff reviewed ALRA Section 2.3.3.A.3 and the UFSAR using the evaluation methodology described in SER Section 2.3. The staff conducted its review in accordance with the guidance described in SRP-LR Section 2.3.

In conducting its review, the staff evaluated the system functions described in the ALRA and UFSAR in accordance with the requirements of 10 CFR 54.4(a) to verify that the applicant had not omitted from the scope of license renewal any components with intended functions delineated under 10 CFR 54.4(a). The staff then reviewed those components that the applicant had identified as being within the scope of license renewal to verify that the applicant had not omitted any passive and long-lived components that should be subject to an AMR in accordance with the requirements of 10 CFR 54.21(a)(1).

During the staff's review of original LRA Section 2.3.3.A.3, the staff identified an area in which additional information was necessary to complete the review of the applicant's scoping and screening results. In RAI 2.3.3.A.3-1, dated November 19, 2004, the staff requested that the applicant identify the portions of the city water system containing components subject to AMR. The applicant response, by letter dated December 22, 2004, has been subsequently incorporated in the ALRA as discussed below.

In its ALRA, dated July 14, 2005, the applicant's ALRA Table 2.3.3.A.3-1 includes a list of the components subject to an AMR and a list of the new LR drawings. The applicant also provided an LR drawing that accurately depicts all the components subject to an AMR, including those subject to an AMR in accordance with 10 CFR 54.4(a)(2). Based on the information submitted in the ALRA, the staff's concern described in RAI 2.3.3.A.3-1 is resolved.

2.3A.3.3.3 Conclusion

The staff reviewed the ALRA and RAI response to determine whether any SSCs that should be within the scope of license renewal had not been identified by the applicant. No omissions were identified. In addition, the staff performed a review to determine whether any components that should be subject to an AMR had not been identified by the applicant. No omissions were identified. On the basis of its review, the staff concludes that there is reasonable assurance that the applicant had adequately identified the city water system components that are within the scope of license renewal, as required by 10 CFR 54.4(a), and the city water system components that are subject to an AMR, as required by 10 CFR 54.21(a)(1).

2.3A.3.4 NMP1 Compressed Air Systems

2.3A.3.4.1 Summary of Technical Information in the Amended Application

In ALRA Section 2.3.3.A.4, the applicant described the compressed air systems. The compressed air systems are designed to provide clean, filtered air to various areas of NMP1. The compressed air systems consist of the house service air system, the instrument air system, and the breathing air system. The house service air system is a NSR system designed to provide a reliable source of clean air for use in maintenance and as a backup to the instrument air system. The instrument air system is designed to provide a source of clean, dry air for use in instruments, controls, and as a backup to the breathing air system. The breathing air system is a NSR system designed to provide a reliable supply of clean, filtered air fit for human breathing.

The compressed air systems contains SR components that are relied upon to remain functional during and following DBEs. The failure of NSR SSCs in the compressed air systems could potentially prevent the satisfactory accomplishment of an SR function. In addition, the compressed air systems performs functions that support EQ and SBO.

The intended functions within the scope of license renewal include the following:

- provides filtration

- provides heat transfer

- provides structural support to NSR components whose failure could prevent accomplishment of SR function(s)

- provides pressure retaining boundary

- maintains mechanical and structural integrity of NSR components that provide structural support to attached SR components

In ALRA Table 2.3.3.A.4-1, the applicant identified the following compressed air systems component types that are within the scope of license renewal and subject to an AMR:

- air dryers: couplings, flanges, heads, nozzles, piping
- air receivers
- bolting
- drain traps
- filters/strainers
- flow gauge
- heat exchangers
- orifices
- piping and fittings
- regulators
- separators
- tanks
- valves

2.3A.3.4.2 Staff Evaluation

The staff reviewed ALRA Section 2.3.3.A.4 and UFSAR Section X.I using the evaluation methodology described in SER Section 2.3. The staff conducted its review in accordance with the guidance described in SRP-LR Section 2.3.

In conducting its review, the staff evaluated the system functions described in the ALRA and UFSAR in accordance with the requirements of 10 CFR 54.4(a) to verify that the applicant had not omitted from the scope of license renewal any components with intended functions delineated under 10 CFR 54.4(a). The staff then reviewed those components that the applicant had identified as being within the scope of license renewal to verify that the applicant had not omitted any passive and long-lived components that should be subject to an AMR in accordance with the requirements of 10 CFR 54.21(a)(1).

The staff's review of original LRA Section 2.3.3.A.4 identified areas in which additional information was necessary to complete the review of the applicant's scoping and screening results. The applicant responded to the staff's RAIs as discussed below.

In RAI 2.3.3.A.4-1, dated November 19, 2004, the staff indicated that the original LRA stated that the compressed air system provides air to inflate the reactor building track bay door seal. The component type inflatable seals are not listed in the original LRA tables as subject to an AMR. The original LRA tables list only the fire protection barrier penetration seals as subject to an AMR. Therefore, the staff requested that the applicant provide the basis for excluding inflatable seals as subject to an AMR.

In its response, by letter dated December 22, 2004, the applicant stated that the reactor building track bay door inflatable seal is within the scope of license renewal and subject to AMR. It is part of the reactor building structure and is covered by the polymer in air component type in original LRA Section 2.4.A.2.

Based on its review, the staff found the applicant's response to RAI 2.3.3.A.4-1 acceptable because the applicant stated that the inflatable seal is within the scope of license renewal and subject to an AMR, and is part of reactor building structure. Therefore, the staff's concern described in RAI 2.3.3.A.4-1 is resolved.

In RAI 2.3.3.A.4-2, dated November 19, 2004, the staff requested that the applicant identify which double acting actuators are included within the scope of license renewal and subject to an AMR. The applicant response, by letter dated December 22, 2004, has been subsequently incorporated in the ALRA as discussed below.

In its ALRA, dated July 14, 2005, the applicant provided the information, including LR drawings requested by this RAI. Based on review of the information submitted in the ALRA, the staff's concern described in RAI 2.3.3.A.4-2 is resolved.

In RAI 2.3.3.A.4-3, dated November 19, 2004, the staff stated that LR drawing LR-22108-0, sheet 34 shows that the air supply tubing and solenoid valves associated with a valve BV-60-13 are not subject to an AMR; however, LR drawing LR-18017-C, sheet 1 shows the air supply piping and solenoid valves associated valve BV-60-13 are subject to an AMR. Therefore, the staff requested that the applicant resolve this inconsistency and provide the basis for the resolution.

In its response, by dated December 22, 2004, the applicant stated that LR drawing LR-18017-C, sheet 1 is incorrect. The applicant stated that, "The air supply piping to valve BV-60-13 is not in-scope for LR. Valve BV-60-13 fails closed on loss of air and is not relied upon for any licensing basis accident mitigation. As such the air supply piping does not perform any intended function for LR."

Based on its review, the staff found the applicant's response to RAI 2.3.3.A.4-3 acceptable because the applicant adequately justified the exclusion of the component in question as not within the scope of license renewal and not not subject to an AMR. Therefore, the staff's concern described in RAI 2.3.3.A.4-3 is resolved.

In RAI 2.3.3.A.4-4, dated November 19, 2004, the staff stated that on several LR drawings (e.g. LR-22111-0, sheet 5) for the compressed air system, the air supply and solenoid valves associated with the SR valves are excluded as subject to an AMR; therefore, the staff requested that the applicant provide the criteria used to exclude some of the compressed air system auxiliaries to SR valves as subject to an AMR.

In its response, by letter dated December 22, 2004, the applicant stated that the SR air supply and solenoid valves identified in the RAI are normally closed fuel pool cooling system isolation valves which fail safe (closed) on loss of air. None of the air system components to these loads are required to be SR for instrument air system integrity or operation. The applicant concluded that, based upon the scoping criteria for license renewal, the subject instrument air valves and piping are not within the scope of license renewal and are not subject to an AMR. The applicant also clarified that, since this system has no liquid-filled components, there are no NSR components within the system that are within the scope of license renewal and subject to an AMR in accordance with 10 CFR 54.4(a)(2).

Based on its review, the staff found the applicant's response to RAI 2.3.3.A.4-4 acceptable because the applicant adequately justified the exclusion of the component types in question from within the scope of license renewal and subject to an AMR. Therefore, the staff's concern described in RAI 2.3.3.A.4-4 is resolved.

2.3A.3.4.3 Conclusion

The staff reviewed the ALRA, RAI responses, and accompanying scoping boundary drawings to determine whether any SSCs that should be within the scope of license renewal had not been identified by the applicant. No omissions were identified. In addition, the staff performed a review to determine whether any components that should be subject to an AMR had not been identified by the applicant. No omissions were identified. On the basis of its review, the staff concludes that there is reasonable assurance that the applicant had adequately identified the compressed air systems components that are within the scope of license renewal, as required by 10 CFR 54.4(a), and the compressed air systems components that are subject to an AMR, as required by 10 CFR 54.21(a)(1).

2.3A.3.5 NMP1 Containment Systems

2.3A.3.5.1 Summary of Technical Information in the Amended Application

In ALRA Section 2.3.3.A.5, the applicant described the containment systems. The containment systems are designed to control and monitor the primary containment environment. The containment systems consist of the combustible gas control system, primary containment area cooling system, containment atmospheric monitoring system, torus temperature monitoring system, torus drain system, and the integrated leak rate monitoring system. The combustible gas control system is designed to prevent a combustible hydrogen-oxygen concentration from accumulating in the primary containment atmosphere immediately following or during a LOCA. The combustible gas control system consists of the containment inerting system and the containment atmosphere dilution system.

The containment inerting system is used to inert and deinert primary containment and to makeup nitrogen as required to maintain low oxygen concentration and containment pressure. The containment atmosphere dilution system is designed to monitor and maintain the oxygen concentration of the primary containment atmosphere to less than four percent during a LOCA.

The primary containment area cooling system is designed to remove and dissipate the primary containment area heat gain. The containment atmospheric monitoring system continuously monitors and provides control room indication of the containment airborne radioactivity level. This provides for detection of leaks of the reactor primary systems. The torus temperature monitoring system provides information on torus temperature, water level and airspace pressure to ensure that the cooling capacity of water maintained in the suppression chamber is available within the TS limits and to ensure that the containment structural integrity is maintained. The torus drain system is used when the reactor is in cold shutdown or refueling condition. It allows the torus to be dewatered to permit maintenance or other activities. The integrated leak rate monitoring system is used to support periodic 10 CFR 50, Appendix J testing for overall leakage from primary containment, which demonstrates the ability of containment to control the spread of radioactivity in the event of an accident.

The containment systems contain SR components that are relied upon to remain functional during and following DBEs. The failure of NSR SSCs in the containment systems could potentially prevent the satisfactory accomplishment of an SR function. In addition, the containment systems perform functions that support fire protection, EQ, and SBO.

The intended functions within the scope of license renewal include the following:

- provides filtration
- provides heat transfer
- maintains mechanical and structural integrity of NSR components to prevent spatial interactions with SR components
- provides structural support to NSR components whose failure could prevent accomplishment of SR function(s)
- provides removal and/or holdup of fission products

- provides pressure retaining boundary
- maintains mechanical and structural integrity of NSR components that provide structural support to attached SR components

In ALRA Table 2.3.3.A.5-1, the applicant identified the following containment systems component types that are within the scope of license renewal and subject to an AMR:

- airborne activity monitor
- blower
- bolting
- ducting
- filters/strainers
- flame arresters
- flow elements
- heat exchangers
- piping and fittings
- pumps
- rupture discs
- tanks
- traps
- valves
- vaporizers

2.3A.3.5.2 Staff Evaluation

The staff reviewed ALRA Section 2.3.3.A.5 and UFSAR Sections VI and VII.G using the evaluation methodology described in SER Section 2.3. The staff conducted its review in accordance with the guidance described in SRP-LR Section 2.3.

In conducting its review, the staff evaluated the system functions described in the ALRA and UFSAR in accordance with the requirements of 10 CFR 54.4(a) to verify that the applicant had not omitted from the scope of license renewal any components with intended functions delineated under 10 CFR 54.4(a). The staff then reviewed those components that the applicant had identified as being within the scope of license renewal to verify that the applicant had not omitted any passive and long-lived components that should be subject to an AMR in accordance with the requirements of 10 CFR 54.21(a)(1).

The staff's review of ALRA Section 2.3.3.A.5 identified an area in which additional information was necessary to complete the review of the applicant's scoping and screening results. The applicant responded to the staff's RAI as discussed below.

In RAI 2.3.3.A.5, dated December 8, 2004, the staff stated that original LRA Tables 2.3.3.A.5-1 and 3.3.2.A-4, and original LRA Section 2.3.3.A.5 for the containment system, do not include piping/fittings and drywell air cooler units for NMP1; however, these items are shown as within the scope of license renewal on LR drawings and are subject to an AMR. Therefore, the staff requested that the applicant provide information on the associated AMR and AMPs in ALRA Tables 2.3.3.A.5-1 and 3.3.2.A-4, if these components are within the scope of license renewal.

The staff requested the applicant to provide justification for the exclusion of these components if they are not within the scope of license renewal.

In its response, by letter dated January 7, 2005, the applicant stated that the piping/fittings and drywell air cooler units in the NMP1 containment system are component types that are within the scope of the license renewal and subject to AMR. The piping and fittings component type is included in the system description portion of ALRA Section 2.3.3.A.5, but was inadvertently omitted from ALRA Tables 2.3.3.A.5-1 and 3.3.2.A-4. The applicant stated that the ALRA tables have been revised to incorporate the requested information regarding the intended function, AMR, and AMPs for these components. With respect to drywell air cooler units, the applicant stated that these are addressed under the component types "Ducting" and "Heat Exchanger," respectively, and are included in ALRA Tables 2.3.3.A.5-1 and 3.3.2.A-4.

Based on it's review, the staff found the applicant's response to RAI 2.3.3.A.5 acceptable because the applicant has included the piping/fittings and drywell air coolers units and associated components within the scope of license renewal subjected to an AMR in accordance with the requirements of 10 CFR 54.21(a)(1). Therefore, the staff's concern described in RAI 2.3.3.A.5 is resolved.

2.3A.3.5.3 Conclusion

The staff reviewed the ALRA, RAI response, and accompanying scoping boundary drawings to determine whether any SSCs that should be within the scope of license renewal had not been identified by the applicant. No omissions were identified. In addition, the staff performed a review to determine whether any components that should be subject to an AMR had not been identified by the applicant. No omissions were identified. On the basis of its review, the staff concludes that there is reasonable assurance that the applicant had adequately identified the containment systems components that are within the scope of license renewal, as required by 10 CFR 54.4(a), and the containment systems components that are subject to an AMR, as required by 10 CFR 54.21(a)(1).

2.3A.3.6 NMP1 Control Room HVAC System

2.3A.3.6.1 Summary of Technical Information in the Amended Application

In ALRA Section 2.3.3.A.6, the applicant described the control room HVAC system. The control room HVAC system provides filtration, pressurization, heating and cooling to the control complex during normal and emergency conditions. The system is also equipped with an independent smoke and heat removal system for the main and auxiliary control rooms and cable spreading room. The control room HVAC system is comprised of three functional systems which are the normal ventilation, emergency ventilation and smoke purge systems. The normal ventilation system provides fresh and recirculated air for heating and cooling the control complex during normal operation. The emergency ventilation system provides clean, filtered fresh air combined with recirculated air for heating and cooling the control complex during emergency conditions. The smoke purge system is a fire protection ventilation system that removes smoke and heat from the main and auxiliary control rooms and cable spreading room.

The control room HVAC system contains SR components that are relied upon to remain functional during and following DBEs. In addition, the control room HVAC system performs functions that support fire protection.

The intended functions within the scope of license renewal include the following:

- provides filtration
- provides rated fire barrier
- provides heat transfer
- provides pressure retaining boundary

In ALRA Table 2.3.3.A.6-1, the applicant identified the following control room HVAC system component types that are within the scope of license renewal and subject to an AMR:

- blowers
- bolting
- ducting
- expansion tank
- filters/strainers
- flow elements
- heat exchangers
- piping and fittings
- pumps
- seals and gaskets
- temperature elements
- valves and dampers

2.3A.3.6.2 Staff Evaluation

The staff reviewed ALRA Section 2.3.3.A.6 and UFSAR Section III.B.2.2 using the evaluation methodology described in SER Section 2.3. The staff conducted its review in accordance with the guidance described in SRP-LR Section 2.3.

In conducting its review, the staff evaluated the system functions described in the ALRA and UFSAR in accordance with the requirements of 10 CFR 54.4(a) to verify that the applicant had not omitted from the scope of license renewal any components with intended functions delineated under 10 CFR 54.4(a). The staff then reviewed those components that the applicant had identified as being within the scope of license renewal to verify that the applicant had not omitted any passive and long-lived components that should be subject to an AMR in accordance with the requirements of 10 CFR 54.21(a)(1).

2.3A.3.6.3 Conclusion

The staff reviewed the ALRA to determine whether any SSCs that should be within the scope of license renewal had not been identified by the applicant. No omissions were identified. In addition, the staff performed a review to determine whether any components that should be subject to an AMR had not been identified by the applicant. No omissions were identified. On the basis of its review, the staff concludes that there is reasonable assurance that the applicant had adequately identified the control room HVAC system components that are within the scope of license renewal, as required by 10 CFR 54.4(a), and the control room HVAC system components that are subject to an AMR, as required by 10 CFR 54.21(a)(1).

2.3A.3.7 NMP1 Diesel Generator Building Ventilation System

2.3A.3.7.1 Summary of Technical Information in the Amended Application

In ALRA Section 2.3.3.A.7, the applicant described the diesel generator building (DGB) ventilation system. The DGB ventilation system is designed to maintain the diesel room temperature below the allowed maximum for continuous operation of the emergency diesel generator. Each diesel generator rooms is equipped with its own ventilation system. The system consists of roof exhaust fans, a roll-up door, electric heaters, and associated controls. The doors operate in conjunction with the room exhaust fan pairs to ensure that the diesel generator room temperature remains below the allowed maximum. The heaters operate to maintain the diesel generator room ambient temperature at or above 50°F.

The DGB ventilation system contains SR components that are relied upon to remain functional during and following DBEs.

The intended function, within the scope of license renewal, is to provide pressure retaining boundary.

In ALRA Table 2.3.3.A.7-1, the applicant identified the blowers component type of the DGB ventilation system as within the scope of license renewal and subject to an AMR.

2.3A.3.7.2 Staff Evaluation

The staff reviewed ALRA Section 2.3.3.A.7 using the evaluation methodology described in SER Section 2.3. The staff conducted its review in accordance with the guidance described in SRP-LR Section 2.3.

In conducting its review, the staff evaluated the system functions described in the ALRA and UFSAR in accordance with the requirements of 10 CFR 54.4(a) to verify that the applicant had not omitted from the scope of license renewal any components with intended functions delineated under 10 CFR 54.4(a). The staff then reviewed those components that the applicant had identified as being within the scope of license renewal to verify that the applicant had not omitted any passive and long-lived components that should be subject to an AMR in accordance with the requirements of 10 CFR 54.21(a)(1).

2.3A.3.7.3 Conclusion

The staff reviewed the ALRA to determine whether any SSCs that should be within the scope of license renewal had not been identified by the applicant. No omissions were identified. In addition, the staff performed a review to determine whether any components that should be subject to an AMR had not been identified by the applicant. No omissions were identified. On the basis of its review, the staff concludes that there is reasonable assurance that the applicant had adequately identified the DGB ventilation system components that are within the scope of license renewal, as required by 10 CFR 54.4(a), and the DGB ventilation system components that are subject to an AMR, as required by 10 CFR 54.21(a)(1).

2.3A.3.8 NMP1 Emergency Diesel Generator System

2.3A.3.8.1 Summary of Technical Information in the Amended Application

In ALRA Section 2.3.3.A.8, the applicant described the emergency diesel generator (EDG) system. The EDG system provides the standby source of electric power for equipment required for mitigation of the consequences of an accident, for safe shutdown and for maintenance of the station in a safe condition under postulated event and accident scenarios. This system consists of two identical, physically separate, and electrically independent standby diesel generators. Each diesel generator has associated subsystems which assist the unit in performing its safety function. The diesel engine subsystem consists of a diesel engine which provides the mechanical power to run the electric generator. The fuel oil subsystem supplies fuel oil for engine combustion and is comprised of the fuel oil storage and handling system and the engine fuel oil system. The air start subsystem supplies high-pressure air to start the diesel engine. The combustion air intake and exhaust subsystem supports the engine combustion process by supplying filtered air to the diesel engine and then discharging the exhaust gases. The lube oil subsystem provides cooling and lubrication for major engine components. The cooling water subsystem removes heat from the diesel engine via the engine cooling system and diesel generator raw water cooling system. The electric generator subsystem provides the electrical output of the diesel generator unit and includes the required controls.

The EDG system contains SR components that are relied upon to remain functional during and following DBEs. The failure of NSR SSCs in the EDG system could potentially prevent the satisfactory accomplishment of an SR function. In addition, the EDG system performs functions that support fire protection and EQ.

The intended functions within the scope of license renewal include the following:

- provides filtration
- provides heat transfer
- provides structural support to NSR components whose failure could prevent accomplishment of SR function(s)
- provides pressure retaining boundary
- maintains mechanical and structural integrity of NSR components that provide structural support to attached SR components
- provides flow restriction

In ALRA Table 2.3.3.A.8-1, the applicant identified the following EDG system component types that are within the scope of license renewal and subject to an AMR:

- air intakes
- air start motors
- bolting

- compressors
- exhausts for EDG
- filters/strainers
- flow elements
- flow glasses
- heat exchangers
- level glasses
- mufflers and silencers
- orifices
- piping and fittings
- pumps
- tanks
- valves

2.3A.3.8.2 Staff Evaluation

The staff reviewed ALRA Section 2.3.3.A.8 and UFSAR Section IX.B.4.1 using the evaluation methodology described in SER Section 2.3. The staff conducted its review in accordance with the guidance described in SRP-LR Section 2.3.

In conducting its review, the staff evaluated the system functions described in the ALRA and UFSAR in accordance with the requirements of 10 CFR 54.4(a) to verify that the applicant had not omitted from the scope of license renewal any components with intended functions delineated under 10 CFR 54.4(a). The staff then reviewed those components that the applicant had identified as being within the scope of license renewal to verify that the applicant had not omitted any passive and long-lived components that should be subject to an AMR in accordance with the requirements of 10 CFR 54.21(a)(1).

The staff's review of original LRA Section 2.3.3.A.8 identified areas in which additional information was necessary to complete the review of the applicant's scoping and screening results. The applicant responded to the staff's RAIs as discussed below.

In RAI 2.3.3.A.8-1, dated November 19, 2004, the staff stated that LR drawing 18026-C, sheet 1 (B-1) for diesel #102 shows that the line leading to the fuel injectors is not subject to an AMR. LR drawing LR-18026-C, sheet 2 (C-1) for diesel #103 shows that the line leading to the injectors is highlighted as subject to an AMR. Therefore, the staff requested that the applicant resolve the apparent discrepancy between the two LR drawings.

In its response by letter dated December 22, 2004, the applicant stated that LR drawing 18026-C sheet 1 is incorrect and does not properly show the components within the scope of license renewal and subject to AMR. The components in question should have been highlighted on the LR drawing showing that they are within the scope of license renewal under 10 CFR 54.4(a) and subject to AMR under 10 CFR 54.21(a) but inadvertently were not highlighted. The components in question have been included within the scope of license renewal and are subject to AMR.

The staff's review found the applicant's response to RAI 2.3.3.A.8-1 acceptable because it adequately explained that the components in question are within the scope of license renewal and subject to AMR but inadvertently were not highlighted on the LR drawing. Therefore, the staff's concern described in RAI 2.3.3.A.8-1 has been resolved.

In RAI 2.3.3.A.8-2, dated November 19, 2004, the staff stated that LR drawing LR-18026-C, sheet 1 and sheet 2 does not show that the pipes and expansion joints leading to the air start motor are subject to an AMR. The staff noted that the pipe and the expansion joints are not shown on sheet 2 of the drawing. Original LRA Table 2.3.3.B.1-1 lists air start motors as subject to an AMR for NMP2; therefore, the staff requested that the applicant provide the basis for not requiring an AMR for these NMP1 components.

In its response, by letter dated December 22, 2004, the applicant stated that the LR drawing is incorrect. It should show the air start motor, associated piping, and expansion joints as subject to AMR. These components have a pressure boundary intended function. The piping and expansion joints are included with the piping and fittings component type. The applicant also stated that the air start motors will be added to the original LRA Table 2.3.3.A.8-1. In addition, the applicant added the air intake silencer, filter, and exhaust muffler on LR drawing LR-18026-C, sheets 1 and 2 should be shown in red to indicate that they are subject to AMR, consistent with original LRA Table 2.3.3.A.8-1.

Based on its review, the staff found the applicant's response to RAI 2.3.3.A.8-2 acceptable because it explained that: (1) the LR drawing depicts the air start motor, associated piping, and expansion joints as not requiring an AMR, but should have been highlighted on the LR drawings; (2) the air start motor is added to Table 2.3.3.A.8-1; and (3) the air intake silencer, filter, and exhaust muffler should be shown on the LR drawings as being subject to an AMR. Therefore, the staff's concern described in RAI 2.3.3.A.8-2 is resolved.

In RAI 2.3.3.A.8-3, dated November 19, 2004, the staff stated that LR drawing 18026-C, sheets 1 and 2 shows that the tubing to the pressure gauges on the air receiver tanks is not highlighted as subject to an AMR. This tubing has a passive pressure boundary function and meets the criteria of 10 CFR 54.4(a)(1). Additionally, a note on the LR drawings indicates that there are root valves for these pressure indicators; therefore, the staff requested that the applicant provide the basis for not requiring an AMR for this tubing and associated root valves.

In its response, by letter dated December 22, 2004, the applicant stated that the LR drawing is incorrect and does not properly show the tubing between the air receiver tanks and the pressure gauges as within the scope of license renewal and subject to an AMR. The components in question should have been included within the scope of license renewal and subject to an AMR but inadvertently were not highlighted. The applicant stated that the original LRA Table 2.3.3.A.8-1 already represents tubing and instrument root valves under the "Piping and Fittings" and "Valves" component types, respectively. The applicant further stated that pressure gauges are active components and, therefore, are not highlighted on the LR drawing as subject to an AMR.

Based on its review, the staff found the applicant's response to RAI 2.3.3.A.8-3 acceptable because it adequately explained that the components in question are within the scope of license renewal and subject to an AMR but were inadvertently left un-highlighted on the LR drawing. Therefore, the staff's concern described in RAI 2.3.3.A.8-3 is resolved.

In RAI 2.3.3.A.8-4, dated November 19, 2004, the staff stated that LR drawing 18026-C, sheets 1 and 2 do not clearly indicate whether two immersion heaters are subject to an AMR. Depending on the heater design, these heaters can have a pressure boundary intended function; therefore, the staff requested that the applicant clarify if the heat exchangers component type original LRA Table 2.3.3.A.8-1 represents these heaters.

In its response, by letter dated December 22, 2004, the applicant stated that drawing LR-18026-C, sheets 1 and 2, are incorrect. Immersion heaters do have a pressure boundary function. Additionally, sheet 2 should look like sheet 1, indicating that there is a chamber around the heating coils. On both sheets, those chambers should be shown in red, indicating that they are within the scope for license renewal and subject to an AMR. The chambers are treated as part of the piping and fittings component type. The heaters themselves are also within the scope of license renewal; however, since they are active components, per Appendix B of NEI 95-10, Revision 3, they are not subject to an AMR.

Based on its review, the staff found the applicant's response to RAI 2.3.3.A.8-4 acceptable because it adequately explained that: (1) although the immersion heaters are within the scope of license renewal, they are active components and do not perform a pressure boundary function, and therefore are not subject to an AMR; and (2) the chambers around the heating coils are also within the scope of license renewal, subject to an AMR, and treated as the piping and fittings component type. Therefore, the staff's concern described in RAI 2.3.3.A.8-4 is resolved.

2.3A.3.8.3 Conclusion

The staff reviewed the ALRA, RAI responses, and accompanying scoping boundary drawings to determine whether any SSCs that should be within the scope of license renewal had not been identified by the applicant. No omissions were identified. In addition, the staff performed a review to determine whether any components that should be subject to an AMR had not been identified by the applicant. No omissions were identified. On the basis of its review, the staff concludes that there is reasonable assurance that the applicant had adequately identified the EDG system components that are within the scope of license renewal, as required by 10 CFR 54.4(a), and the EDG system components that are subject to an AMR, as required by 10 CFR 54.21(a)(1).

2.3A.3.9 NMP1 Fire Detection and Protection System

2.3A.3.9.1 Summary of Technical Information in the Amended Application

In ALRA Section 2.3.3.A.9, the applicant described the fire detection and protection system. The fire detection and protection system is designed to achieve the following objectives:

- provide automatic fire detection in those areas where the danger of fire exists

- provide fire extinguishment by fixed equipment activated automatically or manually for those areas where the danger of fire exists

- provide manually-operated fire extinguishing equipment for use by station personnel at points throughout the property and station

- provide a backup cooling water source for the reactor emergency cooling system in the event of a complete loss of all other sources of condensing water

- provide an emergency source of water for containment and reactor vessel flooding

- provide an emergency source of water to the spent fuel storage pool (hose)

- provide a backup water source for the emergency service water system

- provide an emergency cooling water supply to either diesel generator

These objectives are accomplished by the fire detection and control, fire water, halon suppression, and carbon dioxide (CO_2) suppression systems. The fire detection and control system provides for the identification of a fire, annunciation locally and in the control room, and in certain zones, automatically initiates suppression. The fire water system provides for the extinguishment of fires using water. The halon suppression system provides for the extinguishment of fires using Halon 1301. The CO_2 suppression system provides for the extinguishment of fires using CO_2. Portable fire extinguishers are also provided throughout the station to provide additional protection.

The fire detection and protection system contains SR components that are relied upon to remain functional during and following DBEs. The failure of NSR SSCs in the fire detection and protection system could potentially prevent the satisfactory accomplishment of an SR function. In addition, the fire detection and protection system performs functions that support fire protection and SBO.

The intended functions within the scope of license renewal include the following:

- maintains mechanical and structural integrity of NSR components to prevent spatial interactions with SR components

- provides pressure retaining boundary

- converts liquid into spray

- maintains mechanical and structural integrity of NSR components that provide structural support to attached SR components

In ALRA Table 2.3.3.A.9-1, the applicant identified the following fire detection and protection system component types that are within the scope of license renewal and subject to an AMR:

- bolting
- filters/strainers
- fire hydrants
- flow elements
- gearbox
- heat-actuated devices
- heat exchangers
- orifices
- piping and fittings
- pumps
- silencers

- sluice gate for motor driven fire pump
- spray nozzles
- sprinklers
- tanks and air receivers
- valves

2.3A.3.9.2 Staff Evaluation

The staff reviewed ALRA Section 2.3.3.A.9 and UFSAR Sections X.10A and X.10B using the evaluation methodology described in SER Section 2.3. The staff conducted its review in accordance with the guidance described in SRP-LR Section 2.3.

In conducting its review, the staff evaluated the system functions described in the ALRA and UFSAR in accordance with the requirements of 10 CFR 54.4(a) to verify that the applicant had not omitted from the scope of license renewal any components with intended functions delineated under 10 CFR 54.4(a). The staff then reviewed those components that the applicant had identified as being within the scope of license renewal to verify that the applicant had not omitted any passive and long-lived components that should be subject to an AMR in accordance with the requirements of 10 CFR 54.21(a)(1).

The staff also reviewed approved fire protection safety evaluation (SE) report dated July 26, 1979, and March 21, 1983, for Nine Miles Point Unit 1. This report is referenced directly in the Nine Mile Point Unit 1 fire protection current licensing basis (CLB) and summarize the Fire Protection Program and commitments to 10 CFR 50.48 using the guidance of Appendix A to Branch Technical Position (BTP) Chemical and Mechanical Engineering Branch (CMEB) 9.5-1. The staff then reviewed those components that the applicant identified as being within the scope of license renewal to verify that the applicant did not omit any passive and long-lived components that should be subject to an AMR in accordance with the requirements of 10 CFR 54.21(a)(1).

The staff's review of ALRA Section 2.3.3.A.9 identified areas in which additional information was necessary to complete the evaluation of the applicant's scoping and screening results. The applicant responded to the staff's RAIs as discussed below.

In RAI 2.3.3.A.9-2, dated November 17, 2004, the staff stated that drawing LR-18030-C, sheet 2 shows five foam water systems as within the scope of license renewal and subject to an AMR, including the foam solution supply piping; however, the LR drawing shows the foam tank and pumps as not within the scope of license renewal. Additionally, the foam tanks and pumps are shown disconnected from the foam water system supply piping. The UFSAR does not reference these foam water systems; therefore, the staff requested that the applicant provide the basis for excluding the foam tank and pumps from the scope of license renewal and from being subject to an AMR, since they are necessary for the function of the foam water systems which are shown as within the scope of license renewal.

In its response, by letter dated December 17, 2004, the applicant stated that the ALRA correctly describes the NMP1 fire detection and protection system, as credited for 10 CFR 50.48 and, therefore, 10 CFR 54.4(a). The foam subsystem is not included in this section since it is retired in-place and nonfunctional. The foam subsystem is not within the scope of license renewal for NMP1. The applicant further stated that LR drawing LR-18030-C incorrectly identifies portions

2-73

of the foam subsystem identified as within the scope of license renewal and subject to an AMR. The only portions that are within the scope of license renewal and subject to an AMR are the connections from the fire water headers up to the closed valves to the foam subsystem.

Based on its review, the staff found the applicant's response to RAI 2.3.3.A.9-2 acceptable. The applicant explained that the foam subsystem components in question are not within the scope of license renewal and not subject to an AMR because the foam subsystem is retired in place. The LR drawings inadvertently included highlighted portions of the foam subsystem in error. The staff concludes that there is reasonable assurance that the components were correctly excluded from the scope of license renewal and subject to an AMR. Therefore, the staff's concern described in RAI 2.3.3.A.9-2 is resolved.

In RAI 2.3.3.A.9-3, dated November 17, 2004, the staff stated that LR drawing LR-18030-C, sheet 6 shows that a sprinkler system is within the scope of license renewal and subject to an AMR, except for a portion that services the women's locker room. Areas within the scope of license renewal include administration building, lunch room and wash area, and new locker room and shops. Therefore, the staff requested that the applicant identify the basis for excluding a portion of the sprinkler system from within the scope of license renewal, since the rest of the system is within the scope of license renewal and subject to an AMR.

In its response, by letter dated December 17, 2004, the applicant stated that the portions of the fire water system in the administration building, as depicted on LR drawing LR-18030-C, sheet 6, as within scope and subject to an AMR, are incorrect.

The applicant further stated that original LRA Section 2.3.3.A.9 properly describes the portion of the fire water system as within the scope of license renewal and subject to AMR as "...the connecting fire water supply piping and valves from the pump discharge header to the Reactor Building and Turbine Building fire zones [excluding supplies to non-critical areas, (e.g., storage areas, changing rooms, locker rooms)]." The fire water system in the administration building is provided for commercial purposes and is not credited for compliance with 10 CFR 50.48. As described in UFSAR Section 10A.3.10, the only SR equipment located in the administration building is a DC power board located in the foam room. This area is protected by detection and alarm. There is no fire water suppression to this room. This UFSAR section further states that a fire in the administration building will not result in the loss of capability to achieve safe shutdown and that there are no sources of radioactivity in the building. Therefore, the portion of the fire water system located in the administration building is not within the scope of license renewal is not subject to an AMR.

Based on its review, the staff found the applicant's response to RAI 2.3.3.A.9-3 acceptable because it adequately explained that the administration building fire detection and protection system in the administration building is not credited to meet the requirements of 10 CFR 50.48 and is not part of the plant's license. Therefore, the staff's concern described in RAI 2.3.3.A.9-3 is resolved.

In RAI 2.3.3.A.9-6, dated November 17, 2004, the staff stated that the UFSAR requires at least 1000 gallons of fuel in the fire pump diesel fuel oil storage tank. LR drawing LR-18040-C, sheet 2 shows level instrumentation consisting of air tubing and other components supplying the level indicating instrumentation for a fuel oil storage tank, as excluded from within the scope of license renewal and subject to an AMR. Therefore, the staff asked that the applicant explain the apparent exclusion of these components from within the scope of license renewal and subject to an AMR.

In its response, by letter dated December 17, 2004, the applicant stated that LR drawing LR-18040-C, sheet 2 incorrectly reflects the current plant configuration. The drawing does not reflect the installation of a replacement tank (TANK-82-116) in place of the tank (TANK-88-20) shown on the drawing. TANK-82-116 has new fuel oil level instrumentation (LR-82-113) installed that does not require instrument air as a support system. Instrument LR-82-113 is now used in place of LI-82-28 to verify that the fuel oil supply for the NMP1 diesel fire pump is maintained at greater than or equal to 1000 gallons in compliance with UFSAR Appendix 10A, Section 2.5.2.3.2. Tank level verification is performed on a weekly basis.

Based on its review, the staff found the applicant's response to RAI 2.3.3.A.9-6 acceptable because it adequately explained that the components in question are not within the scope of license renewal and subject to an AMR, but were inadvertently left on the LR drawing that does not reflect the current plant configuration. The staff concludes that there is reasonable assurance that the components were correctly excluded from the scope of license renewal and not subject to an AMR. Therefore, the staff's concern described in RAI 2.3.3.A.9-6 is resolved.

In RAI 2.3.3.A.9-7, dated November 17, 2004, the staff stated that ALRA Table 2.3.3.A.9-1 includes the following component types as subject to an AMR: filters/strainers, flow elements, and orifices; however, the intended function assigned to these components is NSR functional support. ALRA Table 2.0-1 identifies intended functions that are applicable to these component types that are not identified in the ALRA Table 2.3.3.A.9-1. Aging management to ensure that the component level intended functions can be performed is necessary to ensure that the system level intended functions can be maintained. The intended functions include filtration and flow restriction. Therefore, the staff requested that the applicant describe how the intended functions for these components were assigned and evaluated.

In its response, by letter dated December 17, 2004, the applicant stated that a component function would be considered an intended function for license renewal only if failure of that component would cause the failure of a system intended function. Failure of the filtration or flow restriction functions for the above mentioned components would not prevent the NMP1 fire detection and protection system from performing its intended functions. Therefore, the only intended function credited for these components is NSR functional support, as identified in ALRA Table 2.3.3.A.9-1.

In evaluating the applicant's response to RAI 2.3.3.A.9-7, the staff found that it was incomplete and that review of ALRA Section 2.3.3.A.9 could not be completed. The applicant did not explain how the intended function, NSR functional support is applied to the component types in the fire detection and protection system, including piping, valves, strainers, pumps, and orifices, as requested in RAI 2.3.3.A.9-7. Therefore, the staff held a teleconference with the applicant on January 25, 2005, to discuss information necessary to resolve the staff's concern described in RAI 2.3.3.A.9-7. The product of the teleconference was an agreement by the applicant to transmit the required information in a follow-up letter.

By letter dated February 11, 2005, the applicant stated that NSR functional support is a "catch-all" function for NSR components. The applicant also provided a table identifying each component to its intended function, further explaining the use of NSR functional support.

Based on its review, the staff found the applicant's response to RAI 2.3.3.A.9-7, including additional information from the letter dated February 11, 2005, acceptable because they adequately explained what intended functions NSR functional support represent and how it is applied to all the component types in the fire detection and protection system, including piping, valves, strainers, pumps, and orifices. Additionally, the applicant provided a table identifying each component to its intended function, which further explains the use of NSR functional support. Therefore, the staff's concern described in RAI 2.3.3.A.9-7 is resolved.

2.3A.3.9.3 Conclusion

The staff reviewed the ALRA, RAI responses, and accompanying boundary drawings to determine whether any SSCs that should be within the scope of license renewal had not been identified by the applicant. No omissions were identified. In addition, the staff performed a review to determine whether any components that should be subject to an AMR had not been identified by the applicant. No omissions were identified. On the basis of its review, the staff concludes that there is reasonable assurance that the applicant had adequately identified the fire detection and protection system components that are within the scope of license renewal, as required by 10 CFR 54.4(a), and the fire detection and protection system components that are subject to an AMR, as required by 10 CFR 54.21(a)(1).

2.3A.3.10 NMP1 Hydrogen Water Chemistry System

2.3A.3.10.1 Summary of Technical Information in the Amended Application

In ALRA Section 2.3.3.A.10, the applicant described the hydrogen water chemistry system. The hydrogen water chemistry system and noble metal chemical addition systems are designed to mitigate intergranular stress corrosion cracking of the reactor recirculation piping and the RPV internals. The hydrogen water chemistry system injects hydrogen into the FW/HPCI system to suppress the radiolytic generated oxidant concentration in the reactor core regions. This significantly reduces the electrochemical potential of the reactor components and greatly reduces crack initiation and growth. The noble metal chemical addition system includes permanent monitoring equipment as well as connections for periodically injecting a noble metal solution. The hydrogen water chemistry system does not perform any intended functions for license renewal purposes and, therefore, is not described further. The monitoring portion of the noble metal chemical injection system does, however, perform an intended function. The monitoring portion draws a sample from the reactor water cleanup (RWCU) system, analyzes the effectiveness of the noble metal treatment in the durability monitor, and returns the sample to the RWCU system.

The failure of NSR SSCs in the hydrogen water chemistry system could potentially prevent the satisfactory accomplishment of an SR function.

The intended function, within the scope of license renewal, is to maintain mechanical and structural integrity to prevent spatial interactions.

In ALRA Table 2.3.3.A.10-1, the applicant identified the following hydrogen water chemistry system component types that are within the scope of license renewal and subject to an AMR:

- bolting
- flow element
- piping and fittings
- valves

2.3A.3.10.2 Staff Evaluation

The staff reviewed ALRA Section 2.3.3.A.10 and UFSAR Section X.M using the evaluation methodology described in SER Section 2.3. The staff conducted its review in accordance with the guidance described in SRP-LR Section 2.3.

In conducting its review, the staff evaluated the system functions described in the ALRA and UFSAR in accordance with the requirements of 10 CFR 54.4(a) to verify that the applicant had not omitted from the scope of license renewal any components with intended functions delineated under 10 CFR 54.4(a). The staff then reviewed those components that the applicant had identified as being within the scope of license renewal to verify that the applicant had not omitted any passive and long-lived components that should be subject to an AMR in accordance with the requirements of 10 CFR 54.21(a)(1).

2.3A.3.10.3 Conclusion

The staff reviewed the ALRA to determine whether any SSCs that should be within the scope of license renewal had not been identified by the applicant. No omissions were identified. In addition, the staff performed a review to determine whether any components that should be subject to an AMR had not been identified by the applicant. No omissions were identified. On the basis of its review, the staff concludes that there is reasonable assurance that the applicant had adequately identified the hydrogen water chemistry system components that are within the scope of license renewal, as required by 10 CFR 54.4(a), and the hydrogen water chemistry system components that are subject to an AMR, as required by 10 CFR 54.21(a)(1).

2.3A.3.11 NMP1 Liquid Poison System

2.3A.3.11.1 Summary of Technical Information in the Amended Application

In ALRA Section 2.3.3.A.11, the applicant described the liquid poison system. The liquid poison system is a standby, redundant, independent control system that is designed to bring the reactor to a cold shutdown condition in the unlikely event that the control rod system fails to shut down and hold the reactor sub-critical as the reactor cools and xenon decays.

The liquid poison system consists of an ambient pressure tank with immersion heater for low-temperature sodium pentaborate solution storage, two high-pressure positive displacement pumps for injecting the solution into the reactor core, two explosive actuated shear plug valves for isolating the liquid poison from the RPV until required, an in-vessel sparger ring, a test tank, two reactor coolant isolation check valves, pressure relief valves and associated piping, valves, instrumentation and controls.

The liquid poison system contains SR components that are relied upon to remain functional during and following DBEs. The failure of NSR SSCs in the liquid poison system could potentially prevent the satisfactory accomplishment of an SR function. In addition, the liquid poison system performs functions that support fire protection, EQ, and ATWS.

The component types subject to an AMR include the liquid poison tank, the liquid poison accumulators, liquid poison pumps, and the connecting piping, fittings and valves. It also includes the NSR portions starting at the test tank suction lines to the pumps and return lines to the test tank. The demin water makeup to the test tank and flush for the pumps is also subject to an AMR, as are the drains lines from the test tank, pumps, poison tank, and discharge line to the reactor.

The intended functions within the scope of license renewal include the following:

- maintains mechanical and structural integrity of NSR components to prevent spatial interactions with SR components

- provides pressure retaining boundary

- maintains mechanical and structural integrity of NSR components that provide structural support to attached SR components

In ALRA Table 2.3.3.A.11-1, the applicant identified the following liquid poison system component types that are within the scope of license renewal and subject to an AMR:

- bolting
- piping and fittings
- pumps
- tanks
- valves

2.3A.3.11.2 Staff Evaluation

The staff reviewed ALRA Section 2.3.3.A.11 and the UFSAR using the evaluation methodology described in SER Section 2.3. The staff conducted its review in accordance with the guidance described in SRP-LR Section 2.3.

In conducting its review, the staff evaluated the system functions described in the ALRA and UFSAR in accordance with the requirements of 10 CFR 54.4(a) to verify that the applicant had not omitted from the scope of license renewal any components with intended functions delineated under 10 CFR 54.4(a). The staff then reviewed those components that the applicant had identified as being within the scope of license renewal to verify that the applicant had not omitted any passive and long-lived components that should be subject to an AMR in accordance with the requirements of 10 CFR 54.21(a)(1).

The staff's review of original LRA Section 2.3.3.A.11 identified an area in which additional information was necessary to complete the evaluation of the applicant's scoping and screening results. The applicant responded to the staff's RAIs as discussed below.

In RAI 2.3-3, dated October 11, 2005, the staff stated that the liquid poison system contains two elements that monitor the liquid poison storage tank temperature. One of these components is within the scope of license renewal since it is an SR component. The other component is NSR and has no license renewal intended function. The in-scope temperature element itself is an active component and, therefore, not subject to AMR. However, the temperature sensor is housed in a thermowell that is subject to AMR. Therefore, the staff requested that the applicant identify which is the in-scope temperature element, TE 41-35 or TE 41-28.

In its response, by letter dated October 28, 2005, the applicant indicated that the thermowell for temperature element 41-28 is SR and is within the scope of license renewal. Temperature element 41-28 is SR because it monitors the sodium pentaborate solution in the liquid poison storage tank and provides input to the temperature controller that drives the heater that maintains the temperature in the solution within the proper range. Therefore, the staff's concern described in RAI 2.3-3 is resolved.

2.3A.3.11.3 Conclusion

The staff reviewed the ALRA and RAI response to determine whether any SSCs that should be within the scope of license renewal had not been identified by the applicant. No omissions were identified. In addition, the staff performed a review to determine whether any components that should be subject to an AMR had not been identified by the applicant. No omissions were identified. On the basis of its review, the staff concludes that there is reasonable assurance that the applicant had adequately identified the liquid poison system components that are within the scope of license renewal, as required by 10 CFR 54.4(a), and the liquid poison system components that are subject to an AMR, as required by 10 CFR 54.21(a)(1).

2.3A.3.12 NMP1 Miscellaneous Non-Contaminated Vents and Drains System

2.3A.3.12.1 Summary of Technical Information in the Amended Application

In ALRA Section 2.3.3.A.12, the applicant described the miscellaneous non-contaminated vents and drains system. The miscellaneous non-contaminated vents and drains system is designed to route the non-contaminated effluents to floor drains, building sumps, the discharge tunnel, and the turbine building equipment drain tank.

The failure of NSR SSCs in the miscellaneous non-contaminated vents and drains system could potentially prevent the satisfactory accomplishment of an SR function.

The intended function, within the scope of license renewal, is to maintain mechanical and structural integrity to prevent spatial interactions.

In ALRA Table 2.3.3.A.12-1, the applicant identified that the piping and fittings component type of the miscellaneous non-contaminated vents and drains system is within the scope of license renewal and subject to an AMR.

2.3A.3.12.2 Staff Evaluation

The staff reviewed ALRA Section 2.3.3.A.12 and UFSAR Section XII.A.2.2 using the evaluation methodology described in SER Section 2.3. The staff conducted its review in accordance with the guidance described in SRP-LR Section 2.3.

In conducting its review, the staff evaluated the system functions described in the ALRA and UFSAR in accordance with the requirements of 10 CFR 54.4(a) to verify that the applicant had not omitted from the scope of license renewal any components with intended functions delineated under 10 CFR 54.4(a). The staff then reviewed those components that the applicant had identified as being within the scope of license renewal to verify that the applicant had not omitted any passive and long-lived components that should be subject to an AMR in accordance with the requirements of 10 CFR 54.21(a)(1).

2.3A.3.12.3 Conclusion

The staff reviewed the ALRA to determine whether any SSCs that should be within the scope of license renewal had not been identified by the applicant. No omissions were identified. In addition, the staff performed a review to determine whether any components that should be subject to an AMR had not been identified by the applicant. No omissions were identified. On the basis of its review, the staff concludes that there is reasonable assurance that the applicant had adequately identified the miscellaneous non-contaminated vents and drains system components that are within the scope of license renewal, as required by 10 CFR 54.4(a), and the miscellaneous non-contaminated vents and drains system components that are subject to an AMR, as required by 10 CFR 54.21(a)(1).

2.3A.3.13 NMP1 Neutron Monitoring System

2.3A.3.13.1 Summary of Technical Information in the Amended Application

In ALRA Section 2.3.3.A.13, the applicant described the neutron monitoring system. The neutron monitoring system monitors neutron flux level, in the reactor, in three separate ranges: source range, intermediate range, and power range. This system also includes the capability to calibrate the local power range monitors during normal operation.

The source range monitoring and intermediate range monitoring systems are equipped with mechanically retractable detector assemblies which allow the operator to insert the detectors into the reactor core, and then retract the detectors to a low neutron flux region below the core when the proper point in reactor operation is reached. The local power range monitoring detectors are installed at fixed locations in the reactor core. The average power range monitoring system utilizes the signals from the local power range monitoring detectors to provide average power range signals for monitoring.

The neutron monitoring system also includes the traversing in-core probe system which provides the capability to calibrate the local power range monitors during normal operation. The traversing in-core probe system consists of four identical trains, each containing ionization chamber detectors, indexing mechanism, ball valve, shear valve, chamber shield, drive mechanism and drive control unit. The drive mechanism drives the traversing in-core probe detector through the ball and shear valves and indexing mechanism into calibration tubes and then guide tubes located in the reactor core. The ball and shear valves function as reactor coolant isolation valves if a leak were to occur in a calibration or guide tube. The drive mechanisms, indexer mechanisms and calibration and guide tubes are purged continuously with nitrogen gas.

The neutron monitoring system contains SR components that are relied upon to remain functional during and following DBEs.

The components subject to an AMR include the four traversing in-core probe system ball valves and their associated guide tubes from the shear valves to the containment penetration. The dry tubes for source range monitoring and intermediate range monitoring detectors are not included in the system boundary. The dry tubes are included in the RPV internals (Section 2.3.1.A.2).

The intended function, within the scope of license renewal, is to provide pressure retaining boundary.

In ALRA Table 2.3.3.A.13-1, the applicant identified the following neutron monitoring system component types that are within the scope of license renewal and subject to an AMR:

- bolting
- piping
- valves

2.3A.3.13.2 Staff Evaluation

The staff reviewed ALRA Section 2.3.3.A.13 and the UFSAR using the evaluation methodology described in SER Section 2.3. The staff conducted its review in accordance with the guidance described in SRP-LR Section 2.3.

In conducting its review, the staff evaluated the system functions described in the ALRA and UFSAR in accordance with the requirements of 10 CFR 54.4(a) to verify that the applicant had not omitted from the scope of license renewal any components with intended functions delineated under 10 CFR 54.4(a). The staff then reviewed those components that the applicant had identified as being within the scope of license renewal to verify that the applicant had not omitted any passive and long-lived components that should be subject to an AMR in accordance with the requirements of 10 CFR 54.21(a)(1).

2.3A.3.13.3 Conclusion

The staff reviewed the ALRA to determine whether any SSCs that should be within the scope of license renewal had not been identified by the applicant. No omissions were identified. In addition, the staff performed a review to determine whether any components that should be subject to an AMR had not been identified by the applicant. No omissions were identified. On the basis of its review, the staff concludes that there is reasonable assurance that the applicant had adequately identified the neutron monitoring system components that are within the scope of license renewal, as required by 10 CFR 54.4(a), and the neutron monitoring system components that are subject to an AMR, as required by 10 CFR 54.21(a)(1).

2.3A.3.14 NMP1 Process Radiation Monitoring System

2.3A.3.14.1 Summary of Technical Information in the Amended Application

In ALRA Section 2.3.3.A.14, the applicant described the process radiation monitoring system. The process radiation monitoring system is designed to monitor radiation levels of liquid and gaseous processes throughout the plant, assist in controlling the release of radioactive byproducts, and provide for personnel safety by warning of abnormal radiation levels. The process radiation monitoring system consists of the following independent subsystems: main steam line radiation monitoring, air-ejector off-gas radiation monitoring, stack effluent radiation monitoring, process liquid radiation monitoring, reactor building ventilation radiation monitoring, emergency cooling condenser vent monitor, and refueling bridge high radiation monitor. Each of these subsystems consists of an appropriate detector and monitor and provide readouts, alarms and computer points to aide the operator. Only the air-ejector off-gas, stack effluent and process liquid radiation monitors draw a sample from their respective process streams. These subsystems were evaluated and determined to not be within the scope of license renewal. The remaining subsystems measure radiation levels directly on the process piping or local area.

The process radiation monitoring system contains SR components that are relied upon to remain functional during and following DBEs. In addition, the process radiation monitoring system performs functions that support EQ.

The in-scope components for the process radiation monitoring system are active components. Therefore, there are no components requiring an AMR for the process radiation monitoring system.

2.3A.3.14.2 Staff Evaluation

The staff reviewed ALRA Section 2.3.3.A.14 and UFSAR Section VIII.C.3 using the evaluation methodology described in SER Section 2.3. The staff conducted its review in accordance with the guidance described in SRP-LR Section 2.3.

In conducting its review, the staff evaluated the system functions described in the ALRA and UFSAR in accordance with the requirements of 10 CFR 54.4(a) to verify that the applicant had not omitted from the scope of license renewal any components with intended functions delineated under 10 CFR 54.4(a). The staff then reviewed those components that the applicant had identified as being within the scope of license renewal to verify that the applicant had not omitted any passive and long-lived components that should be subject to an AMR in accordance with the requirements of 10 CFR 54.21(a)(1).

2.3A.3.14.3 Conclusion

The staff reviewed the ALRA to determine whether any SSCs that should be within the scope of license renewal had not been identified by the applicant. No omissions were identified. In addition, the staff performed a review to determine whether any components that should be subject to an AMR had not been identified by the applicant. No omissions were identified. On the basis of its review, the staff concludes that there is reasonable assurance that the applicant had adequately identified the process radiation monitoring system components that are within the scope of license renewal, as required by 10 CFR 54.4(a); and the process radiation monitoring system components that are subject to an AMR, as required by 10 CFR 54.21(a)(1).

2.3A.3.15 NMP1 Radioactive Waste Disposal Building HVAC System

2.3A.3.15.1 Summary of Technical Information in the Amended Application

In ALRA Section 2.3.3.A.15, the applicant described the radioactive waste disposal building HVAC system. The radioactive waste disposal building HVAC system provides heating and ventilation for personnel comfort, equipment protection and for controlling possible radioactivity release to the atmosphere. Air is drawn into the system through an inlet louver, filter and heater by two supply fans and distributed throughout the waste building and waste building extension. An air outlet is located in each room and at each piece of equipment where radioactive contamination could be released. The exhaust ductwork leads to two trains of inlet and outlet dampers, roughing and high efficiency filters, and exhaust fans. The discharge from all of the exhaust fans travels through one of three backdraft dampers and exits the station through the vent stack.

The radioactive waste disposal building HVAC system contains SR components that are relied upon to remain functional during and following DBEs.

The intended function, within the scope of license renewal, is to provide pressure retaining boundary.

In ALRA Table 2.3.3.A.15-1, the applicant identified that the dampers component type of the radioactive waste disposal building HVAC system is within the scope of license renewal and subject to an AMR.

2.3A.3.15.2 Staff Evaluation

The staff reviewed ALRA Section 2.3.3.A.15 and UFSAR Section III.C.1.4 using the evaluation methodology described in SER Section 2.3. The staff conducted its review in accordance with the guidance described in SRP-LR Section 2.3.

In conducting its review, the staff evaluated the system functions described in the ALRA and UFSAR in accordance with the requirements of 10 CFR 54.4(a) to verify that the applicant had not omitted from the scope of license renewal any components with intended functions delineated under 10 CFR 54.4(a). The staff then reviewed those components that the applicant had identified as being within the scope of license renewal to verify that the applicant had not omitted any passive and long-lived components that should be subject to an AMR in accordance with the requirements of 10 CFR 54.21(a)(1).

2.3A.3.15.3 Conclusion

The staff reviewed the ALRA to determine whether any SSCs that should be within the scope of license renewal had not been identified by the applicant. No omissions were identified. In addition, the staff performed a review to determine whether any components that should be subject to an AMR had not been identified by the applicant. No omissions were identified. On

the basis of its review, the staff concludes that there is reasonable assurance that the applicant had adequately identified the radioactive waste disposal building HVAC system components that are within the scope of license renewal, as required by 10 CFR 54.4(a), and the radioactive waste disposal building HVAC system components that are subject to an AMR, as required by 10 CFR 54.21(a)(1).

2.3A.3.16 NMP1 Radioactive Waste System

2.3A.3.16.1 Summary of Technical Information in the Amended Application

In ALRA Section 2.3.3.A.16, the applicant described the radioactive waste system. The radioactive waste system is designed to meet the following objectives:

- collect and process all radioactive waste generated without limiting normal station operation

- collect and process radioactive wastes for disposal, or transfer to a vendor for processing and disposal

- release radioactive material to the environment in a controlled manner so that all releases are within the limits of 10 CFR 20 and the TSs

- retain radioactive wastes, if they accidentally leak from the systems, so that they can be recovered and reprocessed

The radioactive waste system consists of the gaseous waste system, liquid waste system, and solid waste system. Gaseous radioactive wastes include airborne particulates as well as gases vented from process equipment. Sources of gaseous waste activity are the offgas system effluent, steam-packing exhauster system effluent, and building ventilation exhausts. The liquid waste system processes the liquids collected in equipment drains and floor drains in areas that are potentially contaminated with radioactive materials. The wastes are collected in the floor drain sumps located within the drywell, the reactor building (RB), the turbine building (TB), the radioactive waste solidification and storage building, the offgas building, and the waste disposal building (WDB). The liquids in these floor drain sumps are pumped into the floor drain collector, waste neutralizer tank, or utility collector tank, which are located in the WDB. The solid waste system processes spent resins, filter sludge, and concentrated waste. It also is designed for collection and shipment of lowlevel solids. Wastes may be processed or solidified onsite, or transferred to a vendor for processing.

The radioactive waste system contains SR components that are relied upon to remain functional during and following DBEs. The failure of NSR SSCs in the radioactive waste system could potentially prevent the satisfactory accomplishment of an SR function. In addition, the radioactive waste system performs functions that support EQ.

The intended functions within the scope of license renewal include the following:

- provides flood protection barrier
- maintains mechanical and structural integrity of NSR components to prevent spatial interactions with SR components

- provides structural support to NSR components whose failure could prevent accomplishment of SR function(s)

- provides pressure retaining boundary

- maintains mechanical and structural integrity of NSR components that provide structural support to attached SR components

In ALRA Table 2.3.3.A.16-1, the applicant identified the following radioactive waste system component types that are within the scope of license renewal and subject to an AMR:

- bolting
- filters/strainers
- flow element
- heat exchangers
- piping and fittings
- pumps
- separator
- tanks
- valves

2.3A.3.16.2 Staff Evaluation

The staff reviewed ALRA Section 2.3.3.A.16 and UFSAR Section XII.A using the evaluation methodology described in SER Section 2.3. The staff conducted its review in accordance with the guidance described in SRP-LR Section 2.3.

In conducting its review, the staff evaluated the system functions described in the ALRA and UFSAR in accordance with the requirements of 10 CFR 54.4(a) to verify that the applicant had not omitted from the scope of license renewal any components with intended functions delineated under 10 CFR 54.4(a). The staff then reviewed those components that the applicant had identified as being within the scope of license renewal to verify that the applicant had not omitted any passive and long-lived components that should be subject to an AMR in accordance with the requirements of 10 CFR 54.21(a)(1).

The staff's review of original LRA Section 2.3.3.A.16 identified areas in which additional information was necessary to complete the review of the applicant's scoping and screening results. The applicant responded to the staff's RAIs as discussed below.

In RAIs 2.3.3.A.16-1 through 2.3.3.A.16-3, dated November 19, 2004, the staff requested that the applicant clarify inconsistencies between the original LRA and LR drawings that the staff encountered in its review. The applicant's response, by letter dated December 22, 2004, has been subsequently incorporated in the ALRA as discussed below.

In its ALRA, dated July 14, 2005, the applicant provided the staff with revised LR drawings correcting the inconsistencies and accurately depicting all the components subject to AMR, including those subject under 10 CFR 54.4(a)(2). Therefore, the staff's concerns described in RAIs 2.3.3.A.16-1 through 2.3.3.A.16-3 are resolved.

In RAI 2.3.3.A.16-4, dated November 19, 2004, the staff stated that LR drawing LR-18006-C, sheet 3 shows piping sleeves for two pipelines as within the scope of license renewal and subject to an AMR. The piping sleeves are passive and long-lived components. Therefore, the staff requested that the applicant clarify if these components are included with a component type which is listed in original LRA Table 2.3.3.A.16-1 or justify the exclusion of these components from subject to an AMR in accordance with the requirements of 10 CFR 54.21(a)(1).

The staff further stated that LR drawing LR-18006-C, sheet 3 shows two pipelines from the N_2 leak test on another LR drawing to the above-mentioned sleeve pipes as subject to an AMR. The AMR boundary flag for one pipeline indicates that this pipeline is included in the containment system. No boundary flag is shown for the other pipeline. Therefore, the staff requested that the applicant clarify whether these pipelines are included in the NMP1 radioactive waste system, or, if not, show the location of the radioactive waste AMR boundary interface with other license renewal systems.

In its response, by letter dated December 22, 2004, the applicant stated that the sleeve pipes represent primary containment penetrations X-25 and X-26, respectively. These penetrations are subject to AMR and are addressed in original LRA Section 2.4.A.1. In addition, the applicant clarified that both these lines should have a containment boundary flag. However, the boundary flag for the latter pipeline was inadvertently omitted.

Based on its review, the staff found the applicant's response to RAI 2.3.3.A.16-4 acceptable because it adequately clarified that the sleeve pipes are part of the primary containment structure penetration and are addressed in original LRA Section 2.4.A.1. Therefore, the staff's concern described in RAI 2.3.3.A.16-4 is resolved.

In RAI 2.3.3.A.16-5, dated November 19, 2004, the staff stated that LR drawing LR 18018-C, sheet 1 shows the pipeline from two shutdown cooling heat exchangers as being subject to an AMR. However, the pipeline from the shutdown cooling heat exchanger is shown as excluded as subject to an AMR, although an AMR boundary flag indicates that this line should be within the scope of the radioactive waste system. Therefore, the staff requested that the applicant clarify whether this is an inadvertent error in highlighting the LR drawing, or, if not, explain how the latter pipeline differs from the former pipelines.

In addition, the staff stated that LR drawing LR 18045-C, sheet 7 shows the shutdown cooling system drains line, which appears to be the continuation of the above-mentioned pipelines to the reactor building equipment drain tank (RBEDT). However, an AMR boundary flag indicates that the portion of this line that is shown on this drawing is within the scope of the compressed air system (CAS). Therefore, the staff requested that the applicant to explain why this line is included in the compressed air system.

In its response, by letter dated, December 22, 2004, the applicant stated that LR drawing LR-18018-C is incorrect for the pipeline from the shutdown cooling heat exchanger to the RBEDT and should have been highlighted in red. The applicant provided corrections to the boundary flag locations and locations on the LR drawings.

Based on its review, the staff found the applicant's response to RAI 2.3.3.A.16-5 acceptable because it stated that the LR drawings are incorrect and identified the required corrections to the drawings. Therefore, the staff's concern described in RAI 2.3.3.A.16-5 is resolved.

In RAI 2.3.3.A.16-6, dated November 19, 2004, the staff stated that LR drawings show the pressure and level instruments' drain lines and their associated components (fittings and valves) tie in to the pipeline which runs to the RBEDT. That pipeline is shown on these drawings as within the scope of license renewal and subject to an AMR. Also, another LR drawing shows a pipeline which connects fuel pool cooling system drains to the reactor building drain tanks as subject to an AMR. However, one of the previous LR drawings shows that pipeline as being excluded as subject to an AMR. Also, this pipeline is not highlighted in red on that LR drawing, although an AMR boundary flag shows it as being within the scope of the RWS. Further, this AMR boundary flag indicates that a portion of the pipeline from the fuel pool cooling drains on an LR drawing is within the scope of the CAS.

Therefore, to resolve the above discrepancies, the staff requested that the applicant:

(a)   Provide drawings or descriptive information that shows how the instrumentation drains header connects to the fuel pool cooling system drains pipeline.

(b)   Provide drawings or descriptive information that clearly identify portions of the radioactive waste system to RBEDT which are within the scope of license renewal and subject to an AMR, and eliminate inconsistencies between the above-mentioned drawings.

In its response to RAI 2.3.3.A.16-6a, by letter dated December 22, 2004, the applicant stated:

The instrument drain headers identified in the RAI do not connect to the fuel pool cooling system drains pipeline. For NMP1, the line identification (i.e., 89-2-C) is not a unique piping component number. Using the line identification legend shown on drawing LR-18000-C, Sheet 1 (location E3), the line identifier "89-2-C" indicates a pipe in system 89 (RWS) that is 2 inches in diameter and made of carbon steel. Therefore, this identification applies to every 2-inch, carbon steel line in system 89 (RWS) regardless of its function. This identification does not, therefore, imply a connection between the identically designated piping segments described in this RAI.

Based on its review, the staff found the applicant's response to RAI 2.3.3.A.16-6a acceptable because the applicant adequately explained why there is no connection between the pipelines in question. Therefore, the staff's concern described in RAI 2.3.3.A.16-6a is resolved.

In its response to RAI 2.3.3.A.16-6b, by letter dated December 22, 2004, the applicant stated:

The depiction of the input lines to the RBEDT on drawing LR-18045-C, Sheets 7 and 7A, that are contrary to the above-referenced description are drafting errors. AMR boundary flag designators contrary to this description, including the ones referencing the CAS, are also drafting errors.

As stated in the response to RAI 2.3.3.A.16-1, drawing LR-18045-C, Sheet 7A, provides no additional information to that shown on Sheet 7 and should, therefore, be disregarded.

Based on it review of the applicant's response, the staff found that it was incomplete and that its review of original LRA Section 2.3.3.A.16 could not be completed. Although the applicant stated that the depiction of the input lines to the RBEDT on drawing LR-18045-C, sheets 7 and 7A, are contrary to the above-referenced description due to drafting errors, it did not adequately identify which of the LR drawings are correct. In addition, the applicant stated that LR drawing LR-18045-C, sheet 7A does not add any information to LR-18045-C, sheet 7 and should be disregarded; however, the applicant did not explain the inconsistency between these two sheets. As a result, the staff held a teleconference with the applicant on January 27, 2005, to discuss information necessary to resolve its concern described in RAI 2.3.3.A.16-6b. The product of the teleconference was an agreement by the applicant to transmit the required information in a follow-up letter.

In its follow-up response, by letter dated February 11, 2005, the applicant provided detailed descriptive information that resolved RAIs 2.3.3.A.16-1 through 2.3.3.A.16-3. That response is also applicable to the concern in 2.3.3.A.16-6b and provides complete resolution.

Based on its review, the staff found the applicant's response to RAI 2.3.3.A.16-6b, including the information in the letter dated February 11, 2005, acceptable because it adequately described all of the portions of the NMP1 radioactive waste system. Additionally, this information described the impact on the original LRA. Therefore, the staff's concern described in RAI 2.3.3.A.16-6b is resolved.

2.3A.3.16.3 Conclusion

The staff reviewed the ALRA, RAI responses, and accompanying scoping boundary drawings to determine whether any SSCs that should be within the scope of license renewal had not been identified by the applicant. No omissions were identified. In addition, the staff performed a review to determine whether any components that should be subject to an AMR had not been identified by the applicant. No omissions were identified. On the basis of its review, the staff concludes that there is reasonable assurance that the applicant had adequately identified the radioactive waste system components that are within the scope of license renewal, as required by 10 CFR 54.4(a), and the radioactive waste system components that are subject to an AMR, as required by 10 CFR 54.21(a)(1).

2.3A.3.17 NMP1 Reactor Building Closed Loop Cooling Water System

2.3A.3.17.1 Summary of Technical Information in the Amended Application

In ALRA Section 2.3.3.A.17, the applicant described the reactor building closed loop cooling (RBCLC) water system. The RBCLC water system is designed to provide demineralized water to cool reactor auxiliary equipment located in the primary containment, RB, TB, and WDB. The closed loop permits isolation of systems containing radioactive liquids from the service water.

The RBCLC water system contains SR components that are relied upon to remain functional during and following DBEs. The failure of NSR SSCs in the RBCLC water system could potentially prevent the satisfactory accomplishment of an SR function. In addition, the RBCLC water system performs functions that support fire protection and EQ.

The intended functions within the scope of license renewal include the following:

- provides filtration

- provides heat transfer

- maintains mechanical and structural integrity of NSR components to prevent spatial interactions with SR components

- provides structural support to NSR components whose failure could prevent accomplishment of SR function(s)

- provides pressure retaining boundary

- maintains mechanical and structural integrity of NSR components that provide structural support to attached SR components

- provides flow restriction

In ALRA Table 2.3.3.A.17-1, the applicant identified the following RBCLC water system component types that are within the scope of license renewal and subject to an AMR:

- actuator
- bolting
- filters/strainers
- flow elements
- heat exchangers
- orifices
- piping and fittings
- pumps
- temperature elements
- valves

2.3A.3.17.2 Staff Evaluation

The staff reviewed ALRA Section 2.3.3.A.17 and UFSAR Section X.D using the evaluation methodology described in SER Section 2.3. The staff conducted its review in accordance with the guidance described in SRP-LR Section 2.3.

In conducting its review, the staff evaluated the system functions described in the ALRA and UFSAR in accordance with the requirements of 10 CFR 54.4(a) to verify that the applicant had not omitted from the scope of license renewal any components with intended functions delineated under 10 CFR 54.4(a). The staff then reviewed those components that the applicant had identified as being within the scope of license renewal to verify that the applicant had not omitted any passive and long-lived components that should be subject to an AMR in accordance with the requirements of 10 CFR 54.21(a)(1).

The staff's review of original LRA Section 2.3.3.A.17 identified areas in which additional information was necessary to complete the review of the applicant's scoping and screening results. The applicant responded to the staff's RAIs as discussed below.

In RAIs 2.3.3.A.17-1 and 2.3.3.A.17-2, dated November 19, 2004, the staff requested that the applicant clarify information given on a license renewal boundary drawing concerning SSC's that are within scope of license renewal in accordance with 10 CFR 54.4(a). The applicant response, by letter dated December 22, 2004, has been subsequently incorporated in the ALRA as discussed below.

In its ALRA, dated July 14, 2005, the applicant provided revised LR drawings which identify SSC's in scope and subject to AMR under 10 CFR 54.4(a)(2). Based on its review of the information submitted in the ALRA, the staff's concerns described in RAIs 2.3.3.A.17-1 and 2.3.3.A.17-2 are resolved.

2.3A.3.17.3 Conclusion

The staff reviewed the ALRA, RAI responses, and accompanying scoping boundary drawings to determine whether any SSCs that should be within the scope of license renewal had not been identified by the applicant. No omissions were identified. In addition, the staff performed a review to determine whether any components that should be subject to an AMR had not been identified by the applicant. No omissions were identified. On the basis of its review, the staff concludes that there is reasonable assurance that the applicant had adequately identified the RBCLC water system components that are within the scope of license renewal, as required by 10 CFR 54.4(a), and the RBCLC water system components that are subject to an AMR, as required by 10 CFR 54.21(a)(1).

2.3A.3.18 NMP1 Reactor Building HVAC System

2.3A.3.18.1 Summary of Technical Information in the Amended Application

In ALRA Section 2.3.3.A.18, the applicant described the reactor building HVAC system. The reactor building HVAC system is designed to control the RB atmosphere within limits during normal and emergency operating conditions. Additionally, the system is an alternative system for venting the primary containment to the atmosphere, if necessary. The reactor building HVAC system consists of the reactor building normal ventilation system and the reactor building emergency ventilation system. The reactor building normal ventilation system provides clean fresh air to the RB, removes air from areas where excessive heat concentration and potential airborn contamination exist, and maintains a negative pressure in the RB relative to the atmosphere by regulating the amount of outside air introduced into the building. The clean air is required to remove air from areas where excessive heat concentration exists. The normal ventilation system automatically isolated upon initiation of the emergency ventilation system. The reactor building emergency ventilation system removes air from areas where excessive heat concentration and potential airborne contamination exists, maintains a negative pressure in the RB relative to atmosphere, and removes and filters contaminated air during accident conditions. The reactor building emergency ventilation system is a standby system consisting of redundant filter trains, which operates in the event of an accident or normal ventilation failure. This system can also be used to process the drywell and torus atmospheres when venting.

The reactor building HVAC system contains SR components that are relied upon to remain functional during and following DBEs. In addition, the reactor building HVAC system performs functions that support EQ.

The intended functions within the scope of license renewal include the following:

- provides filtration
- provides rated fire barrier
- provides pressure retaining boundary

In ALRA Table 2.3.3.A.18-1, the applicant identified the following reactor building HVAC system component types that are within the scope of license renewal and subject to an AMR:

- blowers
- bolting
- ducting
- filters
- flow elements
- piping and fittings
- temperature elements
- valves and dampers

2.3A.3.18.2 Staff Evaluation

The staff reviewed ALRA Section 2.3.3.A.18 and UFSAR Sections VI.E.2 and VII.H using the evaluation methodology described in SER Section 2.3. The staff conducted its review in accordance with the guidance described in SRP-LR Section 2.3.

In conducting its review, the staff evaluated the system functions described in the ALRA and UFSAR in accordance with the requirements of 10 CFR 54.4(a) to verify that the applicant had not omitted from the scope of license renewal any components with intended functions delineated under 10 CFR 54.4(a). The staff then reviewed those components that the applicant had identified as being within the scope of license renewal to verify that the applicant had not omitted any passive and long-lived components that should be subject to an AMR in accordance with the requirements of 10 CFR 54.21(a)(1).

2.3A.3.18.3 Conclusion

The staff reviewed the ALRA to determine whether any SSCs that should be within the scope of license renewal had not been identified by the applicant. No omissions were identified. In addition, the staff performed a review to determine whether any components that should be subject to an AMR had not been identified by the applicant. No omissions were identified. On the basis of its review, the staff concludes that there is reasonable assurance that the applicant had adequately identified the reactor building HVAC system components that are within the scope of license renewal, as required by 10 CFR 54.4(a), and the reactor building HVAC system components that are subject to an AMR, as required by 10 CFR 54.21(a)(1).

2.3A.3.19 NMP1 Reactor Water Cleanup System

2.3A.3.19.1 Summary of Technical Information in the Amended Application

In ALRA Section 2.3.3.A.19, the applicant described the RWCU system. The RWCU system is designed to maintain high reactor water purity in order to: minimize deposits on fuel clad surfaces by reducing the amount of water-borne impurities in the primary system and reduce the secondary sources of beta and gamma radiation resulting from the deposition of corrosion products, fission products, and impurities in the primary system. The RWCU system continuously purifies a portion of the reactor recirculation flow and reactor bottom head drain flow with a minimum of heat loss from the cycle. Water is normally removed at reactor pressure from one of the reactor recirculation loops and the reactor bottom head drain line, cooled in regenerative and non-regenerative heat exchangers, reduced in pressure, filtered, demineralized, and pumped through the shell side of the regenerative heat exchanger to the RPV through the FW/HPCI system. Whenever reactor pressure is insufficient to maintain suction pressure at the main cleanup pumps, an auxiliary pump is used.

The RWCU system contains SR components that are relied upon to remain functional during and following DBEs. The failure of NSR SSCs in the RWCU system could potentially prevent the satisfactory accomplishment of an SR function. In addition, the RWCU system performs functions that support fire protection and EQ.

The intended functions within the scope of license renewal include the following:

- maintains mechanical and structural integrity of NSR components to prevent spatial interactions with SR components

- provides pressure retaining boundary

- maintains mechanical and structural integrity of NSR components that provide structural support to attached SR components

In ALRA Table 2.3.3.A.19-1, the applicant identified the following RWCU system component types that are within the scope of license renewal and subject to an AMR:

- bolting
- heat exchangers
- filters
- flow elements
- flow gauges
- piping and fittings
- pumps
- tanks
- valves

2.3A.3.19.2 Staff Evaluation

The staff reviewed ALRA Section 2.3.3.A.19 and UFSAR Section X.B using the evaluation methodology described in SER Section 2.3. The staff conducted its review in accordance with the guidance described in SRP-LR Section 2.3.

In conducting its review, the staff evaluated the system functions described in the ALRA and UFSAR in accordance with the requirements of 10 CFR 54.4(a) to verify that the applicant had not omitted from the scope of license renewal any components with intended functions delineated under 10 CFR 54.4(a). The staff then reviewed those components that the applicant had identified as being within the scope of license renewal to verify that the applicant had not omitted any passive and long-lived components that should be subject to an AMR in accordance with the requirements of 10 CFR 54.21(a)(1).

The staff's review of original LRA Section 2.3.3.A.19 identified areas in which additional information was necessary to complete the review of the applicant's scoping and screening results. The applicant responded to the staff's RAIs as discussed below.

In RAI 2.3.3.A.19-1, dated November 19, 2004, the staff stated that drawing LR-1809-C, sheet 1 shows oil coolers for the clean-up pumps to be within the "CU" system boundary and requiring an AMR. The original LRA Table 2.3.3.A-19-1 lists heat exchangers as a component type; however, original LRA Table 3.3.2.A-17-1 does not include heat exchangers with a lubricating oil environment and original LRA Section 3.3.2.A.17 does not list lubricating oil as an environment to which the RWCU system is exposed. Therefore, the staff requested that the applicant confirm that the clean-up pump oil coolers have been properly evaluated within the original LRA or justify their exclusion from requiring an AMR.

In its response, by letter dated December 22, 2004, the applicant stated that the LR drawing LR-18009-C, sheet 1, shows the cooling water side of the heat exchangers as being subject to AMR. This is because of the "Pressure Boundary" intended function for the RBCLC system. The shell side of the heat exchanger is not SR so it is not shown as within the scope of license renewal (depicted in black on the drawing), and the heat exchanger itself does not have an LR intended function of heat transfer. Therefore, the drawing boundary flags are incorrect. The "LR-CU" side of each of those flags should be solid blue. Consistent with original LRA Section 2.3.3.A.19, the pump oil coolers are within the scope of license renewal and subject to AMR to meet 10 CFR 54.4(a)(2), since they are NSR equipment containing liquid in the vicinity of SR components. Per LR drawing convention, components within the scope of license renewal and subject to an AMR for criterion 10 CFR 54.4(a)(2) only are not identified in red. The only heat exchanger within the RWCU system that is subject to AMR for criterion 10 CFR 54.4(a)(1) is the non-regenerative heat exchanger, which does not have a lube oil environment.

Based on its review, the staff found that the applicant's response was incomplete and that its review of original LRA Section 2.3.3.A.19 could not be completed because:

- Although the applicant explained that the pump oil cooler is within the scope of license renewal and subject to an AMR in accordance with 10 CFR 54.4(a)(2), original LRA Tables 2.3.3.A.19-1 and 3.3.3.A-17-1 do not include heat exchangers with the intended function to prevent failure from affecting SR equipment in a lubricating oil environment.

2-93

- In its response, the applicant stated that the drawing boundary flags are incorrect and the "LR-CU" side of each of those flags should be solid blue. This does not appear to be correct since the oil cooler tubes are within the scope of license renewal with the pressure boundary intended function for the reactor building closed loop cooling system.

As a result, the staff held a teleconference with the applicant on January 27, 2005 to discuss information necessary to resolve the staff's concern described in RAI 2.3.3.A.19-1. The product of this teleconference was an agreement by the applicant to transmit the required information in a follow-up letter.

In its follow-up response, by letter dated February 11, 2005, the applicant stated that the RWCU pump oil coolers are within the scope of license renewal since they are a pressure boundary for the reactor building closed RBCLC system. As such, only the end covers, tube sheets and tubes exposed to RBCLC water are within the scope of license renewal. The external surface of the tubes exposed to the lubricating oil are within the scope of license renewal, but have no aging effects requiring management.

In the ALRA, submitted by the applicant on July 14, 2005, ALRA Tables 2.3.3.A.19-1 and 3.3.3.A-17-1 are revised and now include the heat exchangers with the intended function to prevent failure from affecting SR equipment in a lubricating oil environment.

Based on its review, the staff found the applicant's response to RAI 2.3.3.A.19-1 acceptable because the applicant stated that the pump oil cooler is within the scope of license renewal and subject to an AMR and the applicant clarified that the drawing boundary flags are incorrect and that the tubes of the RWCU heat exchanger should have been shown within the scope of license renewal and subject to an AMR. Therefore, the staff's concern described in RAI 2.3.3.A.19-1 is resolved.

In RAI 2.3.3.A.19-2, dated November 19, 2004, the staff requested that the applicant clarify the inconsistencies between the original LRA and LR drawings that the staff encountered in its review. The applicant response, by letter dated December 22, 2004, has been subsequently incorporated in the ALRA as discussed below.

In its ALRA, dated July 14, 2005, the applicant provided the staff with revised LR drawing correcting the inconsistencies and accurately depicting all the components subject to AMR, including those subject under criterion 10 CFR 54.4(a)(2). Therefore, the staff's concern described in RAI 2.3.3.A.19-2 is resolved

In RAI 2.3.3.A.19-3, dated November 19, 2004, the staff stated that drawing LR-18009-C, sheet 1 shows piping and penetration downstream of a check valve as not subject to an AMR; however, drawing LR-18006, sheet 2 shows this same piping as subject to an AMR. Therefore, the staff requested that the applicant explain the apparent discrepancy between these drawings and confirm that the piping downstream of the check valve and penetration on the first drawing received an AMR.

In its response, by letter dated December 22, 2004, the applicant stated that drawing LR-18009-C, sheet 1 is incorrect. The applicant explained that the piping and penetration downstream of valve CU-37 (CKV-63.1-02) are subject to an AMR. The penetration itself is part of the primary containment structure, which is addressed in original LRA Sections 2.4.A.1 and

3.5.2.A.1. The penetration piping is covered in the RWCU system, which is covered in original LRA Sections 2.3.3.A.19 and 3.3.2.A.17. In addition, the applicant stated that drawing LR-18006-C, sheet 2, is incorrect at the referenced location. In its response, the applicant provided corrections to the locations of the boundary flags on this drawing and also provided revisions to original LRA Section 2.3.3.A.19.

Based on its review, the staff found the applicant's response to RAI 2.3.3.A.19-3 acceptable because it adequately explained that the LR drawings in question are incorrect, provided corrections to the locations of boundary flags on LR drawings, and provided the revisions required to original LRA 2.3.3.A.19-3. Therefore, the staff's concern described in RAI 2.3.3.A.19-3 is resolved.

In RAI 2.3.3.A.19-4, dated November 19, 2004, the staff stated that drawing LR-18009-C, sheet 1 shows piping upstream of a valve as not subject to an AMR; however, drawing LR-18006-C, sheet 1 shows this same piping as subject to an AMR. Therefore, the staff requested that the applicant explain the apparent discrepancy between these drawings and confirm that the identified piping does not require an AMR, as indicated on drawing LR-18009-C, sheet 1.

In its response, by letter dated December 22, 2004, the applicant stated that with respect to the inconsistency between drawings LR-18009-C, sheet 1 and LR-18006-C, sheet 1, the former drawing is correct. However, as stated in original LRA Section 2.3.3.A.19, the NSR portions of the RWCU system are within the scope of license renewal per 10 CFR 54.4(a)(2) and subject to AMR. Per the convention adopted for the LR drawings, components within the scope of license renewal and subject to an AMR for the 10 CFR 54.4(a)(2) criterion only are not identified in red. Therefore, on LR drawing LR-18009-C, sheet 1, the components shown in red and black are actually in-scope and subject to AMR.

Based on its review, the staff found the applicant's response to RAI 2.3.3.A.19-4 acceptable because the applicant adequately clarified the inconsistency between the two LR drawings. Therefore, the staff's concern described in RAI 2.3.3.A.19-4 is resolved.

2.3A.3.19.3 Conclusion

The staff reviewed the ALRA, RAI responses, and accompanying scoping boundary drawings to determine whether any SSCs that should be within the scope of license renewal had not been identified by the applicant. No omissions were identified. In addition, the staff performed a review to determine whether any components that should be subject to an AMR had not been identified by the applicant. No omissions were identified. On the basis of its review, the staff concludes that there is reasonable assurance that the applicant had adequately identified the RWCU system components that are within the scope of license renewal, as required by 10 CFR 54.4(a), and the RWCU system components that are subject to an AMR, as required by 10 CFR 54.21(a)(1).

2.3A.3.20 NMP1 Sampling System

2.3A.3.20.1 Summary of Technical Information in the Amended Application

In ALRA Section 2.3.3.A.20, the applicant described the sampling system. The sampling system provides for the sampling of liquid, steam and gases from various systems in the plant under all operating modes. Liquid samples can be obtained from the RPV, spent fuel pool, RWCU, core spray, torus, liquid poison, condensate, feedwater, RBCLC, turbine building closed loop cooling (TBCLC), circulating water, radioactive waste disposal and make-up systems. Steam samples from the main steam system are obtainable. Gaseous samples can be obtained from primary containment, vent stack and off gas systems.

The sampling system contains SR components that are relied upon to remain functional during and following DBEs. The failure of NSR SSCs in the sampling system could potentially prevent the satisfactory accomplishment of an SR function. In addition, the sampling system performs functions that support fire protection, EQ, and SBO.

The intended functions within the scope of license renewal include the following:

- maintains mechanical and structural integrity of NSR components to prevent spatial interactions with SR components

- provides pressure retaining boundary

- maintains mechanical and structural integrity of NSR components that provide structural support to attached SR components

In ALRA Table 2.3.3.A.20-1, the applicant identified the following sampling system component types that are within the scope of license renewal and subject to an AMR:

- bolting
- heat exchangers
- piping and fittings
- pumps
- rupture disc
- valves

2.3A.3.20.2 Staff Evaluation

The staff reviewed ALRA Section 2.3.3.A.20 and UFSAR Section VIII.C.3 using the evaluation methodology described in SER Section 2.3. The staff conducted its review in accordance with the guidance described in SRP-LR Section 2.3.

In conducting its review, the staff evaluated the system functions described in the ALRA and UFSAR in accordance with the requirements of 10 CFR 54.4(a) to verify that the applicant had not omitted from the scope of license renewal any components with intended functions delineated under 10 CFR 54.4(a). The staff then reviewed those components that the applicant had identified as being within the scope of license renewal to verify that the applicant had not

omitted any passive and long-lived components that should be subject to an AMR in accordance with the requirements of 10 CFR 54.21(a)(1).

The staff's review of original LRA Section 2.3.3.A.20 identified areas in which additional information was necessary to complete the review of the applicant's scoping and screening results. The applicant responded to the staff's RAIs as discussed below.

In RAI 2.3.3.A.20-1, dated November 19, 2004, the staff requested that the applicant clarify drawing inconsistencies that the staff encountered in its review. The applicant response, by letter dated December 22, 2004, has been subsequently incorporated in the ALRA as discussed below.

In its ALRA, dated July 14, 2005, the applicant provided the staff with revised LR drawings correcting the inconsistencies and accurately depicting all the components subject to AMR, including those subject under criterion 10 CFR 54.4(a)(2). Therefore, the staff's concern described in RAI 2.3.3.A.20-1 is resolved.

In RAI 2.3.3.A.20-2, dated November 19, 2004, the staff stated that drawing LR-18041-C, sheet 1 shows condensate sampling points at BV 110-72 and BV 110-73 as not subject to an AMR. The LR drawing for the condensate system indicates that the condensate line leading to CS 50-233 is within the condensate system AMR boundary flags and subject to an AMR. Therefore, the staff requested that the applicant identify where the AMR boundary exists between the condensate system and the sampling points at BV 110-72 and BV 110-73 and explain the basis for excluding these blocking valves as subject to an AMR.

In its response, by letter dated December 22, 2004, the applicant stated that drawing LR-18041-C, sheet 1 is incorrect. The AMR boundary on this drawing should include valves 110-251, 110-252, and 110-598, and all associated piping up to and including the condensate pumps and the main condenser. The applicant also noted that the component referenced in the RAI should read "CE 50-233" not "CS 50-233."

In addition, the applicant further clarified that per original LRA Section 2.3.4.A.20, the sampling system liquid-filled piping, fittings, equipment on LR drawing LR-18041-C, sheet 1, that are shown in black are also in-scope for LR and subject to AMR for criterion 10 CFR 54.4(a)(2) since they are liquid-filled components in the vicinity of SR components. Per LR drawing convention, only components in-scope for LR and subject to AMR for 10 CFR 54.4(a)(2) are not to be shown in red on LR drawings.

Based on its review, the staff found the applicant's response to RAI 2.3.3.A.20-2 acceptable because it concurred with the applicant that the sampling LR drawing AMR boundary should include valves and all associated piping up to and including condensate pumps, back to the main condensers, but were inadvertently left un-highlighted on the LR drawing. The staff concludes that there is reasonable assurance that the components were correctly included within the scope of license renewal and subject to an AMR. Therefore, the staff's concern described in RAI 2.3.3.A.20-2 is resolved.

In RAIs 2.3.3.A.20-3, 2.3.3.A.20-4, and 2.3.3.A.20-5, dated November 19, 2005, the staff requested that the applicant clarify information given on LR drawings concerning SSC's within the scope of license renewal per 10 CFR 54.4(a). The applicant response, by letter dated December 22, 2004, has been subsequently incorporated in the ALRA as discussed below.

In its ALRA, dated July 14, 2005, the applicant revised the LR drawing which identifies all SSC's within the scope of license renewal and subject to AMR including those in scope under criterion 10 CFR 54.4(a)(2). Therefore, the staff's concerns described in RAIs 2.3.3.A.20-3, 2.3.3.A.20-4, and 2.3.3.A.20-5 are resolved.

In RAI 2.3.3.A.20-6, dated November 19, 2004, the staff indicated that 10 CFR 54.4(a)(2) states that all NSR SSC's whose failure could prevent the satisfactory accomplishment of any of the functions described in 10 CFR 54.4(a)(1) is within the scope of license renewal. Drawing LR-18041-C, sheet 7 shows heat exchangers 122-44 and 122-45 outside the AMR boundary flags of the reactor building closed loop cooling water system and the sampling system. However, these heat exchangers are shown within the reactor building closed loop cooling water system and sampling system AMR boundary on drawing LR-18022-C, sheet 2. Since failure of either the tube side or shell side of these heat exchangers could affect the integrity of the reactor building closed loop cooling water system, the staff requested that the applicant explain the basis for excluding these heat exchangers as subject to an AMR as indicated on drawing LR-18041-C, sheet 7.

In its response, by letter dated December 22, 2004, the applicant stated that the post accident sample system is NSR and therefore, supplying reactor building closed loop cooling water to post accident sample coolers is an NSR function. The reactor building closed loop cooling water system line supplying the post accident sample system includes an excess flow check valve to prevent high flow rates from a downstream break and the return line includes a check valve which prevents back flow from the reactor building closed loop cooling water system.

The applicant further stated that drawing LR-18041-C, sheet 7 correctly shows the boundary at the check valves based on the above evaluation. However, drawing LR-18022-C, sheet 2 incorrectly shows the sample coolers as subject to an AMR. Based on the function of the excess flow check valve and the return line check valve, the NSR portion of the reactor building closed loop cooling water system cannot affect the performance of the required SR function. The boundary at the check valves meets the requirement of 10 CFR 54.4(a)(2). Therefore, drawing LR-18022-C, sheet 2 should be corrected by removing the AMR boundary from the sample coolers and associated piping and placing the reactor building closed loop cooling water system scope boundary flag at the drawing continuation flags. However, the applicant further explained that even though the subject heat exchanger shells should be shown in black, per original LRA Section 2.3.3.A.20, they are still within the scope of license renewal and subject to an AMR in accordance with 10 CFR 54.4(a)(2) since they are liquid-filled components in the vicinity of SR components.

Based on its review, the staff found the applicant's response to RAI 2.3.3.A.20-6 acceptable because it adequately explained that the NSR sample coolers are within the scope of license renewal in accordance with the 10 CFR 54.4(a)(2) only and, as such, were highlighted (red colored) on the LR drawing in error. The applicant also clarified that the AMR boundary on the LR drawing at the check valves meets the requirement of 10 CFR 54.4(a)(2) and the boundary flag on the LR drawing should be moved from the sample coolers to the drawing continuation flag. The staff concludes that there is reasonable assurance that the components were correctly

2-98

included within the scope of license renewal and subject to an AMR. Therefore, the staff's concern described in RAI 2.3.3.A.20-6 is resolved.

In RAI 2.3.3.A.20-7, dated November 19, 2004, the staff stated that drawing LR-18041-C, sheet 2 shows the air supply to AOV 110-83A as subject to an AMR for the NMP1 sampling system. The staff believed that this valve should be evaluated as part of the instrument air system. Therefore, the staff requested that the applicant confirm that this valve and its environment are within the sampling system or explain this apparent discrepancy.

In its response, by letter dated December 22, 2004, the applicant confirmed that the valve operator is not subject to AMR because it is not liquid-filled. Therefore, it is not within the scope of license renewal. The applicant stated that the LR drawing is incorrect and improperly shows the air supply valve as subject to an AMR. The components in question should not have been highlighted on the LR drawing as being subject to an AMR in accordance with 10 CFR 54.21(a) but were highlighted inadvertently.

Based on its review, the staff found the applicant's response to RAI 2.3.3.A.20-7 acceptable because it adequately explained that the air supply valve in question was highlighted (in red color) on the LR drawing in error. This air valve is not necessary for its associated air operated block valve to perform its intended function. Therefore, the air valve in question is not within the scope of license renewal and not subject to an AMR. The staff concludes that there is reasonable assurance that the components were correctly excluded from within the scope of license renewal and from requiring an AMR. Therefore, the staff's concern described in RAI 2.3.3.A.20-7 is resolved.

2.3A.3.20.3 Conclusion

The staff reviewed the ALRA, RAI responses, and accompanying scoping boundary drawings to determine whether any SSCs that should be within the scope of license renewal had not been identified by the applicant. No omissions were identified. In addition, the staff performed a review to determine whether any components that should be subject to an AMR had not been identified by the applicant. No omissions were identified. On the basis of its review, the staff concludes that there is reasonable assurance that the applicant had adequately identified the sampling system components that are within the scope of license renewal, as required by 10 CFR 54.4(a), and the sampling system components that are subject to an AMR, as required by 10 CFR 54.21(a)(1).

2.3A.3.21 NMP1 Service Water System

2.3A.3.21.1 Summary of Technical Information in the Amended Application

In ALRA Section 2.3.3.A.21, the applicant described the service water system. The service water system is designed to provide a reliable supply of cooling water to various safety and NSR components and systems. Systems cooled by the service water system include the RBCLC water system, TBCLC water system, RB HVAC system, TB HVAC system, and radioactive WDB HVAC system. Service water also is supplied to the screenwash pumps, the radwaste solidification and storage building, and the makeup demineralizer. The service water system is injected with chemicals to control biological growth by the chemical injection system.

The service water system contains SR components that are relied upon to remain functional during and following DBEs. The failure of NSR SSCs in the service water system could potentially prevent the satisfactory accomplishment of an SR function. In addition, the service water system performs functions that support fire protection and EQ.

The intended functions within the scope of license renewal include the following:

- provides filtration
- maintains mechanical and structural integrity of NSR components to prevent spatial interactions with SR components
- provides structural support to NSR components whose failure could prevent accomplishment of SR function(s)
- provides pressure retaining boundary
- maintains mechanical and structural integrity of NSR components that provide structural support to attached SR components

In ALRA Table 2.3.3.A.21-1, the applicant identified the following service water system component types that are within the scope of license renewal and subject to an AMR:

- bolting
- filters/strainers
- flow elements
- piping and fittings
- pumps
- valves

2.3A.3.21.2 Staff Evaluation

The staff reviewed ALRA Section 2.3.3.A.21 and UFSAR Section X.F using the evaluation methodology described in SER Section 2.3. The staff conducted its review in accordance with the guidance described in SRP-LR Section 2.3.

In conducting its review, the staff evaluated the system functions described in the ALRA and UFSAR in accordance with the requirements of 10 CFR 54.4(a) to verify that the applicant had not omitted from the scope of license renewal any components with intended functions delineated under 10 CFR 54.4(a). The staff then reviewed those components that the applicant had identified as being within the scope of license renewal to verify that the applicant had not omitted any passive and long-lived components that should be subject to an AMR in accordance with the requirements of 10 CFR 54.21(a)(1).

The staff's review of original LRA Section 2.3.3.A.21 identified areas in which additional information was necessary to complete the review of the applicant's scoping and screening results. The applicant responded to the staff's RAIs as discussed below.

In RAI 2.3.3.A.21-1, dated November 19, 2004, the staff requested that the applicant clarify information given on a license renewal boundary drawing concerning SSC's within the scope of license renewal in accordance with 10 CFR 54.4(a). The applicant response, by letter dated December 22, 2004, has been subsequently incorporated in the ALRA as discussed below.

In its ALRA, dated July 14, 2005, the applicant provided the staff with revised LR drawing which identifies all SSC's in scope and subject to AMR including those under criterion 10 CFR 54.4(a)(2). Based on review of the information submitted in the ALRA the staff's concern described in RAI 2.3.3.A.21-1 is resolved.

In RAI 2.3.3.A.21-2, dated November 19, 2004, the staff requested that the applicant clarify drawing inconsistencies that the staff encountered in its review. The applicant response, by letter dated December 22, 2004, has been subsequently incorporated in the ALRA as discussed below.

In its ALRA, dated July 14, 2005, the applicant provided the staff with revised LR drawings correcting the inconsistencies and accurately depicting all components subject to AMR including those subject under criterion 10 CFR 54.4(a)(2). Based on review of the information submitted in the ALRA the staff's concern described in RAI 2.3.3.A.21-2 is resolved.

2.3A.3.21.3 Conclusion

The staff reviewed the ALRA, RAI responses, and accompanying scoping boundary drawings to determine whether any SSCs that should be within the scope of license renewal had not been identified by the applicant. No omissions were identified. In addition, the staff performed a review to determine whether any components that should be subject to an AMR had not been identified by the applicant. No omissions were identified. On the basis of its review, the staff concludes that there is reasonable assurance that the applicant had adequately identified the service water system components that are within the scope of license renewal, as required by 10 CFR 54.4(a), and the service water system components that are subject to an AMR, as required by 10 CFR 54.21(a)(1).

2.3A.3.22 NMP1 Shutdown Cooling System

2.3A.3.22.1 Summary of Technical Information in the Amended Application

In ALRA Section 2.3.3.A.22, the applicant described the shutdown cooling system. The shutdown cooling system is designed to cool reactor water below temperatures and pressures at which the main condenser may be used as a heat sink following reactor shutdown. This system provides the capability to achieve and maintain a cold shutdown condition by removal of reactor fission product decay heat. The shutdown cooling system consists of reactor coolant isolation valves, three redundant loops each having a pump, heat exchanger and flow control valve, and associated piping, valves, instrumentation and controls. The heater exchangers are cooled by the RBCLC water system.

The shutdown cooling system contains SR components that are relied upon to remain functional during and following DBEs. The failure of NSR SSCs in the shutdown cooling system could potentially prevent the satisfactory accomplishment of an SR function. In addition, the shutdown cooling system performs functions that support fire protection and EQ.

The intended functions within the scope of license renewal include the following:

- provides heat transfer
- maintains mechanical and structural integrity of NSR components to prevent spatial interactions with SR components
- provides pressure retaining boundary
- maintains mechanical and structural integrity of NSR components that provide structural support to attached SR components
- provides flow restriction

In ALRA Table 2.3.3.A.22-1, the applicant identified the following shutdown cooling system component types that are within the scope of license renewal and subject to an AMR:

- bolting
- flow elements
- heat exchangers
- orifices
- piping and fittings
- pumps
- valves

2.3A.3.22.2 Staff Evaluation

The staff reviewed ALRA Section 2.3.3.A.22 and the UFSAR Section X.A using the evaluation methodology described in SER Section 2.3. The staff conducted its review in accordance with the guidance described in SRP-LR Section 2.3.

In conducting its review, the staff evaluated the system functions described in the ALRA and UFSAR in accordance with the requirements of 10 CFR 54.4(a) to verify that the applicant had not omitted from the scope of license renewal any components with intended functions delineated under 10 CFR 54.4(a). The staff then reviewed those components that the applicant had identified as being within the scope of license renewal to verify that the applicant had not omitted any passive and long-lived components that should be subject to an AMR in accordance with the requirements of 10 CFR 54.21(a)(1).

The staff's review of original LRA Section 2.3.3.A.22 identified areas in which additional information was necessary to complete the review of the applicant's scoping and screening results. The applicant responded to the staff's RAIs as discussed below.

In RAI 2.3.3.A.22-1, dated November 19, 2004, the staff requested that the applicant clarify information given on LR drawing 18018-C, sheet 1 concerning SSC's in scope of license renewal per 10 CFR 54.4(a). The applicant response, by letter dated December 22, 2004, has been subsequently incorporated in the ALRA as discussed below.

In its ALRA, dated July 14, 2005, the applicant provided the revised LR drawing which identifies all SSC's in scope of license renewal and subject to AMR including those subject under criterion

10 CFR 54.4(a)(2). Based on review of the information submitted in the ALRA, the staff's concern described in RAI 2.3.3.A.22-1 is resolved.

In RAI 2.3.3.A.22-2, dated November 19, 2004, the staff stated that original LRA Table 2.3.3.A.22-1 does not list temperature elements or thermowells as component types within the shutdown cooling system. Drawing LR-18018-C, sheet 1 shows a temperature element within the AMR boundary of the shutdown cooling system. A general note number on the LR drawing states that all temperature devices including temperature elements have thermowells. Therefore, the staff requested that the applicant explain the basis for excluding temperature elements and/or thermowells (pressure boundary function) as component types in original LRA Table 2.3.3.A.22-1 as subject to an AMR.

In addition the staff stated that the original LRA Table 2.3.3.A.22-1 does not list bolting as a component type within the shutdown cooling system. Bolted connections appear to be used on a number of flow elements within the shutdown cooling system. Therefore, the staff also requested that the applicant explain the basis for excluding bolting as a component type in the original LRA Table 2.3.3.A.22-1 as subject to an AMR.

In its response, by letter dated December 22, 2004, the applicant stated:

> The thermowell for TE-38-115, "Temperature Primary Element - Water To Reactor Recirc Loop," is constructed of the same material as the piping; therefore, there is no reason to create a thermowell subcomponent for TE-38-115 or to list thermowells as a separate component type in [original] LRA Tables 2.3.3.A.22-1 and 3.3.2.A-20. The convention adopted for the [original] LRA was that thermowells made of the same material as the piping were included under the component type of "Piping and Fittings" as being a portion of the pressure boundary of the pipe. Drawing LR-18018-C, Sheet 1, incorrectly highlighted temperature element TE-38-115 as being subject to AMR. It is actually not in-scope for LR. The following additional instruments are also incorrectly shown as being subject to AMR on drawing LR-18018-C, Sheet 1: PT-38-141, PT-38-153, PT-38-148, TE-38-130, and TE-38-136. Pressure transmitters PT-38-141, PT-38-153, and PT-38-148 are in-scope for LR, but are not subject to AMR.

In regard to bolting, the applicant in its ALRA, submitted July 14, 2005, now includes bolting as a component type requiring an AMR in original LRA tables 2.3.3.A.22-1 and 3.3.2.A-20.

Based on its review, the staff found the applicant's response to RAI 2.3.3.A.22-2 acceptable. The applicant adequately clarified that thermowells and bolting are included in the component type of "Piping and Fittings." Additionally, the applicant added carbon steel, "Piping and Fittings," in an air environment with a pressure boundary function to represent the thermowells, and has added bolting to the component types requiring an AMR. The staff agreed that temperature element TE-38-115 is not within the scope of license renewal, and that is because they are active components, the instruments (TEs and PTs) identified in the applicant's response are not subject to AMRs. Therefore, the staff's concern described in RAI 2.3.3.A.22-2 is resolved.

2.3A.3.22.3 Conclusion

The staff reviewed the ALRA, RAI responses, and accompanying scoping boundary drawings to determine whether any SSCs that should be within the scope of license renewal had not been identified by the applicant. No omissions were identified. In addition, the staff performed a review to determine whether any components that should be subject to an AMR had not been identified by the applicant. No omissions were identified. On the basis of its review, the staff concludes that there is reasonable assurance that the applicant had adequately identified the shutdown cooling system components that are within the scope of license renewal, as required by 10 CFR 54.4(a), and the shutdown cooling system components that are subject to an AMR, as required by 10 CFR 54.21(a)(1).

2.3A.3.23 NMP1 Spent Fuel Pool Filtering and Cooling System

2.3A.3.23.1 Summary of Technical Information in the Amended Application

In ALRA Section 2.3.3.A.23, the applicant described the spent fuel pool filtering and cooling system. The spent fuel pool filtering and cooling system is designed to remove decay heat from the spent fuel assemblies and the impurities from the pool water. This system maintains the temperature and purity of the spent fuel pool water at acceptable levels. Cooling water is supplied to the heat exchangers from the RBCLC water system. Makeup water to the spent fuel storage pool is provided by the condensate and condensate transfer system. The spent fuel pool filtering and cooling system is also used after reactor refueling to drain the reactor internals storage pit and head cavity. Alternate lines allow transport of the water to either the main condenser or to the waste disposal system for processing.

The spent fuel pool filtering and cooling system contains SR components that are relied upon to remain functional during and following DBEs. The failure of NSR SSCs in the spent fuel pool filtering and cooling system could potentially prevent the satisfactory accomplishment of an SR function. In addition, the spent fuel pool filtering and cooling system performs functions that support EQ.

The intended functions within the scope of license renewal include the following:

- provides filtration
- provides heat transfer
- maintains mechanical and structural integrity of NSR components to prevent spatial interactions with SR components
- provides structural support to NSR components whose failure could prevent accomplishment of SR function(s)
- provides pressure retaining boundary
- maintains mechanical and structural integrity of NSR components that provide structural support to attached SR components

In ALRA Table 2.3.3.A.23-1, the applicant identified the following spent fuel pool filtering and cooling system component types that are within the scope of license renewal and subject to an AMR:

- bolting
- filters/strainers
- flow elements
- flow gauge
- heat exchangers
- piping and fittings
- pumps
- tanks
- valves

2.3A.3.23.2 Staff Evaluation

The staff reviewed ALRA Section 2.3.3.A.23 and UFSAR Section X.H using the evaluation methodology described in SER Section 2.3. The staff conducted its review in accordance with the guidance described in SRP-LR Section 2.3.

In conducting its review, the staff evaluated the system functions described in the ALRA and UFSAR in accordance with the requirements of 10 CFR 54.4(a) to verify that the applicant had not omitted from the scope of license renewal any components with intended functions delineated under 10 CFR 54.4(a). The staff then reviewed those components that the applicant had identified as being within the scope of license renewal to verify that the applicant had not omitted any passive and long-lived components that should be subject to an AMR in accordance with the requirements of 10 CFR 54.21(a)(1).

The staff's review of original LRA Section 2.3.3.A.23 identified areas in which additional information was necessary to complete the evaluation of the applicant's scoping and screening results. The applicant responded to the staff's RAIs as discussed below.

In RAIs 2.3.3.A.23-1 and -2, dated November 19, 2004, the staff requested that the applicant clarify information given on an LR drawing concerning SSC's in scope of license renewal per 10 CFR 54.4(a). The applicant's response, by letter dated December 22, 2004, has been subsequently incorporated in the ALRA as discussed below.

In its ALRA, dated July 14, 2005, the applicant provided the staff with revised LR drawings which identify all SSC's within the scope of license renewal and subject to AMR including those subject under criterion 10 CFR 54.4(a)(2). Based on review of the information submitted in the amended LRA the staff's concern described in RAI 2.3.3.A.23-1 and RAI 2.3.3.A.23-2 are resolved.

2.3A.3.23.3 Conclusion

The staff reviewed the ALRA, RAI responses, and accompanying scoping boundary drawings to determine whether any SSCs that should be within the scope of license renewal had not been identified by the applicant. No omissions were identified. In addition, the staff performed a review to determine whether any components that should be subject to an AMR had not been identified by the applicant. No omissions were identified. On the basis of its review, the staff concludes that there is reasonable assurance that the applicant had adequately identified the spent fuel pool filtering and cooling system components that are within the scope of license renewal, as required by 10 CFR 54.4(a), and the spent fuel pool filtering and cooling system components that are subject to an AMR, as required by 10 CFR 54.21(a)(1).

2.3A.3.24 NMP1 Technical Support Center HVAC System

2.3A.3.24.1 Summary of Technical Information in the Amended Application

In ALRA Section 2.3.3.A.24, the applicant described the technical support center HVAC system. The technical support center HVAC system is designed to maintain the technical support center temperature and supply tempered, recirculated, and outside air to maintain a suitable environment for emergency response personnel. During the normal mode of operation, air is drawn into the system through a louvered intake, electric heater, filter and cooling coil to the circulating fan. This fan discharges air to the technical support center. Air is exhausted through the exhaust fan to the environment. In the emergency mode, the normal mode flow path isolates and the supply fan draws air through a separate louvered intake. The air is then directed through a prefilter, HEPA filter, charcoal filter and a second HEPA filter to the suction of the circulating fan. There is no direct exhaust path in the emergency mode as the technical support center is maintained at a positive pressure. The HVAC system also has a separate exhaust path for the removal of smoke.

The technical support center HVAC system contains SR components that are relied upon to remain functional during and following DBEs. In addition, the technical support center HVAC system performs functions that support EQ.

The applicant stated that the in-scope components for the technical support center HVAC system are active components. Therefore, there are no components requiring an AMR for the technical support center HVAC system.

2.3A.3.24.2 Staff Evaluation

The staff reviewed ALRA Section 2.3.3.A.24 and UFSAR Section III.E.1.2.2 using the evaluation methodology described in SER Section 2.3. The staff conducted its review in accordance with the guidance described in SRP-LR Section 2.3.

In conducting its review, the staff evaluated the system functions described in the ALRA and UFSAR in accordance with the requirements of 10 CFR 54.4(a) to verify that the applicant had not omitted from the scope of license renewal any components with intended functions delineated under 10 CFR 54.4(a). The staff then reviewed those components that the applicant had identified as being within the scope of license renewal to verify that the applicant had not omitted any passive and long-lived components that should be subject to an AMR in accordance with the requirements of 10 CFR 54.21(a)(1).

2.3A.3.24.3 Conclusion

The staff reviewed the ALRA to determine whether any SSCs that should be within the scope of license renewal had not been identified by the applicant. No omissions were identified. In addition, the staff performed a review to determine whether any components that should be subject to an AMR had not been identified by the applicant. No omissions were identified. On the basis of its review, the staff concludes that there is reasonable assurance that the applicant had adequately identified the technical support center HVAC system components that are within the scope of license renewal, as required by 10 CFR 54.4(a), and the technical support center HVAC system components that are subject to an AMR, as required by 10 CFR 54.21(a)(1).

2.3A.3.25 *NMP1 Turbine Building Closed Loop Cooling Water System*

2.3A.3.25.1 Summary of Technical Information in the Amended Application

In ALRA Section 2.3.3.A.25, the applicant described the TBCLC water system. The TBCLC water system provides demineralized water to cool various NSR auxiliary equipment in the TB in support of power generation. The closed loop provides isolation of systems containing radioactive liquids from the service water, which returns to the lake.

The failure of NSR SSCs in the TBCLC water system could potentially prevent the satisfactory accomplishment of an SR function.

The intended functions within the scope of license renewal include the following:

*   maintains mechanical and structural integrity of NSR components to prevent spatial interactions with SR components

*   provides structural support to NSR components whose failure could prevent accomplishment of SR function(s)

*   maintains mechanical and structural integrity of NSR components that provide structural support to attached SR components

In ALRA Table 2.3.3.A.25-1, the applicant identified the following TBCLC water system component types that are within the scope of license renewal and subject to an AMR:

*   bolting
*   filters/strainers
*   heat exchangers
*   piping and fittings
*   pumps
*   tank
*   valves

2.3A.3.25.2 Staff Evaluation

The staff reviewed ALRA Section 2.3.3.A.25 and UFSAR Section X.E using the evaluation methodology described in SER Section 2.3. The staff conducted its review in accordance with the guidance described in SRP-LR Section 2.3.

In conducting its review, the staff evaluated the system functions described in the ALRA and UFSAR in accordance with the requirements of 10 CFR 54.4(a) to verify that the applicant had not omitted from the scope of license renewal any components with intended functions delineated under 10 CFR 54.4(a). The staff then reviewed those components that the applicant had identified as being within the scope of license renewal to verify that the applicant had not omitted any passive and long-lived components that should be subject to an AMR in accordance with the requirements of 10 CFR 54.21(a)(1).

2.3A.3.25.3 Conclusion

The staff reviewed the ALRA to determine whether any SSCs that should be within the scope of license renewal had not been identified by the applicant. No omissions were identified. In addition, the staff performed a review to determine whether any components that should be subject to an AMR had not been identified by the applicant. No omissions were identified. On the basis of its review, the staff concludes that there is reasonable assurance that the applicant had adequately identified the TBCLC water system components that are within the scope of license renewal, as required by 10 CFR 54.4(a), and the TBCLC water system components that are subject to an AMR, as required by 10 CFR 54.21(a)(1).

2.3A.3.26 NMP1 Turbine Building HVAC System

2.3A.3.26.1 Summary of Technical Information in the Amended Application

In ALRA Section 2.3.3.A.26, the applicant described the turbine building HVAC system. The turbine building HVAC system is designed to provide a continuous flow of fresh tempered air throughout the building, while maintaining a negative atmospheric pressure. This system also has heat and smoke removal capability for three smoke zones and the upper elevation of the TB. The turbine building HVAC system consists of air intakes, filters, electric heating units, flow control dampers, dampers, and ductwork to distribute air to various areas in the TB. Outside air is taken in through louvered, screened penthouses, which supply air to the turbine building HVAC supply fans. The air then passes through filters and heating coils. Exhaust air is directed through a plenum to the stack for discharge and is monitored for radiation. The exhaust system discharges into the plenum, which also receives air from the containment and other buildings. The smoke removal function of the turbine building HVAC system consists of three independent air make-up fans, dampers and ductwork (one for each smoke zone) and automatic isolation dampers and exhaust fans of the normal ventilation system. In addition, there are twelve motor operated roof vents and five sidewall vents.

The turbine building HVAC system contains SR components that are relied upon to remain functional during and following DBEs. In addition, the turbine building HVAC system performs functions that support fire protection.

The intended functions within the scope of license renewal include the following:

- provides path for release of filtered and unfiltered gaseous discharge
- provides pressure retaining boundary

In ALRA Table 2.3.3.A.26-1, the applicant identified the following turbine building HVAC system component types that are within the scope of license renewal and subject to an AMR:

- blowers
- bolting
- ducting
- muffler
- valves and dampers
- vents

2.3A.3.26.2 Staff Evaluation

The staff reviewed ALRA Section 2.3.3.A.26 and UFSAR Section III.A.2.2 using the evaluation methodology described in SER Section 2.3. The staff conducted its review in accordance with the guidance described in SRP-LR Section 2.3.

In conducting its review, the staff evaluated the system functions described in the ALRA and UFSAR in accordance with the requirements of 10 CFR 54.4(a) to verify that the applicant had not omitted from the scope of license renewal any components with intended functions delineated under 10 CFR 54.4(a). The staff then reviewed those components that the applicant had identified as being within the scope of license renewal to verify that the applicant had not omitted any passive and long-lived components that should be subject to an AMR in accordance with the requirements of 10 CFR 54.21(a)(1).

2.3A.3.26.3 Conclusion

The staff reviewed the ALRA to determine whether any SSCs that should be within the scope of license renewal had not been identified by the applicant. No omissions were identified. In addition, the staff performed a review to determine whether any components that should be subject to an AMR had not been identified by the applicant. No omissions were identified. On the basis of its review, the staff concludes that there is reasonable assurance that the applicant had adequately identified the turbine building HVAC system components that are within the scope of license renewal, as required by 10 CFR 54.4(a), and the turbine building HVAC system components that are subject to an AMR, as required by 10 CFR 54.21(a)(1).

2.3A.3.27 NMP1 Electric Steam Boiler System

2.3A.3.27.1 Summary of Technical Information in the Amended Application

In ALRA Section 2.3.3.A.27, the applicant described the electric steam boiler system. The electric steam boiler system is designed to supply saturated steam to the radwaste concentrator #12 heat exchanger to support the processing of radioactive waste, the nitrogen vaporizer to support drywell inerting, and the TB decontamination area to support decontamination activities.

The system includes a condensate receiver which supplies the condensate to the boiler for generation of saturated steam. The steam is routed through steam piping to the above loads.

The failure of NSR SSCs in the electric steam boiler system could potentially prevent the satisfactory accomplishment of an SR function.

The intended function, within the scope of license renewal, is to maintain mechanical and structural integrity to prevent spatial interactions.

In ALRA Table 2.3.3.A.27-1, the applicant identified the following electric steam boiler system component types that are within the scope of license renewal and subject to an AMR:

- boiler
- bolting
- drain trap
- level gauge
- piping and fittings
- strainer
- valves

2.3A.3.27.2 Staff Evaluation

The staff reviewed ALRA Section 2.3.3.A.27 using the evaluation methodology described in SER Section 2.3. The staff conducted its review in accordance with the guidance described in SRP-LR Section 2.3.

In conducting its review, the staff evaluated the system functions described in the ALRA and UFSAR in accordance with the requirements of 10 CFR 54.4(a) to verify that the applicant had not omitted from the scope of license renewal any components with intended functions delineated under 10 CFR 54.4(a). The staff then reviewed those components that the applicant had identified as being within the scope of license renewal to verify that the applicant had not omitted any passive and long-lived components that should be subject to an AMR in accordance with the requirements of 10 CFR 54.21(a)(1).

2.3A.3.27.3 Conclusion

The staff reviewed the ALRA to determine whether any SSCs that should be within the scope of license renewal had not been identified by the applicant. No omissions were identified. In addition, the staff performed a review to determine whether any components that should be subject to an AMR had not been identified by the applicant. No omissions were identified. On the basis of its review, the staff concludes that there is reasonable assurance that the applicant had adequately identified the electric steam boiler system components that are within the scope of license renewal, as required by 10 CFR 54.4(a), and the electric steam boiler system components that are subject to an AMR, as required by 10 CFR 54.21(a)(1).

2.3A.3.28 NMP1 Makeup Demineralizer System

2.3A.3.28.1 Summary of Technical Information in the Amended Application

In ALRA Section 2.3.3.A.28, the applicant described the makeup demineralizer system. The makeup demineralizer system is designed to supply batches of demineralized water to fill the demineralized water makeup tank, the condensate storage tanks, and other reservoirs as necessary. It also provides water directly to the liquid poison system, laboratories and sample sinks, and the stator winding liquid cooling system. Demineralized water from this system can be used as an alternate source for several plant systems, including the RWCU and CRD systems. The makeup demineralizer system utilizes a portable skid- or truck-mounted demineralized water unit to process service water or city water for use in the station. Water is processed through several components acting as tanks (i.e., precipitator, clearwell, filter and purifier) while pumped to the portable demineralized water unit. The discharge of the portable unit is directed to the demineralized water storage tank and then to the condensate storage tanks and/or other system loads.

The failure of NSR SSCs in the makeup demineralizer system could potentially prevent the satisfactory accomplishment of an SR function.

The intended function, within the scope of license renewal, is to maintain mechanical and structural integrity of NSR components to prevent spatial interactions with SR components.

In ALRA Table 2.3.3.A.28-1, the applicant identified the following makeup demineralizer system component types that are within the scope of license renewal and subject to an AMR:

- bolting
- level gauge
- piping and fittings
- pumps
- tanks
- valves

2.3A.3.28.2 Staff Evaluation

The staff reviewed ALRA Section 2.3.3.A.28 and UFSAR Section X.G.1.0 using the evaluation methodology described in SER Section 2.3. The staff conducted its review in accordance with the guidance described in SRP-LR Section 2.3.

In conducting its review, the staff evaluated the system functions described in the ALRA and UFSAR in accordance with the requirements of 10 CFR 54.4(a) to verify that the applicant had not omitted from the scope of license renewal any components with intended functions delineated under 10 CFR 54.4(a). The staff then reviewed those components that the applicant had identified as being within the scope of license renewal to verify that the applicant had not omitted any passive and long-lived components that should be subject to an AMR in accordance with the requirements of 10 CFR 54.21(a)(1).

2.3A.3.28.3 Conclusion

The staff reviewed the ALRA to determine whether any SSCs that should be within the scope of license renewal had not been identified by the applicant. No omissions were identified. In addition, the staff performed a review to determine whether any components that should be subject to an AMR had not been identified by the applicant. No omissions were identified. On

the basis of its review, the staff concludes that there is reasonable assurance that the applicant had adequately identified the makeup demineralizer system components that are within the scope of license renewal, as required by 10 CFR 54.4(a), and the makeup demineralizer system components that are subject to an AMR, as required by 10 CFR 54.21(a)(1).

2.3A.4 Steam and Power Conversion Systems

In ALRA Section 2.3.4.A, the applicant identified the structures and components of the NMP1 steam and power conversion systems that are subject to an AMR for license renewal.

The applicant described the supporting structures and components of the steam and power conversion systems in the following sections of the ALRA:

- 2.3.4.A.1   NMP1 condensate and condensate transfer system
- 2.3.4.A.2   NMP1 condenser air removal and off-gas system
- 2.3.4.A.3   NMP1 feedwater/high pressure coolant injection system
- 2.3.4.A.4   NMP1 main generator and auxiliary system
- 2.3.4.A.5   NMP1 main steam system
- 2.3.4.A.6   NMP1 main turbine and auxiliary systems
- 2.3.4.A.7   NMP1 moisture separator reheater steam system

The staff's review findings regarding ALRA Sections 2.3.4.A.1 – 2.3.4.A.7 are presented in SER Sections 2.3A.4.1 – 2.3A.4.7, respectively.

2.3A.4.1 NMP1 Condensate and Condensate Transfer System

2.3A.4.1.1 Summary of Technical Information in the Amended Application

In ALRA Section 2.3.4.A.1, the applicant described the condensate and condensate transfer system. The condensate system condenses steam exhausted from the lowpressure turbines and the turbine bypass valves. This condensate then becomes the primary water supply to the FW/HPCI system. The main condenser also acts as a collecting basin for various leakage, drains, and relief valve discharges from balance of plant systems. The condensate system also removes impurities from the condensed liquid for re-use as reactor water. The condensate serves as a cooling medium for the off gas system steam jet air ejector condensers, steam entering the condenser when the turbine bypass valves are open, and the turbine exhaust hood spray. Additionally, under emergency conditions such as a small break LOCA, the condensate system supplies water from the main condenser to support the HPCI mode of operation to supply makeup water to the reactor. For license renewal purposes, the condensate system also includes the condensate transfer system. The condensate transfer system supplies various systems and equipment throughout the plant with clean demineralized water. The condensate transfer system takes condensate from the condensate storage tanks (CSTs), which are cross-connected, and delivers the water through one of two redundant pumps.

The condensate and condensate transfer system contains SR components that are relied upon to remain functional during and following DBEs. The failure of NSR SSCs in the condensate and condensate transfer system could potentially prevent the satisfactory accomplishment of an

SR function. In addition, the condensate and condensate transfer system performs functions that support, EQ, and SBO.

The intended functions within the scope of license renewal include the following:

- maintains mechanical and structural integrity of NSR components to prevent spatial interactions with SR components

- provides pressure retaining boundary

- maintains mechanical and structural integrity of NSR components that provide structural support to attached SR components

In ALRA Table 2.3.4.A.1-1, the applicant identified the following condensate and condensate transfer system component types that are within the scope of license renewal and subject to an AMR:

- bolting
- condensate demineralizers
- filters/strainers
- flow elements
- flow gauges
- flow indicators
- flow orifices
- level observation glasses
- main condenser
- piping and fittings
- pumps
- tanks
- valves

2.3A.4.1.2 Staff Evaluation

The staff reviewed ALRA Section 2.3.4.A.1 and UFSAR Section XI.B using the evaluation methodology described in SER Section 2.3. The staff conducted its review in accordance with the guidance described in SRP-LR Section 2.3.

In conducting its review, the staff evaluated the system functions described in the ALRA and UFSAR in accordance with the requirements of 10 CFR 54.4(a) to verify that the applicant had not omitted from the scope of license renewal any components with intended functions delineated under 10 CFR 54.4(a). The staff then reviewed those components that the applicant had identified as being within the scope of license renewal to verify that the applicant had not omitted any passive and long-lived components that should be subject to an AMR in accordance with the requirements of 10 CFR 54.21(a)(1).

The staff's review of original LRA Section 2.3.4.A.1 identified areas in which additional information was necessary to complete the review of the applicant's scoping and screening results. The applicant responded to the staff's RAIs as discussed below.

In RAI 2.3.4.A.1-1, dated November 19, 2004, the staff stated that LR drawing 18003, shows an inter-condenser and after-condenser within the scope of license renewal and subject to an AMR. Additionally, two re-combiner condensers are also shown within scope and subject to an AMR; however, original LRA Table 2.3.4.A.1-1 does not include these heat exchangers individually among the list of components subject to an AMR, nor is the generic component type "Heat Exchanger" included in the table. The applicant was asked to justify exclusion of the heat exchangers from original LRA Table 2.3.4.A.1-1.

In its response, dated December 22, 2004, the applicant stated that original LRA Section 2.3.4.A.1 states that the AMR includes the main flow path from the main condenser to the boundary with the feedwater system. This includes the tube side of the recombiner condensers and the inter-condenser (the after-condenser is retired in place and isolated from the condensate system). The tube sides of these condensers are included in the "Piping and Fittings" component type listed in original LRA Table 2.3.4.A.1-1. The recombiner condensers and inter-condenser are evaluated in the within the scope of license renewal for criteria 10 CFR 54.4(a)(1) or (3). The subject LR drawing incorrectly shows the shell sides of the recombiner condensers and inter-condenser, as well as the entire after-condenser, as within the scope of license renewal and subject to an AMR.

The applicant further stated for clarification that all of the liquid-filled components on the subject LR drawing shown in black are within the scope of license renewal and subject to AMR to meet 10 CFR 54.4(a)(2), since they are in the turbine building and in the vicinity of SR components; however, per the LR drawing convention, components within the scope of license renewal and subject to AMR for 10 CFR 54.4(a)(2) only, are not shown in red on the LR drawings.

Based on its review, the staff found the applicant's response to RAI 2.3.4.A.1-1 acceptable because it adequately explained that the subject LR drawing incorrectly shows the shell sides of the recombiner condensers and inter-condenser, as well as the entire after-condenser, as in-scope for license renewal and subject to AMR. Only the tube side of the recombiner condensers and the inter-condenser are within the scope of license renewal in accordance with 10 CFR 54.4(a) and are subject to an AMR in accordance with 10 CFR 54.21(a). The components in question are included in the "Piping and Fittings" component type listed in original LRA Table 2.3.4.A.1-1. The staff concludes that there is reasonable assurance that the components were correctly included within the scope of license renewal and subject to an AMR. Therefore, the staff's concern described in RAI 2.3.4.A.1-1 is resolved.

In RAI 2.3.4.A.1-2, dated November 19, 2004, the staff stated that LR drawing 18008, sheet 1 shows that a valve labeled CT-38, on line 57-3/4 -B, is outside the scope of license renewal and excluded from requiring an AMR. To ensure that the valve has the capability of isolating this line, the staff believed that it should be within scope of license renewal and subject to an AMR. The applicant was asked to justify exclusion of valve CT-38 from the scope of license renewal and from being subject to an AMR.

In its response, by letter dated December 22, 2004, the applicant stated that the LR drawing is incorrect and does not properly show the component within the scope of license renewal and subject to an AMR. The component in question should have been highlighted on the LR drawing showing that it is within the scope of license renewal and is subject to an AMR. However, the component was inadvertently not highlighted. The component in question has been included within the scope of license renewal and is subject to an AMR.

Based on its review, the staff found the applicant's response to RAI 2.3.4.A.1-2 acceptable because it adequately explained that the component in question is within the scope of license renewal and subject to an AMR but was inadvertently left un-highlighted on the LR drawing. The staff concludes that there is reasonable assurance that the component was correctly included within the scope of license renewal and subject to an AMR. Therefore, the staff's concern described in RAI 2.3.4.A.1-2 is resolved.

In RAI 2.3.4.A.1-3, dated November 19, 2004, the staff requested that the applicant clarify drawing inconsistencies that the staff encountered in its review. The applicant response, by letter dated December 22, 2004, has been subsequently incorporated in the ALRA as discussed below.

In its ALRA, dated July 14, 2005, the applicant provided the staff with revised LR drawings correcting the inconsistencies and accurately depicting the components subject to AMR under criterion 10 CFR 54.4. Therefore, the staff's concern described in RAI 2.3.4.A.1-3 is resolved.

In RAI 2.3.4.A.1-4, dated November 19, 2004, the staff stated that LR drawing 18048, shows line 57-4 -B within the scope of license renewal and ending in a continuation flag labeled "relief to condensate surge and storage tank." This flag shows a continuation to drawing 18003, location G1; however, at this location on drawing 18003, the continuation of line 57-4-B is shown outside the scope of license renewal. There is no license renewal boundary flag on either of these drawings marking this change in classification, nor any valve present that could isolate the in-scope portion from the out-of-scope portion of the line. Therefore, the staff requested that the applicant explain the absence of a boundary flag and an isolation valve separating the in-scope and out-of-scope portions of the abovementioned line.

In its response, dated December 22, 2004, the applicant stated that the LR drawing is incorrect and does not properly show the license renewal boundary for the components within the scope of license renewal and subject to an AMR. The drawing incorrectly shows line 57-4-B beyond valve PSV-57-57 as within the scope of license renewal. There should be a boundary flag on the discharge side of the valve showing solid blue pointing away from the valve and "LR-CS" pointing toward the valve. Line 57-4-B beyond the relief valve should be shown in black.

For clarification the applicant stated, to be consistent with the description in original LRA Section 2.3.4.A.1, all of the liquid-filled components on the subject drawing that are not highlighted are within the scope of license renewal and subject to AMR, since they are in the turbine building and in the vicinity of SR components. Per the LR drawing convention, components within the scope of license renewal and subject to AMR under 10 CFR 54.4(a)(2) only, are not shown in red on the LR drawings.

Based on its review, the staff found the applicant's response to RAI 2.3.4.A.1-4 acceptable because it adequately explained that the components in question beyond the subject valve are not within the scope of license renewal and subject to an AMR but inadvertently were highlighted on the LR drawing. The staff concludes that there is reasonable assurance that the components beyond the valve representing the license renewal boundary were correctly not included within the scope of license renewal and not subject to an AMR. Therefore, the staff's concern described in RAI 2.3.4.A.1-4 is resolved.

In RAI 2.3.4.A.1-5, dated November 19, 2004, the staff stated that the two condensate surge and storage tanks are shown within the scope of license renewal on LR drawing 18003. Original LRA Table 2.3.4.A.1-1 lists "Tanks" as a component type subject to an AMR. Therefore, the staff requested that the applicant confirm that the condensate surge and storage tanks are included in the component type "Tanks" and identify other tanks (if any) belonging to the condensate and condensate transfer system that are within the scope of license renewal and included in the component type "Tanks."

In its response, by letter dated December 22, 2004, the applicant stated that both tanks are within the scope of license renewal, subject to an AMR, and included in the component type "Tanks" in original LRA Table 2.3.4.A.1-1. There are no other tanks within NMP1 condensate and condensate transfer system that are SR and subject to AMR. There are, however, other tanks in the system that are within the scope of license renewal and subject to AMR to meet 10 CFR 54.4(a)(2). These are among the other liquid-filled NSR equipment identified in original LRA Section 2.3.4.A.1 that are located in the reactor building, radwaste solidification and storage building, screen and pump house building, turbine building, or waste disposal building as being in the vicinity of SR components. Per the LR drawing convention, only components within the scope of license renewal and subject to AMR under 10 CFR 54.4(a)(2) are not to be shown in red on the LR drawings.

Based on its review, the staff found the applicant's response acceptable because it adequately explained that the subject tanks in question are within the scope of license renewal in accordance with 10 CFR 54.4(a) and subject to an AMR in accordance with 10 CFR 54.21(a). In addition, the response explained that the subject tanks are included in the component type "tanks" in original LRA Table 2.3.4.A.1-1 and that there are no other tanks within NMP1 condensate and condensate transfer system that are SR and subject to AMR. The response noted, however, that other tanks in the system that are within the scope of license renewal and subject to an AMR under 10 CFR 54.4(a)(2). Therefore, the staff's concern described in RAI 2.3.4.A.1-5 is resolved.

In RAI 2.3.4.A.1-6, dated November 19, 2004, the staff requested information on the intended function of "NSR Functional Support" listed in the original LRA Table 2.3.4.A.1-1. The applicant's response, by letter dated December 22, 2004, is reflected in the ALRA as discussed below.

In its ALRA, dated July 14, 2005, the applicant stated that this intended function is no longer used. Instead, the applicant identified specific NSR intended functions and made them consistent with the standardized list of intended functions in SRP-LR and NEI-95-10. Based on the information submitted in the ALRA the staff's concern described in RAI 2.3.4.A.1-6 is resolved.

In RAI 2.3.4.A.1-7, November 19, 2004, the staff stated that LR drawing 18009, sheet 1 shows that the only component located between valve CT-53 and valves BV 57-103/104 is flow indicator FI 57-168. However, LR drawing 18048 shows that flow gauge FG 57-175 is the only component located between these same valves. The applicant was asked to explain this apparent discrepancy.

In its response, dated December 22, 2004, the applicant stated that it agrees that LR drawing LR-18009-C, sheet 1 (location G5), identifies FI-57-168, a flow indicator for the clean-up demineralizer, as in-scope for license renewal and subject to AMR.

In addition, drawing LR-18048-C (location A6), identifies FG-57-175, a flow gauge for the clean-up demineralizer, as also being within the scope of license renewal and subject to an AMR.

The applicant stated that a review of NMP1 plant drawings C-18009-C, sheet 1, and C-18048-C, revealed the same discrepancy between these drawings as identified in the RAI. The discrepancy is corrected by replacing FG-57-175 on drawing LR-18048-C with FI-57-168, consistent with drawing LR-18009-C, sheet 1, and with the as-built configuration of the plant.

Based on its review, the staff found the applicant's response to RAI 2.3.4.A.1-7 acceptable because it adequately explained that the discrepancy between the two LR drawings was carried over from a discrepancy on two NMP1 P&ID drawings. The as-built configuration of the plant shows FI-57-168 between valve CT-53 and valve BV 57-103/104. LR drawing LR-18009-C, sheet 1 is correct as shown and LR-18048-C should have agreed with it to match the plant as-built configuration. Therefore, the staff's concern described in RAI 2.3.4.A.1-7 is resolved.

2.3A.4.1.3 Conclusion

The staff reviewed the ALRA, RAI responses, and accompanying scoping boundary drawings to determine whether any SSCs that should be within the scope of license renewal had not been identified by the applicant. No omissions were identified. In addition, the staff performed a review to determine whether any components that should be subject to an AMR had not been identified by the applicant. No omissions were identified. On the basis of its review, the staff concludes that there is reasonable assurance that the applicant had adequately identified the condensate and condensate transfer system components that are within the scope of license renewal, as required by 10 CFR 54.4(a), and the condensate and condensate transfer system components that are subject to an AMR, as required by 10 CFR 54.21(a)(1).

2.3A.4.2 NMP1 Condenser Air Removal and Off-Gas System

2.3A.4.2.1 Summary of Technical Information in the Amended Application

In ALRA Section 2.3.4.A.2, the applicant described the condenser air removal and off-gas system. The condenser air removal and off-gas system remove noncondensable radioactive gases that accumulate in the main condenser during plant startup and normal operation. The gases evacuated by this system are mainly concentrated in the condenser, but steam, air, and other gases evacuated by the steam packing exhauster are also discharged to the condenser air removal and off-gas system. The condenser air removal and off-gas system draws a suction from the air volume in the main condenser, processes the gases and exhausts the gases to the main stack. The processing of the non-condensable radioactive gases includes recombining the hydrogen and oxygen gases to form water, removing the moisture content of the gases and providing for radioactive decay so as to minimize the level of radiation exhausted to the main stack. This system also includes equipment to draw the initial vacuum on the main condenser during plant startup. The water removed by the processing of the condenser air is returned to the main condenser.

The condenser air removal and off-gas system contains SR components that are relied upon to remain functional during and following DBEs. The failure of NSR SSCs in the condenser air removal and off-gas system could potentially prevent the satisfactory accomplishment of an SR function.

The intended function, within the scope of license renewal, is to maintain mechanical and structural integrity to prevent spatial interactions.

In ALRA Table 2.3.4.A.2-1, the applicant identified the following condenser air removal and off-gas system component types that are within the scope of license renewal and subject to an AMR:

- air ejectors
- bolting
- heat exchanger
- piping and fittings
- valves

2.3A.4.2.2 Staff Evaluation

The staff reviewed ALRA Section 2.3.4.A.2 and UFSAR Section XI.B.3 using the evaluation methodology described in SER Section 2.3. The staff conducted its review in accordance with the guidance described in SRP-LR Section 2.3.

In conducting its review, the staff evaluated the system functions described in the ALRA and UFSAR in accordance with the requirements of 10 CFR 54.4(a) to verify that the applicant had not omitted from the scope of license renewal any components with intended functions delineated under 10 CFR 54.4(a). The staff then reviewed those components that the applicant had identified as being within the scope of license renewal to verify that the applicant had not omitted any passive and long-lived components that should be subject to an AMR in accordance with the requirements of 10 CFR 54.21(a)(1).

The staff's review of original LRA Section 2.3.4.A.2 identified an area in which additional information was necessary to complete the review of the applicant's scoping and screening results. The applicant responded to the staff's RAI as discussed below.

In RAI 2.3.4.A.2-1, dated November 19, 2004, the staff noted that original LRA Section 2.3.4.A.2 states that the condenser air removal and off-gas system removes and processes non-condensable radioactive gases that accumulate in the main condenser during startup and normal operation. The processing of the radioactive gases includes recombining the hydrogen and oxygen to form water. The original LRA further states that this system is in-scope for performing SR functions per 10 CFR 54.4(a)(1) and that because components within this system are either active or subject to replacement based on qualified life or specified time period no AMR is required. NMP1 UFSAR Section XI, Steam-to-Power Conversion System, B.3.0 (Condenser Air Removal and Offgas System), describes the operation and components of this system. Major components for the system are listed including many which appear to be passive and long-lived. Those components that are described as performing the process include: preheater, recombiner, condenser, drain tank, vent cooler, and 30-min holdup pipe. Therefore, in order for the staff to complete its review, the applicant was requested to confirm that the aforementioned components are not passive or long-lived or otherwise do not perform an intended function identified in original LRA Section 2.3.4.A.2. If found to require an AMR, then the applicant was to identify them on drawing(s) and include them in original LRA Table 2.3.4.A.2.

In its response, by letter dated December 22, 2004, the applicant stated that the intended function of the condenser air removal and off-gas systems is to provide fault protection and isolation for the SR reactor protection system distribution system. This system intended function is performed by active electrical components which are within the scope of license renewal but not subject to an AMR. The functions performed by passive components of the system, such as removing and processing non-condensable radioactive gases that accumulate in the main condenser during startup and normal operation, are not within the scope of license renewal. The information in the response has been subsequently incorporated in the ALRA.

In its ALRA, dated July 14, 2005, the applicant provided the staff with a revised LR drawing which identifies SSC's in scope and subject to an AMR under 10 CFR 54.4(a)(2).

Based on its review, the staff found the applicant's response to RAI 2.3.4.A.2-1 acceptable because it adequately explained that the SR function it performs is accomplished by electrical components that are active and not subject to an AMR in accordance with 10 CFR 54.21(a). Therefore, the staff's concern described in RAI 2.3.4.A.2-1 is resolved

2.3A.4.2.3 Conclusion

The staff reviewed the ALRA, RAI response, and accompanying scoping boundary drawing to determine whether any SSCs that should be within the scope of license renewal had not been identified by the applicant. No omissions were identified. In addition, the staff performed a review to determine whether any components that should be subject to an AMR had not been identified by the applicant. No omissions were identified. On the basis of its review, the staff concludes that there is reasonable assurance that the applicant had adequately identified the condenser air removal and off-gas system components that are within the scope of license renewal, as required by 10 CFR 54.4(a), and the condenser air removal and off-gas system components that are subject to an AMR, as required by 10 CFR 54.21(a)(1).

2.3A.4.3 NMP1 Feedwater/High Pressure Coolant Injection System

2.3A.4.3.1 Summary of Technical Information in the Amended Application

In ALRA Section 2.3.4.A.3, the applicant described the FW/HPCI system. The FW/HPCI system is the main source of processed water to the reactor during normal operation and also is designed to ensure that the core is adequately cooled under small break LOCA conditions, which do not result in a rapid depressurization of the RPV. The primary function of the FW system is to transfer the water from the condensate system to the RPV. The FW system also preheats the feedwater prior to entering the RPV. The HPCI system is an operating mode of the FW system. The purpose of the HPCI system is to provide adequate cooling of the reactor core under abnormal and accident conditions, remove the heat from radioactive decay and residual heat from the reactor core at such a rate that fuel clad melting would be prevented, and provide for continuity of core cooling over the complete range of postulated break sizes in the primary system process barrier. Upon initiation, the HPCI system provides the control functions to deliver water from the condensate storage tanks to the RPV. However, the HPCI system is not an engineered safeguards system and is not considered in any LOCA analyses.

The FW/HPCI system contains SR components that are relied upon to remain functional during and following DBEs. The failure of NSR SSCs in the FW/HPCI system could potentially prevent the satisfactory accomplishment of an SR function. In addition, the FW/HPCI system performs functions that support fire protection, EQ, and SBO.

The intended functions within the scope of license renewal include the following:

- maintains mechanical and structural integrity of NSR components to prevent spatial interactions with SR components

- provides structural support to NSR components whose failure could prevent accomplishment of SR function(s)

- provides removal and/or holdup of fission products

- provides pressure retaining boundary

- maintains mechanical and structural integrity of NSR components that provide structural support to attached SR components

In ALRA Table 2.3.4.A.3-1, the applicant identified the following FW/HPCI system component types that are within the scope of license renewal and subject to an AMR:

- bolting
- feedwater heaters
- filters/strainers
- flow elements
- flow indicators
- flow orifices
- oil coolers
- piping and fittings
- pumps
- valves

2.3A.4.3.2 Staff Evaluation

The staff reviewed ALRA Section 2.3.4.A.3 and UFSAR Sections VII.I and XI.B using the evaluation methodology described in SER Section 2.3. The staff conducted its review in accordance with the guidance described in SRP-LR Section 2.3.

In conducting its review, the staff evaluated the system functions described in the ALRA and UFSAR in accordance with the requirements of 10 CFR 54.4(a) to verify that the applicant had not omitted from the scope of license renewal any components with intended functions delineated under 10 CFR 54.4(a). The staff then reviewed those components that the applicant had identified as being within the scope of license renewal to verify that the applicant had not omitted any passive and long-lived components that should be subject to an AMR in accordance with the requirements of 10 CFR 54.21(a)(1).

The staff's review of original LRA Section 2.3.4.A.3 identified areas in which additional information was necessary to complete the review of the applicant's scoping and screening results. The applicant responded to the staff's RAIs as discussed below.

In RAI 2.3.4.A.3-1, dated November 19, 2004, the staff stated that original LRA Table 2.3.4.A.3-1 includes "oil coolers" as a component type subject to an AMR. However, the staff has been unable to locate oil coolers on the LR drawings referenced in original LRA Section 2.3.4.A.3. Drain coolers, on the other hand, are shown within the scope of license renewal on these drawings and are subject to an AMR, yet have not been included in the table. Therefore, the staff requested the applicant to confirm that "oil coolers" were mistakenly entered in place of "drain coolers" in the table and, if so, make the appropriate corrections. Otherwise, add the component type "drain coolers" to the table and provide drawings showing the subject oil coolers as well as the components they serve.

In its response, by letter dated December 22, 2004, the applicant stated that the subject table is correct and provided the following explanation:

> LR drawing LR-18004-C identifies three drain coolers. They are Drain Cooler 11 (51-04) at location GI; Drain Cooler 12 (51-05) at location G3; and Drain Cooler 13 (51-06) at location G5. These drain coolers are actually the first stage of the feedwater heaters. The shells for these coolers are not safety-related and not subject to AMR; however, the tube sides of these coolers are in-scope for LR and subject to AMR for a feedwater pressure boundary function. Per LRA Section 2.3.4.A.3, as liquid-filled components in the Turbine Building, the shell sides of these coolers are in-scope for LR and subject to AMR for criterion 10 CFR 54.4(a)(2). Consistent with LR drawing convention, the components in-scope for LR and subject to AMR for criterion (a)(2) only are not shown in red on the LR drawings.

> LR drawing LR-18023-C, Sheet 2, which is listed in LRA Section 2.3.4.A.3, identifies two oil coolers, 29-02 and 29-03. HTX-29-02 is the oil cooler for motor-driven Reactor Fedwater Pump 11 and HTX-29-03 is the oil cooler for motor-driven Reactor Feedwater Pump 12. As for the drain coolers above, the shells for these coolers are not safety-related and not subject to AMR for criteria 10 CFR 54.4(a)(1) or (a)(3); however, the tube sides are in-scope for LR and subject to AMR for a lube oil pressure boundary function. As with the drain coolers/feedwater heaters above, since they are liquid-filled components in the Turbine Building, they are in-scope for LR and subject to AMR for criterion 10 CFR 54.4(a)(2).

> As noted in the RAI, LRA Table 2.3.4.A.3-1 includes "Oil Coolers." Since the drain coolers are actually the first stage feedwater heaters, they are included in Table 2.3.4.A.3-1 under the component type "Feedwater Heaters."

Based on its review, the staff found the applicant's response to RAI 2.3.4.A.3-1 acceptable because it adequately explained that the subject drain cooler and oil cooler shell sides as liquid-filled components in the turbine building, are within the scope of license renewal and subject to AMR for criterion 10 CFR 54.4(a)(2). Also, the tube sides are in-scope for license renewal and subject to AMR for a feedwater and lube oil pressure boundary function per criteria 10 CFR 54.4(a)(1). The applicant explains that the drain coolers are the first stage feedwater heater and are included in original LRA Table 2.3.4.A.3-1 under the component type "feedwater heaters." Therefore, the staff's concern described in RAI 2.3.4.A.3-1 is resolved.

In RAI 2.3.4.A.3-2 dated November 19, 2004, the staff stated that original LRA Table 2.3.4.A.3-1 includes the following component types as subject to AMR: filters/strainers, flow elements, flow indicators, and flow orifices; however, the intended function assigned to these components is "NSR Functional Support." original LRA Table 2.0-1 identifies intended functions applicable to these components not identified in original LRA Table 2.3.4.A.3-1. Aging management to ensure that the component level intended functions can be performed is necessary to ensure that the system level intended functions can be maintained. The intended functions include "filtration" and "flow restriction." The applicant was asked to describe how the intended functions for these components are assigned and evaluated.

In its response by letter dated December 22, 2004, the applicant stated that a component's particular function of filtration for a filter or flow restriction for a flow orifice is not an intended function for license renewal but that a component function would be considered an intended function only if failure of that component would cause the failure of a system intended function and failure of the "filtration" or "flow restriction" functions for the components would not prevent the NMP1 FW/HPCI system from performing its intended function; therefore, the only intended function for these components is "NSR Functional Support" as identified in original LRA Table 2.3.4.A.3-1.

In evaluating this response the staff found it incomplete and review of original LRA Section 2.3.4.A.3 could not be completed because the applicant did not explain adequately what intended functions "NSR Functional Support" represents and how that intended function applies to all the component types in the feedwater/high pressure coolant injection system including filters/strainers, flow elements, flow indicators, and flow orifices. The staff held a teleconference with the applicant on January 27, 2005, to discuss information necessary to resolve the concern in RAI 2.3.4.A.3-2. The result of the teleconference was an agreement by the applicant to transmit the required information by a follow-up letter.

By letter dated February 11, 2005, the applicant defined the "NSR Functional Support" function for NSR components. Tables containing clarifying information regarding the intended functions for components in the FW/HPCI system were provided. The tables identify all the functions accomplished by each of the listed component types.

Based on its review, the staff found the applicant's response to RAI 2.3.4.A.3-2 including the information in the letter dated February 11, 2005, acceptable because it adequately explained what intended functions, "NSR Functional Support" represent and how it is applied to all the component types in the feedwater/high pressure coolant injection system including filters/strainers, flow elements, flow indicators, and flow orifices. Therefore, the staff's concern described in RAI 2.3.4.A.3-2 is resolved.

In RAI 2.3.4.A.3-3, dated November 19, 2004, the staff stated that original LRA Table 2.3.4.A.3-1 includes the following component types as being subject to an AMR: flow elements, flow indicators, and flow orifices. However, the drawing legend does not clearly define or distinguish between these components. For example, under "Flow Devices" in the legend, one of the symbols shown is denoted "FE" and defined as "flow element orifice" while another is denoted "FI" and defined as "in-line flow device." The distinction between these three component types and where each appears on the boundary drawings was not clear to the staff.

With respect to the LR drawings the staff asked the applicant to provide examples that clarify the distinction between the above mentioned three component types.

In its response, by letter December 22, 2004, the applicant stated that as an example, drawing LR-18005-C, sheet 1 (location F5), identifies FE-29-113, a flow element in the feedwater pump 13 discharge piping. The component identification is consistent with the drawing legend for flow devices as shown on license drawing LR-18000-C, sheet 1. This component is within the scope of license renewal and subject to AMR.

Another example is drawing LR-18005-C, sheet 1 (location G3), which identifies FI-51-106, a 1-inch flow indicator for feedwater pump 12 seal water. The component identification is consistent with the drawing legend for flow devices as shown on LR drawing LR-18000-C, sheet 1. This component is within the scope of license renewal and subject to AMR.

The last example is drawing LR-18005-C, sheet 1 (location F4), which identifies FOR-29-45, a restricting flow orifice for feedwater pump 12 recirculation. The component identification is consistent with the drawing legend for flow devices as shown on drawing LR-18000-C, sheet 1. This component is within the scope of license renewal and subject to AMR.

The applicant also stated that drawing LR-18000-C, sheet 1 (location A1 to E3 inclusive), provides a comprehensive matrix table that lists numerous component ID prefixes. In the matrix table for these instrument component types, "Flow" is the Measured Variable and the instrument functions are "primary element" (FE), "indicating" (FI), and "orifice restricting" (FOR), respectively. The drawing legend (location G5) is informational only and not meant to be all inclusive.

Based on its review, the staff found the applicant's response to RAI 2.3.4.A.3-3 acceptable because it adequately provided examples of component types flow elements, flow indicators, and flow orifices on license renewal drawings and explains the distinction between the subject three component types. Therefore, the staff's concern described in RAI 2.3.4.A.3-3 is resolved.

In RAI 2.3.4.A.3-4, dated November 19, 2004, the staff stated that LR drawing 18003 shows the symbol "boxed letter B" on the suction side of each of the three feedwater booster pumps. This symbol is not defined in the legend nor is the staff able to determine what it represents. Therefore, the staff requested that the applicant define the above described symbol.

In its response, by letter December 22, 2004, the applicant stated that the symbol "boxed letter B" represents the start-up strainer shell for feedwater booster pumps 11, 12, and 13. The details can be found on license renewal drawing LR-18003-C (location H4). The element has been removed from the strainer.

Based on its review, the staff found the applicant's response to RAI 2.3.4.A.3-4 acceptable because the applicant adequately explained what the "boxed letter B" on the subject LR drawing stands for and how it is not relevant to license renewal. Therefore, the staff's concern described in RAI 2.3.4.A.3-4 is resolved.

2.3A.4.3.3 Conclusion

The staff reviewed the ALRA, RAI responses, and accompanying scoping boundary drawings to determine whether any SSCs that should be within the scope of license renewal had not been identified by the applicant. No omissions were identified. In addition, the staff performed a review to determine whether any components that should be subject to an AMR had not been identified by the applicant. No omissions were identified. On the basis of its review, the staff concludes that there is reasonable assurance that the applicant had adequately identified the FW/HPCI system components that are within the scope of license renewal, as required by 10 CFR 54.4(a), and the FW/HPCI system components that are subject to an AMR, as required by 10 CFR 54.21(a)(1).

2.3A.4.4 NMP1 Main Generator and Auxiliary System

2.3A.4.4.1 Summary of Technical Information in the Amended Application

In ALRA Section 2.3.4.A.4, the applicant described the main generator and auxiliary system. The main generator and auxiliary system consists of the main generator, generator stator cooling water system, hydrogen seal oil system and hydrogen cooling system. The hydrogen cooling system fills the main generator with high-purity hydrogen gas to cool the generator during plant operation. The main generator is filled with hydrogen gas by first purging air with carbon dioxide and then purging the carbon dioxide with hydrogen. The equipment used to supply carbon dioxide to the main generator is the only equipment of the main generator and auxiliary system that is in scope for license renewal.

The main generator and auxiliary system performs functions that support fire protection.

The intended function, within the scope of license renewal, is to provide pressure retaining boundary.

In ALRA Table 2.3.4.A.4-1, the applicant identified the following main generator and auxiliary system component types that are within the scope of license renewal and subject to an AMR:

- bolting
- piping and fittings
- tanks
- valves

2.3A.4.4.2 Staff Evaluation

The staff reviewed ALRA Section 2.3.4.A.4 and UFSAR Section XI.B.1 using the evaluation methodology described in SER Section 2.3. The staff conducted its review in accordance with the guidance described in SRP-LR Section 2.3.

In conducting its review, the staff evaluated the system functions described in the ALRA and UFSAR in accordance with the requirements of 10 CFR 54.4(a) to verify that the applicant had not omitted from the scope of license renewal any components with intended functions delineated under 10 CFR 54.4(a). The staff then reviewed those components that the applicant had identified as being within the scope of license renewal to verify that the applicant had not omitted any passive and long-lived components that should be subject to an AMR in accordance with the requirements of 10 CFR 54.21(a)(1).

2.3A.4.4.3 Conclusion

The staff reviewed the ALRA to determine whether any SSCs that should be within the scope of license renewal had not been identified by the applicant. No omissions were identified. In addition, the staff performed a review to determine whether any components that should be subject to an AMR had not been identified by the applicant. No omissions were identified. On the basis of its review, the staff concludes that there is reasonable assurance that the applicant had adequately identified the main generator and auxiliary system components that are within the scope of license renewal, as required by 10 CFR 54.4(a), and the main generator and auxiliary system components that are subject to an AMR, as required by 10 CFR 54.21(a)(1).

2.3A.4.5 NMP1 Main Steam System

2.3A.4.5.1 Summary of Technical Information in the Amended Application

In ALRA Section 2.3.4.A.5, the applicant described the main steam system. The main steam system supplies dry steam from the RPV to the main turbine and to various support systems. The main steam system consists of two main steam lines, four main steam isolation valves, six electromatic relief valves, four turbine stop valves, four turbine control valves, nine turbine bypass valves, controls, instrumentation, piping, valves and associated equipment. The system extends from the RPV main steam nozzles to the turbine stop, control and bypass valves and to the inlet of the various components to which it supplies steam. The discharge piping and valves from the electromatic relief valves to the torus, including the Y-quenchers, are also included within this system. The electromatic relief valves are also used by the automatic depressurization system to depressurize the RPV during accident conditions.

The main steam system contains SR components that are relied upon to remain functional during and following DBEs. The failure of NSR SSCs in the main steam system could potentially prevent the satisfactory accomplishment of an SR function. In addition, the main steam system performs functions that support fire protection, EQ, and SBO.

The intended functions within the scope of license renewal include the following:

- maintains mechanical and structural integrity of NSR components to prevent spatial interactions with SR components

- provides removal and/or holdup of fission products

- provides pressure retaining boundary

- maintains mechanical and structural integrity of NSR components that provide structural support to attached SR components

In ALRA Table 2.3.4.A.5-1, the applicant identified the following main steam system component types that are within the scope of license renewal and subject to an AMR:

- bolting
- condensing pots
- flow elements
- piping and fittings
- regulator
- valves
- Y-quenchers

2.3A.4.5.2 Staff Evaluation

The staff reviewed ALRA Section 2.3.4.A.5 and UFSAR Sections V.B.1, V.B.5, and XI.B.1 using the evaluation methodology described in SER Section 2.3. The staff conducted its review in accordance with the guidance described in SRP-LR Section 2.3.

In conducting its review, the staff evaluated the system functions described in the ALRA and UFSAR in accordance with the requirements of 10 CFR 54.4(a) to verify that the applicant had not omitted from the scope of license renewal any components with intended functions delineated under 10 CFR 54.4(a). The staff then reviewed those components that the applicant had identified as being within the scope of license renewal to verify that the applicant had not omitted any passive and long-lived components that should be subject to an AMR in accordance with the requirements of 10 CFR 54.21(a)(1).

The staff's review of original LRA Section 2.3.4.A.1 identified areas in which additional information was necessary to complete the review of the applicant's scoping and screening results. The applicant responded to the staff's RAIs as discussed below.

In RAI 2.3.4.A.5-1 dated November 19, 2004, the staff stated that LR Note #1 on drawing LR-18000-C, "License Renewal Boundary Drawing Symbols, Notes, and Acronyms," states that portions of the system subject to AMR are highlighted in red with boundaries indicated by blue flags. The blue flags are described on drawing LR-18000-C as "AMR Boundary Flags." Portions of a license renewal system indicated with solid blue flags may perform intended functions and are within scope of license renewal but are not subject to AMR; however, when an LR drawing is composed of a system diagram that has a continuation on another system diagram not provided by the applicant, the staff is unable to complete its review of whether the license renewal system incorporates all portions necessary to satisfy its plant level system intended functions.

LR-18002-C for the NMP1 main steam system is composed of system diagram C-18002 sheet 1. A portion of the main steam system is depicted on C-18002 sheet 2, which has not been provided. Therefore, the staff requested that the applicant confirm that no portion of the main steam system on C-18002 sheet 2 has SR components or otherwise meets criteria of 10 CFR 54.4(a)(1), (2), or (3), or, if such components exist, identify them and ensure that their component types and intended functions are represented in Table 2.3.4.A.5-1.

In its response dated December 22, 2004, the applicant stated that there are no components subject to AMR under 10 CFR 54.21(a)(1) on the subject system diagram for the main steam system. Therefore, this diagram was not included in the original LRA. The applicant further stated that the main steam system components within the scope of license renewal for 10 CFR 50.54(a)(2) and subject to an AMR are not highlighted on the LR drawings.

Based on its review, the staff found the applicant's response to RAI 2.3.4.A.5-1 acceptable because it adequately explained that there are no components subject to an AMR in accordance with 10 CFR 54.21(a)(1) existing on other system diagrams for the main steam system that are not used as license renewal drawings. Therefore, the staff's concern described in RAI 2.3.4.A.5-1 is resolved.

In RAI 2.3.4.A.5-2 dated November 19, 2004, the staff stated that original LRA Section 2.3.4.A.5 identifies LR drawings depicting components requiring AMR for the NMP1 main steam system. During the review of the original LRA, however, the staff found other LR drawings that have main steam components shown to require AMR not identified in original LRA Section 2.3.4.A.5. These include LR-18006-C, "Drywell and Torus Isolation Valves." For the staff to complete its review of the main steam system the applicant was asked to identify other drawings depicting main steam components requiring AMR.

In its response by letter dated December 22, 2004, the applicant stated that LR drawing LR-18002-C sheet 1 identifies the main steam system components within the scope of license renewal and subject to AMR for 10 CFR 50.54 (a)(1) and (a)(3), that although drawing LR-18006-C sheet 1 also shows main steam system components within the scope of license renewal they are the same as those shown on LR-18002-C, that as such drawing LR-18006-C should have been referenced in original LRA Section 2.3.4.A.5, that LR drawing LR-18006 sheet 1 should have been referenced for the containment spray, liquid poison, emergency cooling, and feedwater systems because components from those systems also appear on the drawing, and that main steam components within scope of license renewal subject to an AMR for 10 CFR 50.54(a)(2) are not redlined on LR drawings; therefore, there are no additional LR drawings that show main steam components subject to an AMR.

Based on its review, the staff found the applicant's response to RAI 2.3.4.A.5-2 acceptable because it adequately explained that although there are some main steam components subject to an AMR in accordance with 10 CFR 54.21(a)(1) that are duplicated on other license renewal drawings, they are highlighted as well in accordance with the license renewal guidance. Therefore, the staff's concern described in RAI 2.3.4.A.5-2 is resolved.

In RAI 2.3.4.A.5-3 dated November 19, 2004, the staff stated that LR drawing 18002 sheet 1 shows the branch line connecting the discharge line of safety relief valve MSER V-2 to temperature element 01-17 omitted from AMR. This branch line forms part of the reactor coolant pressure boundary, and is passive, and long-lived; therefore, it should require an AMR. (Note that the corresponding branch lines for the remaining five safety relief valves are shown correctly as requiring AMR). The staff requested that the applicant justify omission of the branch line from AMR.

In its response by letter dated December 22, 2004, the applicant stated that the LR drawing is incorrect and does not properly show the line within the scope of license renewal and subject to

AMR. The line in question should have been highlighted on the LR drawing showing it within the scope of license renewal and subject to AMR but inadvertently was not highlighted.

Based on its review, the staff found the applicant's response to RAI 2.3.4.A.5-3 acceptable because it adequately explained that the components in question are within the scope of license renewal and subject to an AMR but were inadvertently left un-highlighted on the license renewal drawing. The staff concludes that there is reasonable assurance that the components were correctly included within the scope of license renewal and subject to an AMR. Therefore, the staff's concern described in RAI 2.3.4.A.5-3 is resolved.

In RAI 2.3.4.A.5-4 dated November 19, 2004, the staff stated that LR drawing 18002, sheet 1 (locations A2, A3, D2, D3) shows the discharge line from each of the six safety-relief valves (MSER V-1, 2, 3, 4, 5, and 6) ending at a continuation flag labeled "To Torus" with no continuation drawing specified. At that point there is also a boundary flag showing an interface between the main steam system and the primary containment system. For the staff to determine if all components in this SR system within the scope of license renewal and subject to an AMR have been identified, it must review of the continuation drawing. The applicant was asked to provide a drawing showing the continuation of the safety-relief valve discharge lines to the torus or, if already provided in the original LRA as an LR drawing, identify the drawing number.

In its response by letter dated December 22, 2004, the applicant stated that the safety-relief valve discharge lines are routed through the drywell-to-torus vent lines and terminate in the torus below the water line. The applicant also stated that the continuation of the safety-relief valve discharge lines is as shown on the referenced license renewal drawing to signify that the safety-relief valve discharge line piping continues to the torus, and that there are no other components in those lines.

Based on its review, the staff found the applicant's response to RAI 2.3.4.A.5-4 acceptable because it adequately explained that although the safety relief valve discharge line is routed through the drywell to the torus vent, there are no continuation drawings depicting this piping that contain components other than the piping that is already subject to an AMR. Therefore, the staff's concern described in RAI 2.3.4.A.5-4 is resolved.

In RAI 2.3.4.A.5-5, dated November 19, 2004, the staff requested information on the intended function of "NSR Functional Support" listed in Table 2.3.4.A.5-1 of the original original LRA. The applicant's December 22, 2004, response information to this RAI is reflected in the ALRA as described below.

In its ALRA dated July 14, 2005, the applicant stated that this intended function is no longer used, instead, the applicant identified specific NSR intended functions and made them consistent with the standardized list of intended functions in SRP-LR and NEI-95-10. Based on the information submitted in the ALRA, the staff's concern described in RAI 2.3.4.A.5-5 is resolved.

In RAI 2.3.4.A.5-6, dated November 19, 2004, the staff stated that LR drawing LR-18002-C indicates that bellows expansion joints 66-01R, -02R, -03R, -04R, -05R, and -06R are subject to an AMR in the main steam system. original LRA Table 2.3.4.A.5-1 does not include bellows expansion joints as a "component type" with an intended function; therefore, the staff requested that the applicant justify the omission of the bellows expansion joints from original LRA Table 2.3.4.A.5-1 or revise the table to include this component type.

In its response, dated December 22, 2004, the applicant stated that expansion joints are included under the component type "Piping and Fittings" in the original LRA Table 2.3.4.A.5-1.

Based on its review, the staff found the applicant's response to RAI 2.3.4.A.5-6 acceptable because it adequately explained that the bellows expansion joints in question are within the scope of license renewal and subject to an AMR. Further, the applicant stated that the bellows expansion joints are represented in the original LRA Table. Therefore, the staff's concern described in RAI 2.3.4.A.5-6 is resolved.

2.3A.4.5.3 Conclusion

The staff reviewed the ALRA, RAI responses, and accompanying scoping boundary drawings accompanying scoping boundary drawings to determine whether any SSCs that should be within the scope of license renewal had not been identified by the applicant. No omissions were identified. In addition, the staff performed a review to determine whether any components that should be subject to an AMR had not been identified by the applicant. No omissions were identified. On the basis of its review, the staff concludes that there is reasonable assurance that the applicant had adequately identified the main steam system components that are within the scope of license renewal, as required by 10 CFR 54.4(a), and the main steam system components that are subject to an AMR, as required by 10 CFR 54.21(a)(1).

2.3A.4.6 NMP1 Main Turbine and Auxiliary Systems

2.3A.4.6.1 Summary of Technical Information in the Amended Application

In ALRA Section 2.3.4.A.6, the applicant described the main turbine and auxiliary systems. The main turbine and auxiliary systems converts the thermal energy contained in the steam supplied by the reactor into electrical energy. The turbine is a tandem-compound, 1800 RPM unit with a single admission, double-flow high pressure section and a six-flow low pressure section. A bypass system is provided that allows bypassing excess steam flow to the condenser when the turbine cannot absorb all the generated steam.

The main turbine and auxiliary systems consist of multiple subsystems including the main turbine system, turbine-generator controls system, turbine gland sealing system, turbine oil storage and purification system, turbine protection system and turbine supervisory instruments system. Of these systems, the turbine gland sealing system, turbine oil storage and purification system and the turbine protection system, specifically the turbine overspeed system, contain components that are within the scope of license renewal. The turbine gland sealing system functions to seal the shaft of the main turbine against leakage of steam from the turbine shell to atmosphere as well as leakage of air from atmosphere to the main condenser. The turbine oil storage and purification system supplies purified lubricating and cooling oil to the turbine-generator bearings, shaft-driven feedwater pump and the turbine-generator controls system. The turbine protection system monitors selected parameters, including turbine overspeed, and provides various trips and alarms designed to protect the turbine from damage.

The failure of NSR SSCs in the main turbine and auxiliary systems could potentially prevent the satisfactory accomplishment of an SR function.

The intended function, within the scope of license renewal, is to maintain mechanical and structural integrity to prevent spatial interactions.

In ALRA Table 2.3.4.A.6-1, the applicant identified the following main turbine and auxiliary systems component types that are within the scope of license renewal and subject to an AMR:

- bolting
- heat exchanger
- piping and fittings
- regulator
- valves

2.3A.4.6.2 Staff Evaluation

The staff reviewed ALRA Section 2.3.4.A.6 and UFSAR Sections XI.B.1 and VIII.B.2.3 using the evaluation methodology described in SER Section 2.3. The staff conducted its review in accordance with the guidance described in SRP-LR Section 2.3.

In conducting its review, the staff evaluated the system functions described in the ALRA and UFSAR in accordance with the requirements of 10 CFR 54.4(a) to verify that the applicant had not omitted from the scope of license renewal any components with intended functions delineated under 10 CFR 54.4(a). The staff then reviewed those components that the applicant had identified as being within the scope of license renewal to verify that the applicant had not omitted any passive and long-lived components that should be subject to an AMR in accordance with the requirements of 10 CFR 54.21(a)(1).

2.3A.4.6.3 Conclusion

The staff reviewed the ALRA to determine whether any SSCs that should be within the scope of license renewal had not been identified by the applicant. No omissions were identified. In addition, the staff performed a review to determine whether any components that should be subject to an AMR had not been identified by the applicant. No omissions were identified. On the basis of its review, the staff concludes that there is reasonable assurance that the applicant had adequately identified the main turbine and auxiliary systems components that are within the scope of license renewal, as required by 10 CFR 54.4(a), and the main turbine and auxiliary systems components that are subject to an AMR, as required by 10 CFR 54.21(a)(1).

2.3A.4.7 NMP1 Moisture Separator Reheater Steam System

2.3A.4.7.1 Summary of Technical Information in the Amended Application

In ALRA Section 2.3.4.A.7, the applicant described the moisture separator reheater steam system. The moisture separator reheater steam system removes entrained moisture from the high pressure turbine exhaust and reheats the dried steam to superheated conditions before it passes on to the low pressure turbine.

The failure of NSR SSCs in the moisture separator reheater steam system could potentially prevent the satisfactory accomplishment of an SR function.

The intended function, within the scope of license renewal, is to maintain mechanical and structural integrity to prevent spatial interactions.

In ALRA Table 2.3.4.A.7-1, the applicant identified the following moisture separator reheater steam system component types that are within the scope of license renewal and subject to an AMR:

- bolting
- flow element
- flow orifices
- heat exchanger
- piping and fittings
- separator
- strainer
- tanks
- valves

2.3A.4.7.2 Staff Evaluation

The staff reviewed ALRA Section 2.3.4.A.7 and UFSAR Section XI.B.1.0 using the evaluation methodology described in SER Section 2.3. The staff conducted its review in accordance with the guidance described in SRP-LR Section 2.3.

In conducting its review, the staff evaluated the system functions described in the ALRA and UFSAR in accordance with the requirements of 10 CFR 54.4(a) to verify that the applicant had not omitted from the scope of license renewal any components with intended functions delineated under 10 CFR 54.4(a). The staff then reviewed those components that the applicant had identified as being within the scope of license renewal to verify that the applicant had not omitted any passive and long-lived components that should be subject to an AMR in accordance with the requirements of 10 CFR 54.21(a)(1).

2.3A.4.7.3 Conclusion

The staff reviewed the ALRA to determine whether any SSCs that should be within the scope of license renewal had not been identified by the applicant. No omissions were identified. In addition, the staff performed a review to determine whether any components that should be subject to an AMR had not been identified by the applicant. No omissions were identified. On the basis of its review, the staff concludes that there is reasonable assurance that the applicant had adequately identified the moisture separator reheater steam system components that are within the scope of license renewal, as required by 10 CFR 54.4(a), and the moisture separator reheater steam system components that are subject to an AMR, as required by 10 CFR 54.21(a)(1).

2.3B NMP2 Scoping and Screening Results: Mechanical Systems

2.3B.1 Reactor Vessel, Internals, and Reactor Coolant Systems

In ALRA Section 2.3.1.B, the applicant identified the structures and components of the NMP2 reactor vessel, internals, and reactor coolant systems that are subject to an AMR for license renewal.

The applicant described the supporting structures and components of the reactor vessel, internals, and reactor coolant systems in the following sections of the ALRA:

- 2.3.1.B.1   NMP2 reactor pressure vessel
- 2.3.1.B.2   NMP2 reactor pressure vessel internals
- 2.3.1.B.3   NMP2 reactor pressure vessel instrumentation system
- 2.3.1.B.4   NMP2 reactor recirculation system
- 2.3.1.B.5   NMP2 control rod drive system
- 2.3.1.B.6   NMP2 reactor coolant pressure boundary components in other systems

The staff's review findings regarding ALRA Sections 2.3.1.B.1 through 2.3.1.B.6 are presented in SER Sections 2.3B.1.1 through 2.3B.1.6, respectively.

2.3B.1.1 NMP2 Reactor Pressure Vessel

2.3B.1.1.1 Summary of Technical Information in the Amended Application

In ALRA Section 2.3.1.B.1, the applicant described the RPV. The RPV contains and supports the reactor core, reactor internals, and the reactor coolant/moderator. The RPV forms part of the RCPB and serves as a barrier against leakage of radioactive materials to the drywell. The RPV is a vertical cylindrical pressure vessel of welded construction with hemispherical bottom and top heads. The cylindrical shell and top and bottom heads of the RPV are fabricated of low-alloy steel, the interior of which is clad with stainless steel weld overlay, except for the top head and nozzle and nozzle weld zones. The RPV top head is secured to the RPV by studs and nuts. The RPV flanges are sealed with two concentric metal seal rings designed to permit no detectable leakage through the inner or outer seal at any operating condition. The top head leak detection lines tap off of the vessel head between the seal rings to detect leakage should the inner seal-ring fail. The RPV is penetrated by various nozzles and penetrations. The CRD housings and in-core instrumentation thimbles are welded to the bottom head of the RPV. The concrete and steel vessel support pedestal is constructed as an integral part of the building foundation. Steel anchor bolts, set in the concrete, extend through the bearing plate and secure the flange of the reactor vessel support skirt to the bearing plate, and thus to the support pedestal.

The RPV contains SR components that are relied upon to remain functional during and following DBEs.

The intended functions within the scope of license renewal include the following:

- provides pressure retaining boundary
- provides structural and/or functional support to SR equipment

In ALRA Table 2.3.1.B.1-1, the applicant identified the following RPV component types that are within the scope of license renewal and subject to an AMR:

- bottom head
- nozzles
- nozzle safe ends
- nozzle thermal sleeves
- penetrations: core differential pressure and liquid control, CRD stub tubes, drain lines, in-core instruments, instrumentation
- support skirt
- top head and nozzles
- top head (closure studs and nuts)
- top head (flanges)
- top head (leak detection lines)
- valves
- vessel shells (flange)
- vessel shells: lower intermediate shell, lower shell, upper intermediate shell, upper shell
- vessel welds (including attachment welds)

2.3B.1.1.2 Staff Evaluation

The staff reviewed ALRA Section 2.3.1.B.1 and the USAR using the evaluation methodology described in SER Section 2.3. The staff conducted its review in accordance with the guidance described in SRP-LR Section 2.3.

In conducting its review, the staff evaluated the system functions described in the ALRA and USAR in accordance with the requirements of 10 CFR 54.4(a) to verify that the applicant had not omitted from the scope of license renewal any components with intended functions delineated under 10 CFR 54.4(a). The staff then reviewed those components that the applicant had identified as being within the scope of license renewal to verify that the applicant had not omitted any passive and long-lived components that should be subject to an AMR in accordance with the requirements of 10 CFR 54.21(a)(1).

2.3B.1.1.3 Conclusion

The staff reviewed the ALRA to determine whether any SSCs that should be within the scope of license renewal had not been identified by the applicant. No omissions were identified. In addition, the staff performed a review to determine whether any components that should be subject to an AMR had not been identified by the applicant. No omissions were identified. On the basis of its review, the staff concludes that there is reasonable assurance that the applicant had adequately identified the RPV components that are within the scope of license renewal, as required by 10 CFR 54.4(a), and the RPV components that are subject to an AMR, as required by 10 CFR 54.21(a)(1).

2.3B.1.2 NMP2 Reactor Pressure Vessel Internals

2.3B.1.2.1 Summary of Technical Information in the Amended Application

In ALRA Section 2.3.1.B.2, the applicant described the reactor pressure vessel internals. The reactor pressure vessel internals provide support for the core and other internal components, maintain fuel configuration (coolable geometry) during normal operation and accident conditions, and provide reactor coolant flow distribution through the core. The reactor pressure vessel internals consists of the components internal to the RPV. The main structures within the RPV are the core (fuel, channels, control rods and instrumentation), the core support structure (including the shroud, top guide and core plate), the shroud head and steam separator assembly, the steam dryer assembly, the feedwater spargers, the core spray spargers, and the jet pumps. Except for the Zircaloy used in the fuel assemblies, reactor internals are stainless steel or other corrosion-resistant alloys. The fuel assemblies (which include fuel rods and channel), control rods, in-core instrumentation, shroud head and steam separator assembly, and steam dryers are removable when the reactor vessel is opened for refueling or maintenance.

The reactor pressure vessel internals contain SR components that are relied upon to remain functional during and following DBEs. The failure of NSR SSCs in the reactor pressure vessel internals could potentially prevent the satisfactory accomplishment of an SR function.

The intended functions within the scope of license renewal include the following:

- provides spray shield or curbs for directing flow

- provides structural support to NSR components whose failure could prevent accomplishment of SR function(s)

- provides pressure retaining boundary

- provides structural and/or functional support to SR equipment

In ALRA Table 2.3.1.B.2-1, the applicant identified the following reactor pressure vessel internals component types that are within the scope of license renewal and subject to an AMR:

- access hole covers
- CRD assemblies (includes drive mechanism and housing)
- control rod guide tubes
- core plate, bolts, and supports
- core shroud
- core shroud head bolts
- core/shroud support structures: bolts, brackets, clamps, keepers, supports
- core spray lines and spargers
- differential pressure liquid control line
- flanges
- incore housings
- incore instrumentation dry tubes
- jet pump assemblies
- low pressure coolant injection (LPCI) couplings

- orificed fuel supports
- peripheral fuel supports
- power range detector assemblies
- spray nozzles
- steam dryer assembly
- top guide and supports

2.3B.1.2.2 Staff Evaluation

The staff reviewed ALRA Section 2.3.1.B.2 and the USAR using the evaluation methodology described in SER Section 2.3. The staff conducted its review in accordance with the guidance described in SRP-LR Section 2.3.

In conducting its review, the staff evaluated the system functions described in the ALRA and USAR in accordance with the requirements of 10 CFR 54.4(a) to verify that the applicant had not omitted from the scope of license renewal any components with intended functions delineated under 10 CFR 54.4(a). The staff then reviewed those components that the applicant had identified as being within the scope of license renewal to verify that the applicant had not omitted any passive and long-lived components that should be subject to an AMR in accordance with the requirements of 10 CFR 54.21(a)(1).

The staff's review of original LRA Section 2.3.1.B.2 identified areas in which additional information was necessary to complete the evaluation of the applicant's scoping and screening results. The applicant responded to the staff's RAIs as discussed below.

In RAI-4, dated November 17, 2004, the staff noted that the steam separator assembly consists of a base into which are welded an array of standpipes with a steam separator located at the top of each standpipe. Therefore, the staff requested that the applicant provide justification why these standpipes and steam separators were not included within the scope of license renewal.

In its response by letter dated December 17, 2004, the applicant stated that the steam separators and their standpipes are not included within the scope of license renewal because they are not SR components that perform a license renewal intended function and referred to an evaluation contained in BWRVIP-06-A. Also, the staff's concern about the possibility that failure of these components could prevent the accomplishment of SR functions of nearby components (e.g., the creation of loose parts that might hit and damage SR components). The staff noted that this consideration was also addressed in BWRVIP-06-A and the evaluation was accepted by the staff in letters dated September 15, 1998, and September 16, 2003. Therefore, the staff's concern described in RAI-4 is resolved.

In RAI-5 dated November 17, 2004, the staff requested that the applicant indicate where the feedwater sparger is identified as a vessel internal component requiring an AMR. In its response by letter dated December 17, 2004, the applicant indicated that this sparger also was not included within the scope of license renewal. The applicant stated that, per BWRVIP-06-A, "The sole purpose of the feedwater spargers is to control thermal mixing and extend the life of the vessel and internals. The failure of feedwater spargers or associated brackets would not prevent injection of coolant makeup and are not required to safety shut down the reactor." On this basis the staff found exclusion of the feedwater sparger from within the scope of license renewal acceptable. Therefore, the staff's concern described in RAI-5 is resolved.

In RAI 2.3-4, dated October 11, 2005, the staff requested that the applicant indicate whether the diffuser seal ring and shroud support plate should be identified as an RPV internal component requiring an AMR. In its response, by letter dated October 28, 2005, the applicant indicated that the diffuser seal ring and the shroud support plate are reactor pressure vessel internal components subject to aging management review, that the diffuser seal ring is included under "Jet Pump Assemblies" identified in ALRA Table 2.3.1.B.2-1, and that the shroud support plate is included under "Core Shroud Support Structures" in ALRA Table 2.3.1.B.2-1. Therefore, the staff's concern described in RAI 2.3-4 is resolved.

2.3B.1.2.3 Conclusion

The staff reviewed the ALRA and RAI responses to determine whether any SSCs that should be within the scope of license renewal had not been identified by the applicant. No omissions were identified. In addition, the staff performed a review to determine whether any components that should be subject to an AMR had not been identified by the applicant. No omissions were identified. On the basis of its review, the staff concludes that there is reasonable assurance that the applicant had adequately identified the reactor pressure vessel internals components that are within the scope of license renewal, as required by 10 CFR 54.4(a), and the reactor pressure vessel internals components that are subject to an AMR, as required by 10 CFR 54.21(a)(1).

2.3B.1.3 NMP2 Reactor Pressure Vessel Instrumentation System

2.3B.1.3.1 Summary of Technical Information in the Amended Application

In ALRA Section 2.3.1.B.3, the applicant described the reactor pressure vessel instrumentation system. The reactor pressure vessel instrumentation system provides a means of monitoring and transmitting information concerning key reactor vessel operating parameters during normal and emergency operations. Instrumentation is installed to monitor reactor parameters and indicate these on meters and chart recorders in the control room and remote shutdown panels. The parameters monitored are reactor vessel temperature, water level and pressure, core flow and core plate differential pressure. This system also provides control signals to various systems which, in turn, initiate the appropriate actions required if the monitored parameter exceeds its desired setpoint. Systems receiving control signals from the reactor pressure vessel instrumentation system include reactor protection, primary containment isolation, automatic depressurization, feedwater control, reactor recirculation flow control, redundant reactivity control and residual heat removal (shutdown cooling mode) systems.

The reactor pressure vessel instrumentation system contains SR components that are relied upon to remain functional during and following DBEs. The failure of NSR SSCs in the reactor pressure vessel instrumentation system could potentially prevent the satisfactory accomplishment of an SR function. In addition, the reactor pressure vessel instrumentation system performs functions that support fire protection, EQ, ATWS, and SBO.

The intended functions within the scope of license renewal include the following:

- provides pressure retaining boundary
- provides shielding against radiation

- maintains mechanical and structural integrity of NSR components that provide structural support to attached SR components

In ALRA Table 2.3.1.B.3-1, the applicant identified the following reactor pressure vessel instrumentation system component types that are within the scope of license renewal and subject to an AMR:

- closure bolting
- condensing chambers
- piping and fittings
- radiation collars
- restriction orifices
- vacuum breakers
- valves

2.3B.1.3.2 Staff Evaluation

The staff reviewed ALRA Section 2.3.1.B.3 and the USAR using the evaluation methodology described in SER Section 2.3. The staff conducted its review in accordance with the guidance described in SRP-LR Section 2.3.

In conducting its review, the staff evaluated the system functions described in the ALRA and USAR in accordance with the requirements of 10 CFR 54.4(a) to verify that the applicant had not omitted from the scope of license renewal any components with intended functions delineated under 10 CFR 54.4(a). The staff then reviewed those components that the applicant had identified as being within the scope of license renewal to verify that the applicant had not omitted any passive and long-lived components that should be subject to an AMR in accordance with the requirements of 10 CFR 54.21(a)(1).

The staff's review of original LRA Section 2.3.1.B.3 identified an area in which additional information was necessary to complete the evaluation of the applicant's scoping and screening results. The applicant responded to the staff's RAIs as discussed below.

In RAI 2.3-2 dated October 11, 2005, the staff requested that the applicant indicate whether temperature equalizing columns should be identified as RPV instrumentation system components requiring an AMR.

In its response by letter dated October 28, 2005, the applicant indicated that the NMP2 RPV instrumentation system does not utilize temperature equalizing columns. For NMP2 the type of level measurement system makes no correction for changes in reactor vessel or reference leg water temperature or density and is termed "non-compensated." Each instrument is calibrated at the vessel pressure and drywell temperature at which the instrument is normally used. Therefore, there are no temperature equalizing columns listed for review in the rector vessel instrumentation system section.

Based on this review, the staff found the applicant's assessment acceptable. Therefore, the staff's concern described in RAI 2.3-2 is resolved.

2.3B.1.3.3 Conclusion

The staff reviewed the ALRA and RAI response to determine whether any SSCs that should be within the scope of license renewal had not been identified by the applicant. No omissions were identified. In addition, the staff performed a review to determine whether any components that should be subject to an AMR had not been identified by the applicant. No omissions were identified. On the basis of its review, the staff concludes that there is reasonable assurance that the applicant had adequately identified the reactor pressure vessel instrumentation system components that are within the scope of license renewal, as required by 10 CFR 54.4(a), and the reactor pressure vessel instrumentation system components that are subject to an AMR, as required by 10 CFR 54.21(a)(1).

2.3B.1.4 NMP2 Reactor Recirculation System

2.3B.1.4.1 Summary of Technical Information in the Amended Application

In ALRA Section 2.3.1.B.4, the applicant described the reactor recirculation system. The reactor recirculation system is designed to provide a variable reactor coolant flow in order to control reactor power levels. The reactor recirculation system is part of the RCPB and consists of two external loops. Each loop contains a pump, flow control valve, two blocking valves, piping and associated controls and instrumentation. Coolant flow is from the RPV annulus region, through a recirculation pump and flow control valve, into an external manifold from which individual recirculation inlet lines are routed to the jet pump risers within the RPV. The jet pumps are evaluated as part of the reactor pressure vessel internals. The recirculation pumps operate at two speeds with power coming from either the low frequency motor generator set (25 percent) or a 60-Hz power source (100 percent). The flow control valves are controlled by two separate sets of control system components, one for each valve.

The reactor recirculation system contains SR components that are relied upon to remain functional during and following DBEs. The failure of NSR SSCs in the reactor recirculation system could potentially prevent the satisfactory accomplishment of an SR function. In addition, the reactor recirculation system performs functions that support fire protection, EQ, ATWS, and SBO.

The intended functions within the scope of license renewal include the following:

- provides heat transfer
- maintains mechanical and structural integrity of NSR components to prevent spatial interactions with SR components
- provides pressure retaining boundary
- provides shielding against radiation
- maintains mechanical and structural integrity of NSR components that provide structural support to attached SR components
- provides flow restriction

In ALRA Table 2.3.1.B.4-1, the applicant identified the following reactor recirculation system component types that are within the scope of license renewal and subject to an AMR:

- closure bolting
- piping and fittings
- pumps
- radiation collars
- restriction orifices
- seal coolers
- valves

2.3B.1.4.2 Staff Evaluation

The staff reviewed ALRA Section 2.3.1.B.4 and the USAR using the evaluation methodology described in SER Section 2.3. The staff conducted its review in accordance with the guidance described in SRP-LR Section 2.3.

In conducting its review, the staff evaluated the system functions described in the ALRA and USAR in accordance with the requirements of 10 CFR 54.4(a) to verify that the applicant had not omitted from the scope of license renewal any components with intended functions delineated under 10 CFR 54.4(a). The staff then reviewed those components that the applicant had identified as being within the scope of license renewal to verify that the applicant had not omitted any passive and long-lived components that should be subject to an AMR in accordance with the requirements of 10 CFR 54.21(a)(1).

2.3B.1.4.3 Conclusion

The staff reviewed the ALRA to determine whether any SSCs that should be within the scope of license renewal had not been identified by the applicant. No omissions were identified. In addition, the staff performed a review to determine whether any components that should be subject to an AMR had not been identified by the applicant. No omissions were identified. On the basis of its review, the staff concludes that there is reasonable assurance that the applicant had adequately identified the reactor recirculation system components that are within the scope of license renewal, as required by 10 CFR 54.4(a), and the reactor recirculation system components that are subject to an AMR, as required by 10 CFR 54.21(a)(1).

2.3B.1.5 NMP2 Control Rod Drive System

2.3B.1.5.1 Summary of Technical Information in the Amended Application

In ALRA Section 2.3.1.B.5, the applicant described the CRD system. The CRD system is designed to change core reactivity by changing the position of control rods within the reactor core in response to manual control signals and to scram the reactor in response to manual or automatic signals. The system also provides water to the nuclear boiler instrumentation system reference leg backfill injection lines and the reactor water cleanup (RWCU) and reactor recirculation pump seals. The control rod drive system consists of two redundant pumps, filters, control valves, hydraulic control units, control rod drive mechanisms, scram discharge volume and associated piping, valves, controls and instrumentation. The normal water supply for the pumps is the condensate system with a backup supply from the condensate storage tank (CST). The discharge of each pump provides water to the nuclear boiler instrumentation system reference leg backfill injection lines, RWCU and reactor recirculation pump seals and through filters and pressure and control valves to several portions of the system. These portions are cooling water to the control rod drive mechanisms, charging water to the hydraulic control units, and drive water to the directional control valves. Following a reactor scram the exhaust water from the control rod drive mechanisms is collected in the scram discharge volume.

The CRD system contains SR components that are relied upon to remain functional during and following DBEs. The failure of NSR SSCs in the CRD system could potentially prevent the satisfactory accomplishment of an SR function. In addition, the CRD system performs functions that support fire protection, EQ, ATWS, and SBO.

The intended functions within the scope of license renewal include the following:

- maintains mechanical and structural integrity of NSR components to prevent spatial interactions with SR components

- provides pressure retaining boundary

- maintains mechanical and structural integrity of NSR components that provide structural support to attached SR components

- provides flow restriction

In ALRA Table 2.3.1.B.5-1, the applicant identified the following CRD system component types that are within the scope of license renewal and subject to an AMR:

- accumulators
- closure bolting
- CRD hydraulic control units
- filters
- flow elements
- flow indicators
- flow orifices
- piping and fittings
- pumps
- rupture discs
- valves

2.3B.1.5.2 Staff Evaluation

The staff reviewed ALRA Section 2.3.1.B.5 and the USAR using the evaluation methodology described in SER Section 2.3. The staff conducted its review in accordance with the guidance described in SRP-LR Section 2.3.

In conducting its review, the staff evaluated the system functions described in the ALRA and USAR in accordance with the requirements of 10 CFR 54.4(a) to verify that the applicant had not omitted from the scope of license renewal any components with intended functions delineated under 10 CFR 54.4(a). The staff then reviewed those components that the applicant had identified as being within the scope of license renewal to verify that the applicant had not omitted any passive and long-lived components that should be subject to an AMR in accordance with the requirements of 10 CFR 54.21(a)(1).

2.3B.1.5.3 Conclusion

The staff reviewed the ALRA to determine whether any SSCs that should be within the scope of license renewal had not been identified by the applicant. No omissions were identified. In addition, the staff performed a review to determine whether any components that should be subject to an AMR had not been identified by the applicant. No omissions were identified. On the basis of its review, the staff concludes that there is reasonable assurance that the applicant had adequately identified the CRD system components that are within the scope of license renewal, as required by 10 CFR 54.4(a), and the CRD system components that are subject to an AMR, as required by 10 CFR 54.21(a)(1).

2.3B.1.6 NMP2 Reactor Coolant Pressure Boundary Components in Other Systems

2.3B.1.6.1 Summary of Technical Information in the Amended Application

RCPB components in other systems are evaluated in the GALL Report as part of the reactor vessel, internals, and reactor coolant system. In ALRA Section 2.3 RCPB components requiring an AMR have been maintained in the plant system to which they are normally assigned rather than grouped with other RCPB components in the reactor vessel, internals, and reactor coolant system. These NMP2 plant systems are listed below:

- NMP2 feedwater system (SER Section 2.3B.4.3)
- NMP2 floor and equipment drains system (SER Section 2.3B.3.14)
- NMP2 high-pressure core spray system (SER Section 2.3B.2.3)
- NMP2 low-pressure core spray system (SER Section 2.3B.2.4)
- NMP2 main steam system (SER Section 2.3B.4.4)
- NMP2 reactor core isolation cooling system (SER Section 2.3B.2.6)
- NMP2 reactor water cleanup system (SER Section 2.3B.3.25)
- NMP2 residual heat removal system (SER Section 2.3B.2.7)
- NMP2 standby liquid control system (SER Section 2.3B.3.31)

2.3B.1.6.2 Staff Evaluation

The staff's evaluations of each plant system is contained in the SER sections listed above.

2.3B.1.6.3 Conclusion

The staff's conclusions for each plant system are provided in the SER sections listed above.

2.3B.2 Engineered Safety Features Systems

In ALRA Section 2.3.2.B, the applicant identified the structures and components of the NMP2 ESF systems that are subject to an AMR for license renewal.

The applicant described the supporting structures and components of the ESF systems in the following sections of the ALRA:

- 2.3.2.B.1   NMP2 automatic depressurization system
- 2.3.2.B.2   NMP2 hydrogen recombiner system
- 2.3.2.B.3   NMP2 high pressure core spray system
- 2.3.2.B.4   NMP2 low pressure core spray system
- 2.3.2.B.5   NMP2 primary containment isolation system
- 2.3.2.B.6   NMP2 reactor core isolation cooling system
- 2.3.2.B.7   NMP2 residual heat removal system
- 2.3.2.B.8   NMP2 standby gas treatment system

The staff's review findings regarding ALRA Sections 2.3.2.B.1 through 2.3.2.B.8 are presented in SER Sections 2.3B.2.1 through 2.3B.2.8, respectively.

2.3B.2.1 NMP2 Automatic Depressurization System

2.3B.2.1.1 Summary of Technical Information in the Amended Application

In ALRA Section 2.3.2.B.1, the applicant described the automatic depressurization system. The purpose of the automatic depressurization system is to reduce reactor pressure following small line breaks in the event of high-pressure core spray (HPCS) failure. When reactor vessel pressure is reduced to within the capacity of the low-pressure systems [LPCI (described in the residual heat removal (RHR) system) and low pressure core spray (LPCS) systems] these systems provide inventory makeup to maintain acceptable post-accident temperatures.

The automatic depressurization system contains SR components that are relied upon to remain functional during and following DBEs. In addition, the automatic depressurization system performs functions that support fire protection and EQ.

The component types subject to an AMR that perform the system intended functions for the automatic depressurization system are part of and evaluated in the main steam system (SER Section 2.3B4.4). No additional components within the automatic depressurization system are subject to aging management review.

2.3B.2.1.2 Staff Evaluation

The staff reviewed ALRA Section 2.3.2.B.1 and the USAR using the evaluation methodology described in SER Section 2.3. The staff conducted its review in accordance with the guidance described in SRP-LR Section 2.3.

In conducting its review, the staff evaluated the system functions described in the ALRA and USAR in accordance with the requirements of 10 CFR 54.4(a) to verify that the applicant had not omitted from the scope of license renewal any components with intended functions delineated under 10 CFR 54.4(a). The staff then reviewed those components that the applicant had identified as being within the scope of license renewal to verify that the applicant had not omitted any passive and long-lived components that should be subject to an AMR in accordance with the requirements of 10 CFR 54.21(a)(1).

2.3B.2.1.3 Conclusion

The staff reviewed the ALRA to determine whether any SSCs that should be within the scope of license renewal had not been identified by the applicant. No omissions were identified. In addition, the staff performed a review to determine whether any components that should be subject to an AMR had not been identified by the applicant. No omissions were identified. On the basis of its review, the staff concludes that there is reasonable assurance that the applicant had adequately identified the automatic depressurization system components that are within the scope of license renewal, as required by 10 CFR 54.4(a), and the automatic depressurization system components that are subject to an AMR, as required by 10 CFR 54.21(a)(1).

2.3B.2.2 NMP2 Hydrogen Recombiner System

2.3B.2.2.1 Summary of Technical Information in the Amended Application

In ALRA Section 2.3.2.B.2, the applicant described the hydrogen recombiner system. The purpose of the hydrogen recombiner system is to process the hydrogen and oxygen released to the primary containment during a LOCA. The hydrogen recombiner system takes suction from the drywell and suppression pool, recombines the hydrogen and oxygen gases, and returns the resulting water vapor and other gases to the suppression pool.

The hydrogen recombiner system contains SR components that are relied upon to remain functional during and following DBEs. The failure of NSR SSCs in the hydrogen recombiner system could potentially prevent the satisfactory accomplishment of an SR function. In addition, the hydrogen recombiner system performs functions that support EQ.

The intended functions within the scope of license renewal include the following:

- provides pressure retaining boundary
- maintains mechanical and structural integrity of NSR components that provide structural support to attached SR components

In ALRA Table 2.3.2.B.2-1, the applicant identified the following hydrogen recombiner system component types that are within the scope of license renewal and subject to an AMR:

- blowers
- bolting
- filters/strainers
- flow elements
- hydrogen recombiners
- piping and fittings
- valves

2.3B.2.2.2 Staff Evaluation

The staff reviewed ALRA Section 2.3.2.B.2 and USAR Section 6.2.5.2.2 using the evaluation methodology described in SER Section 2.3. The staff conducted its review in accordance with the guidance described in SRP-LR Section 2.3.

In conducting its review, the staff evaluated the system functions described in the ALRA and USAR in accordance with the requirements of 10 CFR 54.4(a) to verify that the applicant had not omitted from the scope of license renewal any components with intended functions delineated under 10 CFR 54.4(a). The staff then reviewed those components that the applicant had identified as being within the scope of license renewal to verify that the applicant had not omitted any passive and long-lived components that should be subject to an AMR in accordance with the requirements of 10 CFR 54.21(a)(1).

2.3B.2.2.3 Conclusion

The staff reviewed the ALRA to determine whether any SSCs that should be within the scope of license renewal had not been identified by the applicant. No omissions were identified. In addition, the staff performed a review to determine whether any components that should be subject to an AMR had not been identified by the applicant. No omissions were identified. On the basis of its review, the staff concludes that there is reasonable assurance that the applicant had adequately identified the hydrogen recombiner system components that are within the scope of license renewal, as required by 10 CFR 54.4(a), and the hydrogen recombiner system components that are subject to an AMR, as required by 10 CFR 54.21(a)(1).

2.3B.2.3 NMP2 High Pressure Core Spray System

2.3B.2.3.1 Summary of Technical Information in the Amended Application

In ALRA Section 2.3.2.B.3, the applicant described the HPCS system. The purpose of the HPCS system is to maintain RPV coolant inventory after small breaks that do not depressurize the RPV. The HPCS system also provides spray cooling heat transfer during breaks in which core uncovery is calculated.

The HPCS system contains SR components that are relied upon to remain functional during and following DBEs. The failure of NSR SSCs in the HPCS system could potentially prevent the satisfactory accomplishment of an SR function. In addition, the HPCS system performs functions that support fire protection, EQ, and ATWS.

The intended functions within the scope of license renewal include the following:

- maintains mechanical and structural integrity of NSR components to prevent spatial interactions with SR components

- provides pressure retaining boundary

- maintains mechanical and structural integrity of NSR components that provide structural support to attached SR components

In ALRA Table 2.3.2.B.3-1, the applicant identified the following HPCS system component types that are within the scope of license renewal and subject to an AMR:

- bolting
- filters/strainers
- flow elements
- piping and fittings
- pumps
- restriction orifices
- valves

2.3B.2.3.2 Staff Evaluation

The staff reviewed ALRA Section 2.3.2.B.3 and USAR Sections 6.3.1.2.1 and 6.3.2.2.1 using the evaluation methodology described in SER Section 2.3. The staff conducted its review in accordance with the guidance described in SRP-LR Section 2.3.

In conducting its review, the staff evaluated the system functions described in the ALRA and USAR in accordance with the requirements of 10 CFR 54.4(a) to verify that the applicant had not omitted from the scope of license renewal any components with intended functions delineated under 10 CFR 54.4(a). The staff then reviewed those components that the applicant had identified as being within the scope of license renewal to verify that the applicant had not omitted any passive and long-lived components that should be subject to an AMR in accordance with the requirements of 10 CFR 54.21(a)(1).

2.3B.2.3.3 Conclusion

The staff reviewed the ALRA to determine whether any SSCs that should be within the scope of license renewal had not been identified by the applicant. No omissions were identified. In addition, the staff performed a review to determine whether any components that should be subject to an AMR had not been identified by the applicant. No omissions were identified. On the basis of its review, the staff concludes that there is reasonable assurance that the applicant had adequately identified the HPCS system components that are within the scope of license renewal, as required by 10 CFR 54.4(a), and the HPCS system components that are subject to an AMR, as required by 10 CFR 54.21(a)(1).

2.3B.2.4 NMP2 Low Pressure Core Spray System

2.3B.2.4.1 Summary of Technical Information in the Amended Application

In ALRA Section 2.3.2.B.4, the applicant described the low pressure core spray (LPCS) system. The purpose of the LPCS system is to provide RPV coolant inventory makeup and spray cooling during large breaks in which the core is calculated to uncover. Also, following a small break and automatic depressurization system initiation, the LPCS system provides coolant inventory makeup.

The LPCS system contains SR components that are relied upon to remain functional during and following DBEs. The failure of NSR SSCs in the LPCS system could potentially prevent the satisfactory accomplishment of an SR function. In addition, the LPCS system performs functions that support fire protection and EQ.

The intended functions within the scope of license renewal include the following:

- maintains mechanical and structural integrity of NSR components to prevent spatial interactions with SR components

- provides pressure retaining boundary

- maintains mechanical and structural integrity of NSR components that provide structural support to attached SR components

In ALRA Table 2.3.2.B.4-1, the applicant identified the following LPCS system component types that are within the scope of license renewal and subject to an AMR:

- bolting
- filters/strainers
- flow elements
- piping and fittings
- pumps
- restriction orifices
- valves

2.3B.2.4.2 Staff Evaluation

The staff reviewed ALRA Section 2.3.2.B.4 and the USAR Sections 6.3.1.2.2 and 6.3.2.2.3 using the evaluation methodology described in SER Section 2.3. The staff conducted its review in accordance with the guidance described in SRP-LR Section 2.3.

In conducting its review, the staff evaluated the system functions described in the ALRA and USAR in accordance with the requirements of 10 CFR 54.4(a) to verify that the applicant had not omitted from the scope of license renewal any components with intended functions delineated under 10 CFR 54.4(a). The staff then reviewed those components that the applicant had identified as being within the scope of license renewal to verify that the applicant had not omitted any passive and long-lived components that should be subject to an AMR in accordance with the requirements of 10 CFR 54.21(a)(1).

2.3B.2.4.3 Conclusion

The staff reviewed the ALRA to determine whether any SSCs that should be within the scope of license renewal had not been identified by the applicant. No omissions were identified. In addition, the staff performed a review to determine whether any components that should be subject to an AMR had not been identified by the applicant. No omissions were identified. On the basis of its review, the staff concludes that there is reasonable assurance that the applicant had adequately identified the LPCS system components that are within the scope of license renewal, as required by 10 CFR 54.4(a), and the LPCS system components that are subject to an AMR, as required by 10 CFR 54.21(a)(1).

2.3B.2.5 NMP2 Primary Containment Isolation System

2.3B.2.5.1 Summary of Technical Information in the Amended Application

In ALRA Section 2.3.2.B.5, the applicant described the primary containment isolation system. The purpose of the primary containment isolation system is to provide protection against a release of radioactive materials to the environment from accidents occurring to the RCPB, lines connected to the RCPB, or lines that penetrate the primary containment. This is accomplished by automatic isolation valve closure of appropriate lines that penetrate the primary containment system. The primary containment isolation system consists of automatic isolation valves and associated piping for lines that penetrate the primary containment.

The primary containment isolation system contains SR components that are relied upon to remain functional during and following DBEs. In addition, the primary containment isolation system performs functions that support SBO.

The component types requiring an AMR for the primary containment isolation system are evaluated in their respective systems.

2.3B.2.5.2 Staff Evaluation

The staff reviewed ALRA Section 2.3.2.B.5 and USAR Section 6.2.4 using the evaluation methodology described in SER Section 2.3. The staff conducted its review in accordance with the guidance described in SRP-LR Section 2.3.

In conducting its review, the staff evaluated the system functions described in the ALRA and USAR in accordance with the requirements of 10 CFR 54.4(a) to verify that the applicant had not omitted from the scope of license renewal any components with intended functions delineated under 10 CFR 54.4(a). The staff then reviewed those components that the applicant had identified as being within the scope of license renewal to verify that the applicant had not omitted any passive and long-lived components that should be subject to an AMR in accordance with the requirements of 10 CFR 54.21(a)(1).

2.3B.2.5.3 Conclusion

The staff reviewed the ALRA to determine whether any SSCs that should be within the scope of license renewal had not been identified by the applicant. No omissions were identified. In addition, the staff performed a review to determine whether any components that should be subject to an AMR had not been identified by the applicant. No omissions were identified. On the basis of its review, the staff concludes that there is reasonable assurance that the applicant had adequately identified the primary containment isolation system components that are within the scope of license renewal, as required by 10 CFR 54.4(a), and the primary containment isolation system components that are subject to an AMR, as required by 10 CFR 54.21(a)(1).

2.3B.2.6 NMP2 Reactor Core Isolation Cooling System

2.3B.2.6.1 Summary of Technical Information in the Amended Application

In ALRA Section 2.3.2.B.6, the applicant described the reactor core isolation cooling (RCIC) system. The purpose of the RCIC system is to assure that sufficient reactor water inventory is maintained in the reactor vessel to permit adequate core cooling following those events in which the normal feedwater supply is unavailable. This system can be used for accident and nonaccident conditions.

The RCIC system contains SR components that are relied upon to remain functional during and following DBEs. The failure of NSR SSCs in the RCIC system could potentially prevent the satisfactory accomplishment of an SR function. In addition, the RCIC system performs functions that support fire protection, EQ, and SBO.

The intended functions within the scope of license renewal include the following:

- provides heat transfer

- maintains mechanical and structural integrity of NSR components to prevent spatial interactions with SR components

- provides pressure retaining boundary

- maintains mechanical and structural integrity of NSR components that provide structural support to attached SR components

In ALRA Table 2.3.2.B.6-1, the applicant identified the following RCIC system component types that are within the scope of license renewal and subject to an AMR:

- blower
- bolting
- condensing chambers
- drain pots
- filters/strainers
- flow elements
- heat exchangers
- piping and fittings
- pumps
- restriction orifices
- rupture discs
- terry turbine
- valves

2.3B.2.6.2 Staff Evaluation

The staff reviewed ALRA Section 2.3.2.B.6 and the USAR using the evaluation methodology described in SER Section 2.3. The staff conducted its review in accordance with the guidance described in SRP-LR Section 2.3.

In conducting its review, the staff evaluated the system functions described in the ALRA and USAR in accordance with the requirements of 10 CFR 54.4(a) to verify that the applicant had not omitted from the scope of license renewal any components with intended functions delineated under 10 CFR 54.4(a). The staff then reviewed those components that the applicant had identified as being within the scope of license renewal to verify that the applicant had not omitted any passive and long-lived components that should be subject to an AMR in accordance with the requirements of 10 CFR 54.21(a)(1).

2.3B.2.6.3 Conclusion

The staff reviewed the ALRA to determine whether any SSCs that should be within the scope of license renewal had not been identified by the applicant. No omissions were identified. In addition, the staff performed a review to determine whether any components that should be subject to an AMR had not been identified by the applicant. No omissions were identified. On the basis of its review, the staff concludes that there is reasonable assurance that the applicant had adequately identified the RCIC system components that are within the scope of license renewal, as required by 10 CFR 54.4(a), and the RCIC system components that are subject to an AMR, as required by 10 CFR 54.21(a)(1).

2.3B.2.7 NMP2 Residual Heat Removal System

2.3B.2.7.1 Summary of Technical Information in the Amended Application

In ALRA Section 2.3.2.B.7, the applicant described the residual heat removal system. The residual heat removal system is composed of three independent loops, each of which contains a motor-driven pump, piping, valves, instrumentation, and controls. Each loop has a suction source from the suppression pool and is capable of discharging water to either the reactor vessel via a separate nozzle or back to the suppression pool via a full-flow test line. The A and B loops have heat exchangers cooled by service water. Loops A and B can also take suction from the reactor recirculation system suction and can discharge into the reactor recirculation discharge or to the suppression pool and drywell spray spargers. The A and B loops also have connections to reactor steam via the RCIC system steam line (Section 2.3.2.B.6) and can discharge the resultant condensate to the RCIC system pump suction or to the suppression pool. In addition, Loops A and B take suction from the fuel pool and discharge to the fuel pool cooling discharge.

The three loops of the RHR system combine to fulfill five modes of operation. Each mode has its own functional requirements and is presented separately as follows:

(1) low pressure coolant injection mode–following a small break and automatic depressurization system initiation this mode provides coolant inventory makeup

(2) suppression pool cooling mode–ensures that the suppression pool temperature does not exceed design limits following a reactor vessel blowdown or isolation event

(3) containment spray cooling mode–provides two redundant means to spray the drywell and suppression pool to reduce internal pressure to below design limits

(4) reactor steam-condensing mode–provides in conjunction with the RCIC turbine the capability to condense all of the steam generated 1-½ hours after a reactor scram

(5)   shutdown cooling mode–provides the capability to remove decay and sensible heat from the reactor primary system so that the cold shutdown condition can be achieved and maintained

The RHR system contains SR components that are relied upon to remain functional during and following DBEs. The failure of NSR SSCs in the RHR system could potentially prevent the satisfactory accomplishment of an SR function. In addition, the RHR system performs functions that support fire protection, EQ, and SBO.

The intended functions within the scope of license renewal include the following:

- provides heat transfer
- maintains mechanical and structural integrity of NSR components to prevent spatial interactions with SR components
- provides pressure retaining boundary
- maintains mechanical and structural integrity of NSR components that provide structural support to attached SR components

In ALRA Table 2.3.2.B.7-1, the applicant identified the following RHR system component types that are within the scope of license renewal and subject to an AMR:

- "T" quenchers
- bolting
- condensing chambers
- filters/strainers
- flow elements
- heat exchangers
- level elements
- piping and fittings
- pumps
- restriction orifices
- temperature elements
- valves

2.3B.2.7.2 Staff Evaluation

The staff reviewed ALRA Section 2.3.2.B.7 and the USAR Section 5.4.7 using the evaluation methodology described in SER Section 2.3. The staff conducted its review in accordance with the guidance described in SRP-LR Section 2.3.

In conducting its review, the staff evaluated the system functions described in the ALRA and USAR in accordance with the requirements of 10 CFR 54.4(a) to verify that the applicant had not omitted from the scope of license renewal any components with intended functions delineated under 10 CFR 54.4(a). The staff then reviewed those components that the applicant had identified as being within the scope of license renewal to verify that the applicant had not omitted any passive and long-lived components that should be subject to an AMR in accordance with the requirements of 10 CFR 54.21(a)(1).

2-150

2.3B.2.7.3 Conclusion

The staff reviewed the ALRA to determine whether any SSCs that should be within the scope of license renewal had not been identified by the applicant. No omissions were identified. In addition, the staff performed a review to determine whether any components that should be subject to an AMR had not been identified by the applicant. No omissions were identified. On the basis of its review, the staff concludes that there is reasonable assurance that the applicant had adequately identified the RHR system components that are within the scope of license renewal, as required by 10 CFR 54.4(a), and the RHR system components that are subject to an AMR, as required by 10 CFR 54.21(a)(1).

2.3B.2.8 NMP2 Standby Gas Treatment System

2.3B.2.8.1 Summary of Technical Information in the Amended Application

In ALRA Section 2.3.2.B.8, the applicant described the standby gas treatment system (SGTS). The purpose of the SGTS is to limit the release of radioactive gases from the RB to the environment within the guidelines of 10 CFR 100 in the event of a LOCA and to maintain a negative pressure in the RB under accident conditions. It is also used to provide charcoal filtration of the primary containment atmosphere when inerting, deinerting or controlling primary containment pressure.

The SGTS contains SR components that are relied upon to remain functional during and following DBEs. The failure of NSR SSCs in the SGTS could potentially prevent the satisfactory accomplishment of an SR function. In addition, the SGTS performs functions that support EQ.

The intended functions within the scope of license renewal include the following:

- provides filtration

- provides pressure retaining boundary

- maintains mechanical and structural integrity of NSR components that provide structural support to attached SR components

In ALRA Table 2.3.2.B.8-1, the applicant identified the following SGTS component types that are within the scope of license renewal and subject to an AMR:

- actuator
- blowers
- bolting
- filters/strainers
- flow elements
- heaters
- piping and fittings
- restriction orifices
- tanks
- valves

2.3B.2.8.2 Staff Evaluation

The staff reviewed ALRA Section 2.3.2.B.8 and USAR Section 6.5.1 using the evaluation methodology described in SER Section 2.3. The staff conducted its review in accordance with the guidance described in SRP-LR Section 2.3.

In conducting its review, the staff evaluated the system functions described in the ALRA and USAR in accordance with the requirements of 10 CFR 54.4(a) to verify that the applicant had not omitted from the scope of license renewal any components with intended functions delineated under 10 CFR 54.4(a). The staff then reviewed those components that the applicant had identified as being within the scope of license renewal to verify that the applicant had not omitted any passive and long-lived components that should be subject to an AMR in accordance with the requirements of 10 CFR 54.21(a)(1).

2.3B.2.8.3 Conclusion

The staff reviewed the ALRA to determine whether any SSCs that should be within the scope of license renewal had not been identified by the applicant. No omissions were identified. In addition, the staff performed a review to determine whether any components that should be subject to an AMR had not been identified by the applicant. No omissions were identified. On the basis of its review, the staff concludes that there is reasonable assurance that the applicant had adequately identified the SGTS components that are within the scope of license renewal, as required by 10 CFR 54.4(a), and the SGTS components that are subject to an AMR, as required by 10 CFR 54.21(a)(1).

2.3B.3 Auxiliary Systems

In ALRA Section 2.3.3.B, the applicant identified the structures and components of the NMP2 auxiliary systems that are subject to an AMR for license renewal.

The applicant described the supporting structures and components of the auxiliary systems in the following sections of the ALRA:

- 2.3.3.B.1   NMP2 air startup-standby diesel generator system
- 2.3.3.B.2   NMP2 alternate decay heat removal system
- 2.3.3.B.3   NMP2 auxiliary service building HVAC system
- 2.3.3.B.4   NMP2 chilled water ventilation system (removed)
- 2.3.3.B.5   NMP2 compressed air systems
- 2.3.3.B.6   NMP2 containment atmosphere monitoring system
- 2.3.3.B.7   NMP2 containment leakage monitoring system
- 2.3.3.B.8   NMP2 control building chilled water system
- 2.3.3.B.9   NMP2 control building HVAC system
- 2.3.3.B.10   NMP2 diesel generator building ventilation system
- 2.3.3.B.11   NMP2 domestic water system
- 2.3.3.B.12   NMP2 engine-driven fire pump fuel oil system
- 2.3.3.B.13   NMP2 fire detection and protection system
- 2.3.3.B.14   NMP2 floor and equipment drains system
- 2.3.3.B.15   NMP2 generator standby lube oil system

- 2.3.3.B.16 NMP2 glycol heating system (removed)
- 2.3.3.B.17 NMP2 hot water heating system
- 2.3.3.B.18 NMP2 makeup water system
- 2.3.3.B.19 NMP2 neutron monitoring system
- 2.3.3.B.20 NMP2 primary containment purge system
- 2.3.3.B.21 NMP2 process sampling system
- 2.3.3.B.22 NMP2 radiation monitoring system
- 2.3.3.B.23 NMP2 reactor building closed loop cooling water system
- 2.3.3.B.24 NMP2 reactor building HVAC system
- 2.3.3.B.25 NMP2 reactor water cleanup system
- 2.3.3.B.26 NMP2 seal water system (removed)
- 2.3.3.B.27 NMP2 service water system
- 2.3.3.B.28 NMP2 spent fuel pool cooling and cleanup system
- 2.3.3.B.29 NMP2 standby diesel generator fuel oil system
- 2.3.3.B.30 NMP2 standby diesel generator protection (generator) system
- 2.3.3.B.31 NMP2 standby liquid control system
- 2.3.3.B.32 NMP2 yard structures ventilation system
- 2.3.3.B.33 NMP2 auxiliary boiler system
- 2.3.3.B.34 NMP2 circulating water system
- 2.3.3.B.35 NMP2 makeup water treatment system
- 2.3.3.B.36 NMP2 radioactive liquid waste management system
- 2.3.3.B.37 NMP2 roof drainage system
- 2.3.3.B.38 NMP2 sanitary drains and disposal system
- 2.3.3.B.39 service water chemical treatment system
- 2.3.3.B.40 NMP2 turbine building closed loop cooling water system

The staff's review findings regarding ALRA Sections 2.3.3.B.1 through 2.3.3.B.40 are presented in SER Sections 2.3B.3.1 through 2.3B.3.40, respectively.

2.3B.3.1 NMP2 Air Startup-Standby Diesel Generator System

2.3B.3.1.1 Summary of Technical Information in the Amended Application

In ALRA Section 2.3.3.B.1, the applicant described the air startup-standby diesel generator system. The air startup-standby diesel generator system includes the diesel generator combustion air intake and exhaust system. It is designed to provide: (1) a sufficient volume and pressure of compressed air to enable the EDG to start within the required times and (2) reliable combustion air intake and exhaust paths that supply clean air for combustion and a means to discharge exhaust gases outside the diesel generator building. Each standby diesel generator has redundant air starting systems, either of which is capable of starting the engine. To supply combustion air and an exhaust path, fresh air is drawn from outside and passes through an intake filter and an intake silencer located just inside the DGB. The air then passes through the overspeed trip valve, an exhaust driven turbocharger, through a pair of combination intercooler-heaters and then is distributed to each cylinder bank through the engine intake manifolds. Exhaust gases are discharged to the atmosphere above the DGB.

The air startup-standby diesel generator system contains SR components that are relied upon to remain functional during and following DBEs. The failure of NSR SSCs in the air startup-standby diesel generator system could potentially prevent the satisfactory accomplishment of an SR function. In addition, the air startup-standby diesel generator system performs functions that support fire protection.

The intended functions within the scope of license renewal include the following:

- provides pressure retaining boundary
- maintains mechanical and structural integrity of NSR components that provide structural support to attached SR components

In ALRA Table 2.3.3.B.1-1, the applicant identified the following air startup-standby diesel generator system component types that are within the scope of license renewal and subject to an AMR:

- bolting
- diesel engine air start motors
- expansion joints
- filters/strainers
- moisture air separators
- mufflers
- piping and fittings
- starting air lubricator
- tanks
- valves

2.3B.3.1.2 Staff Evaluation

The staff reviewed ALRA Section 2.3.3.B.1 and USAR Sections 9.5.6 and 9.5.8 using the evaluation methodology described in SER Section 2.3. The staff conducted its review in accordance with the guidance described in SRP-LR Section 2.3.

In conducting its review, the staff evaluated the system functions described in the ALRA and USAR in accordance with the requirements of 10 CFR 54.4(a) to verify that the applicant had not omitted from the scope of license renewal any components with intended functions delineated under 10 CFR 54.4(a). The staff then reviewed those components that the applicant had identified as being within the scope of license renewal to verify that the applicant had not omitted any passive and long-lived components that should be subject to an AMR in accordance with the requirements of 10 CFR 54.21(a)(1).

The staff's review of original LRA Section 2.3.3.B.1 identified areas in which additional information was necessary to complete the evaluation of the applicant's scoping and screening results. The applicant responded to the staff's RAIs as discussed below.

In RAI 2.3.3.B.1-1, dated November 19, 2004, the staff stated that drawing LR-104A-0 shows that the license renewal boundary for the air startup standby diesel generator system stops at two valves. Downstream piping and equipment have an SR pressure boundary function. Original LRA Table 2.3.3.B-1 lists the starting air lubricators as subject to an AMR. The drawing shows that two lubricators are not part of the boundary. Therefore, the staff requested that the applicant provide the basis for not including this piping and the associated equipment within the license renewal boundary.

In its response by letter dated December 22, 2004, the applicant stated that the boundary flag upstream of air startup lubricators 2EGA*LU325A and 2EGA*LU325B on drawing LR-104A-0 is incorrect, that these components are SR and subject to an AMR, and that the boundary flag should be downstream of these components but upstream of the high pressure core spray diesel generator 2EGS*EG2. The applicant further clarified that the high pressure core spray diesel generator remains within the scope of license renewal but is not subject to an AMR since it is an active component and that there is no piping or equipment in this system within the scope of license renewal and subject to an AMR under 10 CFR 54.4(a)(2) because there is no liquid in the system.

Based on its review, the staff found the applicant's response to RAI 2.3.3.B.1-1 acceptable because it adequately explained that the components in question are within the scope of license renewal in accordance with 10 CFR 54.4(a) and subject to an AMR in accordance with 10 CFR 54.21(a), but were inadvertently left un-highlighted on the LR drawing. Therefore, the staff's concern described in RAI 2.3.3.B.1-1 is resolved.

In RAI 2.3.3.B.1-2 dated November 19, 2004, the staff stated that USAR Section 9.5.8 states that turbocharger and intercooler heaters are part of the combustion air intake and exhaust system. These components are not listed in the original LRA Table 2.3.3.B-1. These components have a pressure boundary function. Therefore, the staff requested that the applicant provide the basis for excluding these components as subject to an AMR.

In its response by letter dated December 22, 2004, the applicant stated that the turbocharger is an engine-mounted subcomponent of the Division I and II standby diesel generators and that the Division III HPCS diesel generator also has an engine-mounted turbocharger. The applicant explained that the Division I and II and HPCS diesel generators are SR and within the scope of license renewal but are not subject to an AMR as they are active components. Thus, according to the applicant, their turbocharger subcomponents are also within the scope of license renewal but as part of the engine not subject to an AMR. The description of these subcomponents can be found in USAR Section 9.5.8.2.

Based on its review the staff found the applicant's response to RAI 2.3.3.B.1-2 acceptable because it adequately explained that the turbocharger is an engine-mounted subcomponent of the Division I and II standby diesel generators. The Division III HPCS diesel generator also has an engine-mounted turbocharger. The Division I and II and HPCS diesel generators are SR and within the scope of license renewal but are not subject to an AMR because they are active components. For license renewal purposes, the intercooler heater is part of the NMP2 standby generator protection system and listed in original LRA Table 2.3.3.B.30-1 as a heat exchanger. The intercooler heater is passive and long-lived and within the scope of license renewal under 10 CFR 54.4(a) and subject to an AMR under 10 CFR 54.21(a). The intercooler heater is shown in red on the LR drawing for the jacket water subsystem of the standby diesel generator protection system. Therefore, the staff's concern described in RAI 2.3.3.B.1-2 is resolved.

In RAI 2.3.3.B.1-3 dated November 19, 2004, the staff requested that the applicant clarify information given on the related boundary drawing and in original LRA Table 2.3.3.B.1-1 concerning SSCs in scope of license renewal per 10 CFR 54.4(a). In original LRA

Table 2.3.3.B.1 the component type "Moisture Separator" is not listed as being subject to an AMR. This exclusion is inconsistent with the original LRA drawing LRA-104A-0 showing the moisture separator within the license renewal boundary.

In its response by letter dated December 22, 2004, the applicant stated that original LRA Table 2.3.3.B.1-1 does list the component type "Air Separators," an abbreviated form of the description "Moisture Air Separator" listed for components 2EGA*SP 1A, 2EGA*SP1B, 2EGA*SP2A, and 2EGA*SP2B. These air separators are subject to an AMR as shown in original LRA Table 3.3.2.B-1 (page 3.3-191). Therefore, the applicant stated that the original LRA drawing LR-104A-0 is correct as drawn with respect to the moisture separators. The applicant will revise the original LRA Table 2.3.3.B.l-1 to call these components "Moisture Air Separators" and this response has been incorporated in the ALRA as discussed below.

In its ALRA dated July 14, 2005, the applicant provided the staff with the revised LR drawing identifying SSCs within the scope of license renewal and subject to an AMR under 10 CFR 54.4(a)(2). The applicant also provided clarification on the component groups included in original LRA Table 2.3.3.B.1-1.

Based on review of the information submitted in the amended LRA the staff's concern described in RAI 2.3.3.B.1-3 is resolved.

2.3B.3.1.3 Conclusion

The staff reviewed the ALRA, RAI responses, and accompanying scoping boundary drawings to determine whether any SSCs that should be within the scope of license renewal had not been identified by the applicant. No omissions were identified. In addition, the staff performed a review to determine whether any components that should be subject to an AMR had not been identified by the applicant. No omissions were identified. On the basis of its review, the staff concludes that there is reasonable assurance that the applicant had adequately identified the air startup-standby diesel generator system components that are within the scope of license renewal, as required by 10 CFR 54.4(a), and the air startup-standby diesel generator system components that are subject to an AMR, as required by 10 CFR 54.21(a)(1).

2.3B.3.2 NMP2 Alternate Decay Heat Removal System

2.3B.3.2.1 Summary of Technical Information in the Amended Application

In ALRA Section 2.3.3.B.2, the applicant described the alternate decay heat removal system. The alternate decay heat removal system in conjunction with natural circulation is designed to remove the decay heat released from the spent fuel pool, reactor core, reactor internals, storage pool, and cavity during refueling outages to maintain reactor coolant temperatures suitable for refueling. The alternate decay heat removal system accomplishes its design function by utilizing a primary loop for removing decay heat from the spent fuel pool and the reactor core and a secondary loop to transfer the decay heat to the atmosphere.

The alternate decay heat removal system contains SR components that are relied upon to remain functional during and following DBEs. The failure of NSR SSCs in the alternate decay heat removal system could potentially prevent the satisfactory accomplishment of an SR function. In addition, the alternate decay heat removal system performs functions that support EQ.

The intended functions within the scope of license renewal include the following:

- maintains mechanical and structural integrity of NSR components to prevent spatial interactions with SR components

- provides pressure retaining boundary

- maintains mechanical and structural integrity of NSR components that provide structural support to attached SR components

In ALRA Table 2.3.3.B.2-1, the applicant identified the following alternate decay heat removal system component types that are within the scope of license renewal and subject to an AMR:

- bolting
- flow elements
- heat exchangers
- piping and fittings
- pumps
- valves

2.3B.3.2.2 Staff Evaluation

The staff reviewed ALRA Section 2.3.3.B.2 and the USAR Section 9.1.6 using the evaluation methodology described in SER Section 2.3. The staff conducted its review in accordance with the guidance described in SRP-LR Section 2.3.

In conducting its review, the staff evaluated the system functions described in the ALRA and USAR in accordance with the requirements of 10 CFR 54.4(a) to verify that the applicant had not omitted from the scope of license renewal any components with intended functions delineated under 10 CFR 54.4(a). The staff then reviewed those components that the applicant had identified as being within the scope of license renewal to verify that the applicant had not omitted any passive and long-lived components that should be subject to an AMR in accordance with the requirements of 10 CFR 54.21(a)(1).

2.3B.3.2.3 Conclusion

The staff reviewed the ALRA to determine whether any SSCs that should be within the scope of license renewal had not been identified by the applicant. No omissions were identified. In addition, the staff performed a review to determine whether any components that should be subject to an AMR had not been identified by the applicant. No omissions were identified. On the basis of its review, the staff concludes that there is reasonable assurance that the applicant had adequately identified the alternate decay heat removal system components that are within the scope of license renewal, as required by 10 CFR 54.4(a), and the alternate decay heat removal system components that are subject to an AMR, as required by 10 CFR 54.21(a)(1).

2.3B.3.3 NMP2 Auxiliary Service Building HVAC System

2.3B.3.3.1 Summary of Technical Information in the Amended Application

In ALRA Section 2.3.3.B.3, the applicant described the auxiliary service building HVAC system. The auxiliary service building HVAC system is designed to provide an environment that ensures habitability of the areas served consistent with personnel comfort and optimum performance of equipment. The system also supplies filtered and tempered outdoor air for all air conditioned areas.

The auxiliary service building HVAC system performs functions that support fire protection.

The intended function within the scope of license renewal is to provide pressure retaining boundary.

In ALRA Table 2.3.3.B.3-1 the applicant identified that the "Fire Dampers" component type of the auxiliary service building HVAC system is within the scope of license renewal and subject to an AMR.

2.3B.3.3.2 Staff Evaluation

The staff reviewed ALRA Section 2.3.3.B.3 and USAR Section 9.4.9 using the evaluation methodology described in SER Section 2.3. The staff conducted its review in accordance with the guidance described in SRP-LR Section 2.3.

In conducting its review, the staff evaluated the system functions described in the ALRA and USAR in accordance with the requirements of 10 CFR 54.4(a) to verify that the applicant had not omitted from the scope of license renewal any components with intended functions delineated under 10 CFR 54.4(a). The staff then reviewed those components that the applicant had identified as being within the scope of license renewal to verify that the applicant had not omitted any passive and long-lived components that should be subject to an AMR in accordance with the requirements of 10 CFR 54.21(a)(1).

2.3B.3.3.3 Conclusion

The staff reviewed the ALRA to determine whether any SSCs that should be within the scope of license renewal had not been identified by the applicant. No omissions were identified. In addition, the staff performed a review to determine whether any components that should be subject to an AMR had not been identified by the applicant. No omissions were identified. On the basis of its review, the staff concludes that there is reasonable assurance that the applicant had adequately identified the auxiliary service building HVAC system components that are within the scope of license renewal, as required by 10 CFR 54.4(a), and the auxiliary service building HVAC system components that are subject to an AMR, as required by 10 CFR 54.21(a)(1).

2.3B.3.4 NMP2 Chilled Water Ventilation System (Removed)

In ALRA Section 2.3.3.B.4, the applicant stated the chilled water ventilation system has been removed from the scope of license renewal since it has been determined that it does not meet any of the criteria of 10 CFR 54.4. The original LRA included this system within the scope of license renewal in accordance with 10 CFR 54.4(a)(2). However, based upon the applicant's detailed evaluations this system is not credited for mitigation of any CLB event, contains no SR/NSR interfaces, nor introduces any spatial interactions with SR SSCs. Therefore, the chilled water ventilation system is not within the scope of license renewal.

2.3B.3.5 NMP2 Compressed Air Systems

2.3B.3.5.1 Summary of Technical Information in the Amended Application

In ALRA Section 2.3.3.B.5, the applicant described the compressed air systems. The compressed air systems are designed to provide clean, filtered air to various areas of NMP2. The compressed air systems consist of the instrument air system, service air system, breathing air system, and the primary containment ventilation, purge, and nitrogen system. The instrument air system is designed to supply clean, dry, and oil-free air to plant instrumentation and control systems that require an air supply. The service air system is designed to distribute service air to the plant systems that require air as a motive force or for mixing. The breathing air system is designed to provide a reliable supply of clean, filtered air for human breathing. It also supplies clean dry air for use of instruments. The primary containment ventilation, purge, and nitrogen system is used in conjunction with the standby gas treatment system to inert and de-inert the primary containment as required. Functions of the primary containment ventilation, purge, and nitrogen system include providing a dedicated source of nitrogen gas for the operation of the automatic depressurization system relief valves, providing a primary source of instrument nitrogen for the operation of gas-operated valves in primary containment, providing containment isolation and containment bypass leakage control.

The compressed air systems contains SR components that are relied upon to remain functional during and following DBEs. The failure of NSR SSCs in the compressed air systems could potentially prevent the satisfactory accomplishment of an SR function. In addition, the compressed air systems performs functions that support fire protection, EQ, ATWS, and SBO.

The intended functions within the scope of license renewal include the following:

- provides removal and/or holdup of fission products
- provides pressure retaining boundary
- provides shielding against radiation
- maintains mechanical and structural integrity of NSR components that provide structural support to attached SR components
- provides flow restriction

In ALRA Table 2.3.3.B.5-1, the applicant identified the following compressed air systems component types that are within the scope of license renewal and subject to an AMR:

- bolting
- filter
- piping and fittings
- orifices
- radiation collars
- rupture discs
- tanks and receivers
- valves

2.3B.3.5.2 Staff Evaluation

The staff reviewed ALRA Section 2.3.3.B.5 and the USAR Section 9.3.1 using the evaluation methodology described in SER Section 2.3. The staff conducted its review in accordance with the guidance described in SRP-LR Section 2.3.

In conducting its review, the staff evaluated the system functions described in the ALRA and USAR in accordance with the requirements of 10 CFR 54.4(a) to verify that the applicant had not omitted from the scope of license renewal any components with intended functions delineated under 10 CFR 54.4(a). The staff then reviewed those components that the applicant had identified as being within the scope of license renewal to verify that the applicant had not omitted any passive and long-lived components that should be subject to an AMR in accordance with the requirements of 10 CFR 54.21(a)(1).

The staff's review of LRA Section 2.3.3.B.5 identified areas in which additional information was necessary to complete the review of the applicant's scoping and screening results. The applicant responded to the staff's RAIs as discussed below.

In RAIs 2.3.3.B.5-1 and 2.3.3.B.5-4 dated November 19, 2004, the staff requested that the applicant clarify information given on LR boundary drawings concerning SSCs in scope of license renewal per 10 CFR 54.4(a). In RAI 2.3.3.B.5-1 the staff stated that LR drawing LR-006 Sheet A shows air operated control valves FV2A, 2B, and 2C (locations F-3, F-7 and F-10) as subject to an AMR. Therefore, the staff requested that the applicant provide the basis for excluding compressed air system auxiliaries to these valves from requiring an AMR. In RAI 2.3.3.B.5-4 the staff requested that the applicant provide the basis for excluding the double acting cylinders from AMR, noting that the LR drawings do not show the air cylinders as part of the license renewal boundary. This RAI was based on the assumption that the valves will go to their fail-safe position on loss of air pressure. This assumption would be true for single-acting air cylinders with springs but double-acting cylinders require air pressure for one of the cylinders to effect valve repositioning to its fail-safe position. Therefore, the double acting cylinders have a pressure boundary function.

In its response to 2.3.3.B.5-1 by letter dated December 22, 2004, the applicant stated that drawing LR-6A-0 is incorrect, that air-operated valves 2FWR-FV2A, 2FWR-FV2B, and 2FWRFV2C are not within the scope of license renewal per 10 CFR 54.4(a)(1) or (3) as these valves are NSR and part of the feedwater pump recirculation balance drum leakoff system which supports feedwater control and main feedwater in their functions to maintain reactor

2-160

water level during normal plant operation and as such the system and the subject valves are not credited for safe shutdown of the plant. Therefore, the applicant explained that the air supply tubing and the solenoid valves are not within the scope of license renewal. Per LRA Section 2.3.4.B.3, however, the referenced FWR valves and their associated piping are in-scope for LR and subject to an AMR per 10 CFR 54.4(a)(2) as they are liquid-filled components located in the turbine building in the vicinity of SR components. In its response to RAI 2.3.3.B.5-4 by letter dated December 22, 2004, the applicant agreed that SR double-acting actuators are in-scope for LR and subject to an AMR for a "Pressure Boundary" intended function. The applicant stated that revisions to incorporate the AMR results and other associated LRA changes will be submitted to the staff.

In its ALRA dated July 14, 2005, the applicant provided the staff with revised LR drawings which identify all SSCs within the scope of license renewal and subject to an AMR including those subject under 10 CFR 54.4(a)(2).

Based on review of the information submitted in the ALRA the staff's concerns described in RAIs 2.3.3.B.5-1 and 2.3.3.B.5-4 are resolved.

In RAI 2.3.3.B.5-2 dated November 19, 2004, the staff stated that drawing LR-013 sheet E shows a fail closed valve as subject to an AMR. However, the air supply tubing and solenoid valves are not shown as subject to an AMR. Therefore, the staff requested that the applicant provide the basis for excluding the compressed air system auxiliaries to this valve as subject to an AMR.

In its response by letter dated December 22, 2004, the applicant stated:

> Drawing LR-13E-0 is correct for the air supply tubing to valve 2CCP*AOV37B (location D7), but incorrect for the air supply tubing to valve 2CCP*AOV38B. The tubing to the actuator for 2CCP*AOV38B should also be shown in red as safety-related and subject to AMR. The actuators for both of these fail-safe, air-operated valves (AOV) are safety-related and in-scope for LR but are not subject to AMR since they are active components per NEI 95-10, Revision 3, Appendix B. As such, the actuators are correctly colored black; however, both actuators should have a boundary flag at the actuator air inlet with an indicating arrow "LR-CAS" towards the air supply side of the actuator and a solid blue arrow towards the actuator itself.
>
> Similar corrections apply on drawing LR-13E-0 for the air supplies to valves 2CCP*AOV37A and 2CCP*AOV38A (coordinated K2 and K4, respectively).

Based on its review, the staff found the applicant's response to RAI 2.3.3.B.5-2 acceptable because the air supply tubing and solenoid valves to the valve in question are within the scope of license renewal and subject to an AMR. Therefore, the staff's concern described in RAI 2.3.3.B.5-2 is resolved.

In RAI 2.3.3.B.5-3 dated November 19, 2004, the staff stated that on drawing LR-019 sheets L and M main steam isolation valves are shown as subject to an AMR. However, the air supply tubing and solenoid valves are not shown on detail A of drawing LR-019 sheet L as requiring an

AMR. Therefore, the staff requested that the applicant provide the basis for excluding the compressed air system auxiliaries to these valves as subject to an AMR.

In its response by letter dated December 22, 2004, the applicant stated that detail A of drawing LR-19L-0 is incorrect. The applicant explained that the instrument air tubing and solenoid valves should all be highlighted in red up to the operator for valve 2MSS*AOV6C as SR within the scope of license renewal and subject to an AMR consistent with the indication for these lines and components at the other locations for the AOV-6s on this drawing. Since the MSS*AOV-6s are SR valves the supply air tubing and in-line components are also SR.

Based on its review, the staff found the applicant's response to RAI 2.3.3.B.5-3 acceptable because it stated that the air tubing and solenoid valves to the operator for valve AOV6C are SR within the scope of license renewal and subject to an AMR in accordance with 10 CFR 54.4(a)(1) and 10 CFR 54.21(a)(1), respectively. Therefore, the staff's concern described in RAI 2.3.3.B.5-3 is resolved.

2.3B.3.5.3 Conclusion

The staff reviewed the ALRA, RAI responses, and accompanying scoping boundary drawings to determine whether any SSCs that should be within the scope of license renewal had not been identified by the applicant. No omissions were identified. In addition, the staff performed a review to determine whether any components that should be subject to an AMR had not been identified by the applicant. No omissions were identified. On the basis of its review, the staff concludes that there is reasonable assurance that the applicant had adequately identified the compressed air systems components that are within the scope of license renewal, as required by 10 CFR 54.4(a), and the compressed air systems components that are subject to an AMR, as required by 10 CFR 54.21(a)(1).

2.3B.3.6 NMP2 Containment Atmosphere Monitoring System

2.3B.3.6.1 Summary of Technical Information in the Amended Application

In ALRA Section 2.3.3.B.6, the applicant described the containment atmosphere monitoring system. The containment atmosphere monitoring system is designed to supply information concerning containment parameters during normal and post-accident conditions. Monitored drywell parameters are pressure, air temperature, hydrogen, and oxygen concentration along with gaseous and particulate radiation levels. Monitored suppression chamber parameters are pressure, air temperature, hydrogen and oxygen concentration, suppression pool level, and temperature. In addition, drywell and suppression chamber humidity are monitored during containment leak rate testing. The containment atmosphere monitoring system consists of radiation and hydrogen/oxygen monitoring lines. Each line penetrates the primary containment and monitors the radiation level and hydrogen/oxygen concentration during normal operation so they are equipped with containment isolation valves.

The containment atmosphere monitoring system contains SR components that are relied upon to remain functional during and following DBEs. The failure of NSR SSCs in the containment atmosphere monitoring system could potentially prevent the satisfactory accomplishment of an

SR function. In addition, the containment atmosphere monitoring system performs functions that support fire protection, EQ, and SBO.

The intended functions within the scope of license renewal include the following:

- provides pressure retaining boundary
- maintains mechanical and structural integrity of NSR components that provide structural support to attached SR components

In ALRA Table 2.3.3.B.6-1, the applicant identified the following containment atmosphere monitoring system component types that are within the scope of license renewal and subject to an AMR:

- bolting
- condensing chambers
- piping and fittings
- pumps
- valves

2.3B.3.6.2 Staff Evaluation

The staff reviewed ALRA Section 2.3.3.B.6 and USAR Sections 6.2.1.7 and 6.2.4.3.2 using the evaluation methodology described in SER Section 2.3. The staff conducted its review in accordance with the guidance described in SRP-LR Section 2.3.

In conducting its review, the staff evaluated the system functions described in the ALRA and USAR in accordance with the requirements of 10 CFR 54.4(a) to verify that the applicant had not omitted from the scope of license renewal any components with intended functions delineated under 10 CFR 54.4(a). The staff then reviewed those components that the applicant had identified as being within the scope of license renewal to verify that the applicant had not omitted any passive and long-lived components that should be subject to an AMR in accordance with the requirements of 10 CFR 54.21(a)(1).

2.3B.3.6.3 Conclusion

The staff reviewed the ALRA to determine whether any SSCs that should be within the scope of license renewal had not been identified by the applicant. No omissions were identified. In addition, the staff performed a review to determine whether any components that should be subject to an AMR had not been identified by the applicant. No omissions were identified. On the basis of its review, the staff concludes that there is reasonable assurance that the applicant had adequately identified the containment atmosphere monitoring system components that are within the scope of license renewal, as required by 10 CFR 54.4(a), and the containment atmosphere monitoring system components that are subject to an AMR, as required by 10 CFR 54.21(a)(1).

2.3B.3.7 NMP2 Containment Leakage Monitoring System

2.3B.3.7.1 Summary of Technical Information in the Amended Application

In ALRA Section 2.3.3.B.7, the applicant described the containment leakage monitoring system. The containment leakage monitoring system is designed to provide a means of monitoring the drywell area pressure and the suppression chamber pressure during periodic leak rate testing. Two independent pressure sensing lines penetrate the primary containment and connect to instrumentation outside the drywell during testing. The system also continuously monitors the drywell electrical penetrations to detect leakage past the sealing mechanism.

The containment leakage monitoring system contains SR components that are relied upon to remain functional during and following DBEs. The failure of NSR SSCs in the containment leakage monitoring system could potentially prevent the satisfactory accomplishment of an SR function. In addition, the containment leakage monitoring system performs functions that support EQ.

The intended functions within the scope of license renewal include the following:

- ·provides pressure retaining boundary
- maintains mechanical and structural integrity of NSR components that provide structural support to attached SR components

In ALRA Table 2.3.3.B.7-1, the applicant identified the following containment leakage monitoring system component types that are within the scope of license renewal and subject to an AMR:

- bolting
- piping and fittings
- valves

2.3B.3.7.2 Staff Evaluation

The staff reviewed ALRA Section 2.3.3.B.7 and USAR Section 6.2.6 using the evaluation methodology described in SER Section 2.3. The staff conducted its review in accordance with the guidance described in SRP-LR Section 2.3.

In conducting its review, the staff evaluated the system functions described in the ALRA and USAR in accordance with the requirements of 10 CFR 54.4(a) to verify that the applicant had not omitted from the scope of license renewal any components with intended functions delineated under 10 CFR 54.4(a). The staff then reviewed those components that the applicant had identified as being within the scope of license renewal to verify that the applicant had not omitted any passive and long-lived components that should be subject to an AMR in accordance with the requirements of 10 CFR 54.21(a)(1).

2.3B.3.7.3 Conclusion

The staff reviewed the ALRA to determine whether any SSCs that should be within the scope of license renewal had not been identified by the applicant. No omissions were identified. In addition, the staff performed a review to determine whether any components that should be subject to an AMR had not been identified by the applicant. No omissions were identified. On the basis of its review, the staff concludes that there is reasonable assurance that the applicant had adequately identified the containment leakage monitoring system components that are within the scope of license renewal, as required by 10 CFR 54.4(a), and the containment leakage monitoring system components that are subject to an AMR, as required by 10 CFR 54.21(a)(1).

2.3B.3.8 NMP2 Control Building Chilled Water System

2.3B.3.8.1 Summary of Technical Information in the Amended Application

In ALRA Section 2.3.3.B.8, the applicant described the control building chilled water system. The control building chilled water system is designed to provide chilled water to the air conditioning units that provide cooling for personnel and equipment in the control room, relay room, remote shutdown room, and computer room. This system is designed to perform during normal operation, plant shutdown, or accident conditions without loss of function. The control building chilled water system is a closed loop piping system consisting of two independent, redundant chilled water loops.

The control building chilled water system contains SR components that are relied upon to remain functional during and following DBEs. The failure of NSR SSCs in the control building chilled water system could potentially prevent the satisfactory accomplishment of an SR function. In addition, the control building chilled water system performs functions that support fire protection and EQ.

The intended functions within the scope of license renewal include the following:

- provides heat transfer
- maintains mechanical and structural integrity of NSR components to prevent spatial interactions with SR components
- provides pressure retaining boundary
- maintains mechanical and structural integrity of NSR components that provide structural support to attached SR components

In ALRA Table 2.3.3.B.8-1, the applicant identified the following control building chilled water system component types that are within the scope of license renewal and subject to an AMR:

- bolting
- chillers
- flow elements
- piping and fittings
- pumps

- tanks
- valves

2.3B.3.8.2 Staff Evaluation

The staff reviewed ALRA Section 2.3.3.B.8 and the USAR Sections 7.3.1.1.11 and 9.4.10.1 using the evaluation methodology described in SER Section 2.3. The staff conducted its review in accordance with the guidance described in SRP-LR Section 2.3.

In conducting its review, the staff evaluated the system functions described in the ALRA and USAR in accordance with the requirements of 10 CFR 54.4(a) to verify that the applicant had not omitted from the scope of license renewal any components with intended functions delineated under 10 CFR 54.4(a). The staff then reviewed those components that the applicant had identified as being within the scope of license renewal to verify that the applicant had not omitted any passive and long-lived components that should be subject to an AMR in accordance with the requirements of 10 CFR 54.21(a)(1).

2.3B.3.8.3 Conclusion

The staff reviewed the ALRA to determine whether any SSCs that should be within the scope of license renewal had not been identified by the applicant. No omissions were identified. In addition, the staff performed a review to determine whether any components that should be subject to an AMR had not been identified by the applicant. No omissions were identified. On the basis of its review, the staff concludes that there is reasonable assurance that the applicant had adequately identified the control building chilled water system components that are within the scope of license renewal, as required by 10 CFR 54.4(a), and the control building chilled water system components that are subject to an AMR, as required by 10 CFR 54.21(a)(1).

2.3B.3.9 NMP2 Control Building HVAC System

2.3B.3.9.1 Summary of Technical Information in the Amended Application

In ALRA Section 2.3.3.B.9, the applicant described the control building HVAC system. The control building HVAC system provides filtration, pressurization, heating, and cooling to the control building envelope during normal and emergency operations by operating in normal, smoke purge, and emergency modes. Outdoor air is supplied to the control building through missile and tornado protected air intakes. From the intakes the air is drawn into large duct chases by the four air conditioning units. The air is heated or cooled by cooling coils in the air conditioning units or by heaters in the ductwork and force circulated by the air conditioning unit fans throughout the control building envelope. Natural exhaust ventilation is provided through return registers back to the duct chases where most of the air is then recirculated. In the emergency mode the system will divert the intake air through special filters under certain conditions. The filter trains are normally bypassed and automatically come on line on either a supply air radiation monitor trip system signal or a LOCA signal. They would then provide filtered air to the control, relay, and computer rooms. The system is equipped with a special smoke removal system for use post-fire. It removes smoke and heat from the control building using special supply and exhaust fans, dampers, and controls.

The control building HVAC system contains SR components that are relied upon to remain functional during and following DBEs. In addition, the control building HVAC system performs functions that support fire protection, EQ, SBO.

The intended functions within the scope of license renewal include the following:

- provides filtration
- provides heat transfer
- provides pressure retaining boundary

In ALRA Table 2.3.3.B.9-1, the applicant identified the following control building HVAC system component types that are within the scope of license renewal and subject to an AMR:

- air handling unit
- blowers
- bolting
- ducting
- filters/strainers
- flow elements
- heat exchangers
- piping and fittings
- radiation sample point
- valves and dampers (includes fire dampers)

2.3B.3.9.2 Staff Evaluation

The staff reviewed ALRA Section 2.3.3.B.9 and USAR Section 9.4.1 using the evaluation methodology described in SER Section 2.3. The staff conducted its review in accordance with the guidance described in SRP-LR Section 2.3.

In conducting its review, the staff evaluated the system functions described in the ALRA and USAR in accordance with the requirements of 10 CFR 54.4(a) to verify that the applicant had not omitted from the scope of license renewal any components with intended functions delineated under 10 CFR 54.4(a). The staff then reviewed those components that the applicant had identified as being within the scope of license renewal to verify that the applicant had not omitted any passive and long-lived components that should be subject to an AMR in accordance with the requirements of 10 CFR 54.21(a)(1).

2.3B.3.9.3 Conclusion

The staff reviewed the ALRA to determine whether any SSCs that should be within the scope of license renewal had not been identified by the applicant. No omissions were identified. In addition, the staff performed a review to determine whether any components that should be subject to an AMR had not been identified by the applicant. No omissions were identified. On the basis of its review, the staff concludes that there is reasonable assurance that the applicant had adequately identified the control building HVAC system components that are within the scope of license renewal, as required by 10 CFR 54.4(a), and the control building HVAC system components that are subject to an AMR, as required by 10 CFR 54.21(a)(1).

2.3B.3.10 NMP2 Diesel Generator Building Ventilation System

2.3B.3.10.1 Summary of Technical Information in the Amended Application

In ALRA Section 2.3.3.B.10, the applicant described the diesel generator building ventilation system. The diesel generator building ventilation system is designed to provide heating and outside air ventilation to the diesel rooms and diesel generator control rooms. Each diesel generator room is equipped with its own ventilation system. Additionally, the diesel generator building ventilation HVAC system is designed with unit coolers to maintain habitable conditions for personnel comfort within the diesel generator control rooms. The diesel generator building ventilation system performs the following functions: normal heating, normal ventilation, control room cooling, and general area emergency ventilation. The normal duty heating function maintains the diesel generator rooms above 65°F during the winter. The normal ventilation function maintains the diesel generator rooms adequately ventilated and exhausts room air to the atmosphere. The control room cooling function maintains the diesel generator rooms below the maximum design temperature via unit coolers which is provided from the service water system (SER Section 2.3.3B.27). The general area emergency ventilation function establishes a ventilating flow of outside air through the diesel generator rooms to ensure that the space temperatures remain below 125°F outside the control room or 104°F inside the control room for efficient equipment operation.

The diesel generator building ventilation system contains SR components that are relied upon to remain functional during and following DBEs. In addition, the diesel generator building ventilation system performs functions that support fire protection and EQ.

The intended function within the scope of license renewal is to provide pressure retaining boundary.

In ALRA Table 2.3.3.B.10-1, the applicant identified the following diesel generator building ventilation system component types that are within the scope of license renewal and subject to an AMR:

- blowers
- dampers (includes fire dampers)
- ducting
- unit coolers

2.3B.3.10.2 Staff Evaluation

The staff reviewed ALRA Section 2.3.3.B.10 and USAR Section 9.4.6 using the evaluation methodology described in SER Section 2.3. The staff conducted its review in accordance with the guidance described in SRP-LR Section 2.3.

In conducting its review, the staff evaluated the system functions described in the ALRA and USAR in accordance with the requirements of 10 CFR 54.4(a) to verify that the applicant had not omitted from the scope of license renewal any components with intended functions delineated under 10 CFR 54.4(a). The staff then reviewed those components that the applicant had identified as being within the scope of license renewal to verify that the applicant had not

omitted any passive and long-lived components that should be subject to an AMR in accordance with the requirements of 10 CFR 54.21(a)(1).

2.3B.3.10.3 Conclusion

The staff reviewed the ALRA to determine whether any SSCs that should be within the scope of license renewal had not been identified by the applicant. No omissions were identified. In addition, the staff performed a review to determine whether any components that should be subject to an AMR had not been identified by the applicant. No omissions were identified. On the basis of its review, the staff concludes that there is reasonable assurance that the applicant had adequately identified the diesel generator building ventilation system components that are within the scope of license renewal, as required by 10 CFR 54.4(a), and the diesel generator building ventilation system components that are subject to an AMR, as required by 10 CFR 54.21(a)(1).

2.3B.3.11 NMP2 Domestic Water System

2.3B.3.11.1 Summary of Technical Information in the Amended Application

In ALRA Section 2.3.3.B.11, the applicant described the domestic water system. The domestic water system is designed to provide sufficient domestic water from an existing city main to various areas of the plant including the makeup water treatment system (SER Section 2.3.3B.35) and the fire protection system (SER Section 2.3.3B.13). Additionally the domestic water system ensures minimization of flooding potential by providing isolation capabilities of the control building from domestic water supply should piping within the building rupture during a seismic event. Domestic water is supplied to various buildings throughout the plant including the control building, TB, and the auxiliary building. The domestic water system also provides makeup water to various systems including the fire protection system and the filtered water tank.

The domestic water system contains SR components that are relied upon to remain functional during and following DBEs. The failure of NSR SSCs in the domestic water system could potentially prevent the satisfactory accomplishment of an SR function.

The intended functions within the scope of license renewal include the following:

- maintains mechanical and structural integrity of NSR components to prevent spatial interactions with SR components
- provides pressure retaining boundary

In ALRA Table 2.3.3.B.11-1, the applicant identified the following domestic water system component types that are within the scope of license renewal and subject to an AMR:

- bolting
- piping and fittings
- tanks
- valves

2.3B.3.11.2 Staff Evaluation

The staff reviewed ALRA Section 2.3.3.B.11 and USAR Sections 1.2.10.10 and 9.2.4 using the evaluation methodology described in SER Section 2.3. The staff conducted its review in accordance with the guidance described in SRP-LR Section 2.3.

In conducting its review, the staff evaluated the system functions described in the ALRA and USAR in accordance with the requirements of 10 CFR 54.4(a) to verify that the applicant had not omitted from the scope of license renewal any components with intended functions delineated under 10 CFR 54.4(a). The staff then reviewed those components that the applicant had identified as being within the scope of license renewal to verify that the applicant had not omitted any passive and long-lived components that should be subject to an AMR in accordance with the requirements of 10 CFR 54.21(a)(1).

2.3B.3.11.3 Conclusion

The staff reviewed the ALRA to determine whether any SSCs that should be within the scope of license renewal had not been identified by the applicant. No omissions were identified. In addition, the staff performed a review to determine whether any components that should be subject to an AMR had not been identified by the applicant. No omissions were identified. On the basis of its review, the staff concludes that there is reasonable assurance that the applicant had adequately identified the domestic water system components that are within the scope of license renewal, as required by 10 CFR 54.4(a), and the domestic water system components that are subject to an AMR, as required by 10 CFR 54.21(a)(1).

2.3B.3.12 NMP2 Engine-Driven Fire Pump Fuel Oil System

2.3B.3.12.1 Summary of Technical Information in the Amended Application

In ALRA Section 2.3.3.B.12, the applicant described the engine-driven fire pump fuel oil system. The engine-driven fire pump fuel oil system is designed to supply fuel oil to the diesel engine-driven fire pump. The electric-driven fire pump and diesel engine-driven fire pump are located in separate rooms within the SWB. The fuel oil storage tank for the diesel fire pump is located in the diesel fire pump room above a sump. Fuel is gravity fed to the engine and excess fuel supplied to the engine by its fuel pump is recirculated to the tank.

The engine-driven fire pump fuel oil system performs functions that support fire protection.

The intended function within the scope of license renewal is to provide pressure retaining boundary.

In ALRA Table 2.3.3.B.12-1, the applicant identified the following engine-driven fire pump fuel oil system component types that are within the scope of license renewal and subject to an AMR:

- bolting
- piping and fittings
- tank
- valves

2.3B.3.12.2 Staff Evaluation

The staff reviewed ALRA Section 2.3.3.B.12 and USAR Sections 9.5.1.2.2 and 9A.3.1.2.5.6 using the evaluation methodology described in SER Section 2.3. The staff conducted its review in accordance with the guidance described in SRP-LR Section 2.3.

In conducting its review, the staff evaluated the system functions described in the ALRA and USAR in accordance with the requirements of 10 CFR 54.4(a) to verify that the applicant had not omitted from the scope of license renewal any components with intended functions delineated under 10 CFR 54.4(a). The staff then reviewed those components that the applicant had identified as being within the scope of license renewal to verify that the applicant had not omitted any passive and long-lived components that should be subject to an AMR in accordance with the requirements of 10 CFR 54.21(a)(1).

The staff also reviewed approved fire protection safety evaluation report NUREG-1047, dated February 1985 (and Supplements 1 through 6) for Unit 2. This report is referenced directly in the Nine Mile Point Unit 2 fire protection CLB and summarize the Fire Protection Program and commitments to 10 CFR 50.48 using the guidance of Appendix A to BTP CMEB 9.5-1. The staff then reviewed those components that the applicant identified as being within the scope of license renewal to verify that the applicant did not omit any passive and long-lived components that should be subject to an AMR in accordance with the requirements of 10 CFR 54.21(a)(1).

On the basis of its review, the staff found that the applicant has identified those portions of the engine driven fire pump fuel oil system that meet the scoping requirements of 10 CFR 54.4(a) and has included them within the scope of license renewal in ALRA Section 2.3.3.B.12. The applicant has also included engine driven fire pump fuel oil system components that are subject to an AMR in accordance with the requirements of 10 CFR 54.4(a) and 10 CFR 54.21(a)(1) in LRA Table 2.3.3.B.12-1. The staff did not identify any omissions.

2.3B.3.12.3 Conclusion

The staff reviewed the ALRA to determine whether any SSCs that should be within the scope of license renewal had not been identified by the applicant. No omissions were identified. In addition, the staff performed a review to determine whether any components that should be subject to an AMR had not been identified by the applicant. No omissions were identified. On the basis of its review, the staff concludes that there is reasonable assurance that the applicant had adequately identified the engine-driven fire pump fuel oil system components that are within the scope of license renewal, as required by 10 CFR 54.4(a), and the engine-driven fire pump fuel oil system components that are subject to an AMR, as required by 10 CFR 54.21(a)(1).

2.3B.3.13 NMP2 Fire Detection and Protection System

2.3B.3.13.1 Summary of Technical Information in the Amended Application

In ALRA Section 2.3.3.B.13, the applicant described the fire detection and protection system. The fire detection and protection system is designed for detecting, alarming, isolating, and suppressing fires in the plant. The fire protection system consists in part of a reliable freshwater supply, one electric motor-driven fire pump and one diesel engine-driven fire pump, two pressure maintenance fire pumps, one pressure maintenance pump supply tank, one hydropneumatic tank, fire water yard mains, hydrants, standpipes, hose stations, sprinkler, water spray, preaction and deluge systems, foamwater deluge systems, low-pressure CO_2 systems, Halon 1301 systems, and a detection and signaling system. These components in the fire detection and protection system are further divided into the fire protection foam system, the fire protection halon system, the Cardox fire protection system, the fire detection system, and the fire protection water system. The collective capability of the fire suppression systems is adequate to minimize potential damage to SR equipment and is a major element in the facility Fire Protection Program. The fire protection foam system provides fire suppression through blanketing affected areas with dense foam provided by mixing of fire system water, foam concentrate, and air. The fire protection halon system is designed to suppress cable fires in the floor sections of the computer room, relay room, control room, and the radwaste control room. The Cardox fire protection system is designed to supply CO_2 to fixed and hose reel stations for the purpose of extinguishing fires. The fire detection system is designed to provide early detection, annunciation, and actuation of suppression systems in the event of a fire. The thermal and smoke detection systems function to detect products of combustion, alarming both locally and in the main control room. The fire protection water system is designed to provide a reliable, readily available source of water for controlling and extinguishing fires. Additionally the fire protection water system provides control room indication and may be used as an alternative injection/spray source into the RPV or as primary containment by cross-connecting fire protection water to the RHR system. The water source for the fire protection system is Lake Ontario, which is considered to be unlimited.

The failure of NSR SSCs in the fire detection and protection system could potentially prevent the satisfactory accomplishment of an SR function. The fire detection and protection system also performs functions that support fire protection.

The intended functions within the scope of license renewal include the following:

- maintains mechanical and structural integrity of NSR components to prevent spatial interactions with SR components

- provides pressure retaining boundary

- provides over-pressure protection

- converts liquid into spray

- maintains mechanical and structural integrity of NSR components that provide structural support to attached SR components

In ALRA Table 2.3.3.B.13-1 the applicant identified the following fire detection and protection system component types that are within the scope of license renewal and subject to an AMR:

- bolting
- fire hydrants
- flow elements
- halon tank flex hoses
- heat exchangers
- hose reels

2-172

- manifold
- nozzles
- odorizers
- orifices
- piping and fittings
- pumps
- ratio flow proportioner
- rupture discs
- silencer
- strainers
- tanks
- temperature indicators
- valves

2.3B.3.13.2 Staff Evaluation

The staff reviewed LRA and ALRA Section 2.3.3.B.13 and USAR Sections 9.5.1, 9A.3.1.2.5.4, and 9A.3.6 using the evaluation methodology described in SER Section 2.3. The staff conducted its review in accordance with the guidance described in SRP-LR Section 2.3.

The staff also reviewed approved fire protection safety evaluation report NUREG-1047, dated February 1985 (and Supplements 1 through 6) for Unit 2. This report is referenced directly in the Nine Mile Point Unit 2 fire protection CLB and summarize the Fire Protection Program and commitments to 10 CFR 50.48 using the guidance of Appendix A to BTP CMEB 9.5-1. The staff then reviewed those components that the applicant identified as being within the scope of license renewal to verify that the applicant did not omit any passive and long-lived components that should be subject to an AMR in accordance with the requirements of 10 CFR 54.21(a)(1).

In reviewing LRA Section 2.3.3.B.13, the staff identified areas in which additional information was necessary to complete the evaluation of the applicant's scoping and screening results. Therefore, by letter to the applicant dated November 17, 2004, the staff issued a request for RAI concerning the specific issues to determine whether the applicant has properly applied the scoping criteria of 10 CFR 54.4(a) and the screening criteria of 10 CFR 54.21(a)(1). The applicant responded to the staff's RAIs as discussed below.

In RAI 2.3.3.B.13-1 dated November 17, 2004, the staff stated that GALL Report Section XI.27 describes the requirements for aging management of the fire protection water system. It requires that an AMP be established to evaluate the aging effects of microbiologically induced corrosion (MIC) and biofouling of carbon steel and cast iron components in fire protection systems exposed to water.

The LRA discusses requirements for the fire detection and protection program but does not mention trash racks and traveling screens for the fire pump suction water supply. Trash racks and traveling screens are mentioned in the LRA section for the service water system but are not listed in the associated LRA table for that system containing the list of components that require aging management. The components are not mentioned in the LRA.

The USAR states that the trash racks and traveling screens are located upstream of the fire pump suctions to remove any major debris from the water. Therefore, the staff requested that the applicant explain the apparent exclusion of the trash racks and traveling screens located upstream of the fire pump suctions from the scope of license renewal and from being subject to an AMR.

In its response by letter dated December 17, 2004, the applicant stated that although the trash racks and traveling screens are addressed in USAR Section 9.2.5 as preventing large debris from reaching the service water pumps and therefore the fire pumps as well the collection of debris on the trash racks and/or the traveling screens such that blockage could occur is not a license renewal intended function under 10 CFR 50.48. If such a blockage occurred bypass valves would open automatically to bypass the blockage and continue to supply water to the pump suctions. Additionally the fire pump suction headers have their own strainers in line so the loss of the trash racks or traveling screens would not affect the operation of these pumps until repair/replacement of the damaged component could be performed.

The applicant further stated that the supports of the trash racks are within the scope of license renewal and subject to an AMR.

In evaluating this response the staff found that it was incomplete and that review of LRA Section 2.3.3.B.13 could not be completed. The response explains that the trash racks and traveling screens are addressed in the USAR but that they perform no intended function. The staff found this explanation contrary to the USAR, which includes the original NMP2 fire protection SE as CLB. As a result the staff held a teleconference with the applicant on January 25, 2005, to discuss information necessary to resolve the concern in RAI 2.3.3.B.13-1. The product of the teleconference was an agreement by the applicant to transmit the required information by a follow-up letter.

By letter dated February 11, 2005, the applicant provided references to the USAR that describes the bypass valves that provide a traveling water screen bypass flow path to the service water pumps and the fire water pumps. This bypass operates automatically using safety-related, seismically-qualified components, thereby assuring sufficient service water suction bay water level and adequate fire pump suction supply in the event of blockage of the trash rakes or traveling screens. Thus, the applicant stated that the trash rakes and traveling screens do not perform or support any fire protection intended functions.

Based on its review, the staff found the applicant's response to RAI 2.3.3.B.13-1, including the information in the teleconference and letter dated February 11, 2005, acceptable because it adequately described the intended function supporting the fire pump suction supply as accomplished by the automatic bypass valves. Further, the bypass valves and their controls are SR and are within the scope of license renewal under 10 CFR 54.4(a)(1). Therefore, the staff's concern described in RAI 2.3.3.B.13-1 is resolved.

In RAI 2.3.3.B.13-2 dated November 17, 2004, the staff noted that the GALL Report Section XI.27, "Fire Water System," states the requirements for sprinklers, including the inspection frequency recommended by the National Fire Protection Association 25 leading to their eventual replacement. LRA Table 2.3.3.B.13-1 does not include sprinklers within the scope of license renewal and subject to an AMR. The applicant was asked to verify that sprinklers in the fire detection and protection system are within the scope of license renewal and to indicate where they are identified in the LRA.

In its response by letter dated December 17, 2004, the applicant stated that sprinklers are included within the scope of license renewal and subject to an AMR under the component type "Nozzles" in the LRA table associated with the fire detection and protection system and that the corresponding AMR summary is addressed in LRA Section 3.3.2.B.13 and in Table 3.3.2.B-13.

Based on its review, the staff found the applicant's response to RAI 2.3.3.B.13-2 acceptable because it adequately explained that the sprinklers in question are within the scope of license renewal under 10 CFR 54.4(a) and subject to an AMR under 10 CFR 54.21(a). Further, the applicant stated that the sprinklers are represented in the LRA table by the component type "Nozzles." Therefore, the staff's concern described in RAI 2.3.3.B.13-2 is resolved.

In RAI 2.3.3.B.13-3 dated November 17, 2004, the staff noted that LRA Section 2.3.3.B.13 states that the fire protection foam subsystem components subject to an AMR consist of one water header valve. LRA Table 3.3.2.B-13 lists valve environments as air, dried air or gas, and raw water, low flow. Also, the water supply portion of the fire protection foam subsystem includes two foam concentrate storage tanks, four foam concentrate pumps, a foam water ratio flow proportioner, numerous valves exposed to foam concentrate and valves exposed to foam-water mixture, and a piping distribution system exposed to both foam concentrate and foam-water mixture. Therefore, the staff requested that the applicant identify where the components in question were shown in the LRA.

In its response by letter dated December 17, 2004, the applicant stated that the LRA section for the fire detection and protection system identifying the foam subsystem components subject to an AMR consisting of one water header valve is in error and that the components subject to an AMR for the fire protection foam subsystem are shown on its LR drawing and include the foam concentrate tanks, foam concentrate pumps, valves, ratio flow proportioner, and piping as indicated on the drawing. The applicant revised LRA Sections 2.3.3.B.13 and 3.3.2.B.13, and Table 3.3.2.B-13 to address the fire protection foam subsystem properly. The applicant revised LRA Section 2.3.3.B.13 to include the two foam concentrate tanks, the four foam concentrate pumps, the ratio foam proportioner, and the associated piping, fittings, and valves connecting these components which make up the foam distribution system as components subject to an AMR.

LRA Section 3.3.2.B.13 was revised to add liquid foam concentrate and liquid foam concentrate/raw water, low flow environments. LRA Table 3.3.2.B-13 was revised by the applicant to add AMR for the subject foam system components in these environments. The applicant stated that LRA scoping and screening Table 2.3.3.B.13-1 did not require revision because it already generically included the components in question.

Based on its review, the staff found the applicant's response to RAI 2.3.3.B.13-3 acceptable because it adequately explained that the LRA erroneously excluded components in the foam subsystem of the fire detection and protection system and that the associated AMR table and LRA sections have been revised to add the subject components and environments. Therefore, the staff's concern described in RAI 2.3.3.B.13-3 is resolved.

In RAI 2.3.3.B.13-4 dated November 17, 2005, the staff stated that the NMP2 USAR states that fusible link-actuated heat vents are provided in the turbine building roof. The fusible links are set high enough to preclude release due to a steam leak. These vents reduce the possibility of roof collapse in the event of a fire on the operating level.

Fusible links are not described in the fire detection and protection section of the LRA. Fire dampers are described in another part of the LRA and identify a one-time inspection to manage their aging effects.

Heat-sensitive fusible links are composed of heat-sensitive solder and are long-lived passive components that should be within the scope of license renewal and subject to an AMR. Therefore, the staff requested that the applicant explain the apparent exclusion of the heat-sensitive fusible links in the turbine building heat removal system from requiring an AMR and how a one-time inspection would adequately manage the aging effects of fire dampers which utilize fusible links as their actuating devices.

In its response by letter dated December 17, 2004, the applicant stated that because the fusible links require a change in state to perform their function they are considered an active component. Therefore, although they are within scope of license renewal they are not subject to an AMR. Further, the applicant explained that the NMP2 turbine building roof vent housings inadvertently were excluded from scope and AMR. The applicant stated that these components fall into a component type already addressed in the LRA. In LRA Table 2.4.B.13-1 the roof vent housings fall under the last component type in the table, i.e., "Structural Steel (Carbon and Low Alloy Steel) in Air." Their intended function is "Structural Support for NSR." Their AMR is addressed in LRA Section 3.5.2.B.13 and Table 3.5.2.B-13. These components will be managed for aging consistent with the information presented in the referenced LRA locations.

As to aging management of the fire dampers included in LRA Table 3.3.2.B-9, because of their fabrication material, the environment to which they are exposed, and plant operating history the fire dampers are not expected to experience loss of material. For that reason the one-time inspection program has been deemed to be adequate for aging management of these dampers. If the aging effect is discovered as a function of that inspection the inspection scope will be expanded consistent with the program requirements described in LRA Section B2.1.20.

Based on its review, the staff found the applicant's response to RAI 2.3.3.B.13-4 acceptable because it adequately explained that although the fusible links in question are within the scope of license renewal under 10 CFR 54.4(a) they are active components and are not subject to an AMR under 10 CFR 54.21(a). In addition the applicant explained that due to their fabrication material, the environment to which they are exposed, and plant operating history the fire dampers are not expected to experience loss of material and the one-time inspection program is adequate for managing the aging of these components. Therefore, the staff's concern described in RAI 2.3.3.B.13-4 is resolved.

In RAI 2.3.3.B.13-5 dated November 17, 2004, the staff stated that LR drawing LR-43A-O for the diesel engine-driven fire pump identifies the fuel oil supply piping immediately upstream of the diesel engine as within scope of license renewal and subject to an AMR. Additionally the LR drawing indicates that the engine should be subject to an AMR because it is enclosed within "flags." The engine itself, however, is not shown as highlighted for AMR. The applicant was asked to explain the apparent discrepancy between the LR drawing and the AMR boundary flags.

In its response by letter dated December 17, 2004, the applicant explained that drawing LR-43A-0 is incorrect and does not properly show the components within the scope of license renewal and subject to an AMR. The applicant further stated that the engine is not subject to an AMR but its fuel oil supply piping is.

Based on its review, the staff found the applicant's response to RAI 2.3.3.B.13-5 acceptable because it adequately explained that the components in question are within the scope of license renewal in accordance with 10 CFR 54.4(a) and subject to an AMR in accordance with 10 CFR 54.21(a), but were inadvertently left un-highlighted on the LR drawing, except for the engine which was correctly identified as not being subject to an AMR. Therefore, the staff's concern described in RAI 2.3.3.B.13-5 is resolved.

In RAI 2.3.3.B.13-6 dated November 17, 2005, the staff stated that LR drawing LR-43B-0 for the fire detection and protection system identifies piping upstream of a valve leading to the radwaste building via the yard as within the scope of license renewal and subject to an AMR. However, the yard piping between the valve in question and the radwaste building appears to be excluded from requiring an AMR. The applicant was asked to explain the apparent exclusion of this portion of the fire detection and protection system from requiring an AMR.

In its response by letter dated December 17, 2004, the applicant stated that the LR drawing in question, LR-43B-0, was in error and that the yard piping and components in question should have been highlighted on the LR drawing showing them within the scope of license renewal and subject to an AMR but inadvertently were not highlighted. The components in question have been included within the scope of license renewal and subject to an AMR.

Based on its review, the staff found the applicant's response to RAI 2.3.3.B.13-6 acceptable because it adequately explained that the components in question are within the scope of license renewal under 10 CFR 54.4(a) and subject to an AMR under 10 CFR 54.21(a) but were inadvertently left un-highlighted on the LR drawing. The staff concludes that there is reasonable assurance that the components were correctly included within the scope of license renewal and subject to an AMR. Therefore, the staff's concern described in RAI 2.3.3.B.13-6 is resolved.

In RAI 2.3.3.B.13-7 dated November 17, 2004, the staff stated that although LR drawing LR-43E-0 for the fire detection and protection shows a valve installed in piping within the scope of license renewal and subject to an AMR the valve itself is not. The applicant was asked to explain the apparent exclusion of this valve from being subject to an AMR.

In its response by letter dated December 17, 2004, the applicant stated that drawing LR-43E-0 is incorrect and does not properly show the component within the scope of license renewal and subject to an AMR. The component in question should have been highlighted on the LR drawing showing it within the scope of license renewal and subject to an AMR but inadvertently was not highlighted. The component in question has been included within the scope of license renewal and is subject to an AMR.

Based on its review, the staff found the applicant's response to RAI 2.3.3.B.13-7 acceptable because it adequately explained that the component in question is within the scope of license renewal in accordance with 10 CFR 54.4(a) and subject to an AMR in accordance with 10 CFR 54.21(a), but inadvertently was not highlighted on the LR drawing. The staff concludes that there is reasonable assurance that the components were included correctly within the scope of license renewal and subject to an AMR. Therefore, the staff's concern described in RAI 2.3.3.B.13-7 is resolved.

In RAI 2.3.3.B.13-8 dated November 17, 2004, the staff stated that although drawing LR-43E-0 depicts the water spray system for one of the reactor feedwater pumps in scope of license renewal and subject to an AMR two other reactor feedwater pumps do not require AMR. The applicant was asked to explain the apparent exclusion of this portion of the fire protection system from being subject to an AMR.

In its response by letter dated December 17, 2004, the applicant stated that LR drawing LR-43E-0 is incorrect and does not properly show the components within the scope of license renewal and subject to an AMR. The components in question should have been highlighted on the LR drawing showing that they are within the scope of license renewal and subject to an AMR but inadvertently were not highlighted. The components in question have been included within the scope of license renewal and are subject to an AMR.
Based on its review, the staff found the applicant's response to RAI 2.3.3.B.13-8 acceptable because it adequately explained that the components in question are within the scope of license renewal in accordance with 10 CFR 54.4(a) and subject to an AMR in accordance with 10 CFR 54.21(a) but inadvertently were not highlighted on the LR drawing. The staff concludes that there is reasonable assurance that the components were included correctly within the scope of license renewal and subject to an AMR. Therefore, the staff's concern described in RAI 2.3.3.B.13-8 is resolved.

In RAI 2.3.3.B.13-10 dated November 17, 2004, the staff stated that drawing LR-43F-0 depicts several portions of piping zones within the scope of license renewal and subject to an AMR but not other portions of the same piping zone. Therefore, the staff requested that the applicant explain the apparent exclusion of these portions of the fire detection and protection system from requiring an AMR.

In its response by letter dated December 17, 2004, the applicant stated that drawing LR-43F-0 is incorrect and does not properly show the components within the scope of license renewal and subject to an AMR, that the components in question should have been highlighted on the LR drawing showing them within the scope of license renewal and subject to an AMR but inadvertently were not highlighted. The components in question have been included within the scope of license renewal and subject to an AMR.

Based on its review, the staff found the applicant's response to RAI 2.3.3.B.13-10 acceptable because it adequately explained that the components in question are within the scope of license renewal under 10 CFR 54.4(a) and subject to an AMR under 10 CFR 54.21(a) but inadvertently were not highlighted on the LR drawing. The staff concludes that there is reasonable assurance that the components were included correctly within the scope of license renewal and subject to an AMR. Therefore, the staff's concern described in RAI 2.3.3.B.13-10 is resolved.

In RAI 2.3.3.B.13-11 dated November 17, 2004, the staff stated that although drawing LR-43G-0 at location E-4 depicts piping adjacent to a specific valve as within the scope of license renewal and subject to an AMR the piping to this valve is not. The applicant was asked to explain the apparent exclusion of these portions of the fire detection and protection system from requiring an AMR.

In its response by letter dated December 17, 2004, the applicant stated that drawing LR-43G-0 is incorrect and does not properly show the components within the scope of license renewal and subject to an AMR. The components in question should have been highlighted on the LR drawing showing them within the scope of license renewal and subject to an AMR but inadvertently were not highlighted. The components in question have been included within the scope of license renewal and subject to an AMR.

Based on its review, the staff found the applicant's response to RAI 2.3.3.B.13-11 acceptable because it adequately explained that the components in question are within the scope of license renewal under 10 CFR 54.4(a) and subject to an AMR under 10 CFR 54.21(a) but inadvertently were not highlighted on the LR drawing. The staff concludes that there is reasonable assurance that the components were included correctly within the scope of license renewal and subject to an AMR. Therefore, the staff's concern described in RAI 2.3.3.B.13-11 is resolved.

In RAI 2.3.3.B.13-12 dated November 17, 2004, the staff stated that drawing LR-43H-0 depicts piping to some pressure switches in the system as within the scope of license renewal and subject to an AMR but not other piping to similar pressure switches. Therefore, the staff requested that the applicant explain the apparent exclusion of this portion of the fire detection and protection system from requiring an AMR.

In its response by letter dated December 17, 2004, the applicant stated that drawing LR-43H-0 is incorrect and does not properly show the components within the scope of license renewal and subject to an AMR. The components in question should have been highlighted on the LR drawing showing them within the scope of license renewal and subject to an AMR but inadvertently were not highlighted. The components in question have been included within the scope of license renewal and subject to an AMR.

Based on its review, the staff found the applicant's response to RAI 2.3.3.B.13-12 acceptable because it adequately explained that the components in question are within the scope of license renewal under 10 CFR 54.4(a) and subject to an AMR under 10 CFR 54.21(a) but inadvertently were not highlighted on the LR drawing. The staff concludes that there is reasonable assurance that the components were correctly included within the scope of license renewal and subject to an AMR. Therefore, the staff's concern described in RAI 2.3.3.B.13-12 is resolved.

In RAI 2.3.3.B.13-13 dated November 17, 2005, the staff stated that drawing LR-43H-0 depicts piping to some valve tamper switches in the system as within the scope of license renewal and subject to an AMR but not other piping to similar valve tamper switches. Therefore, the staff requested that the applicant explain the apparent exclusion of this portion of the fire detection and protection system from requiring an AMR.

In its response by letter dated December 17, 2004, the applicant stated that drawing LR-43H-0 is incorrect and does not properly show the components within the scope of license renewal and subject to an AMR. The components in question should not have been highlighted on the LR drawing showing that they are within the scope of license renewal and are subject to an AMR but were highlighted. inadvertently. The components in question should not have been included within the scope of license renewal and are not subject to an AMR.

Based on its review, the staff found the applicant's response to RAI 2.3.3.B.13-13 acceptable because it adequately explained that the components in question are not within the scope of license renewal under 10 CFR 54.4(a) and subject to an AMR under 10 CFR 54.21(a) but were highlighted inadvertently on the LR drawing. The staff concludes that there is reasonable assurance that the components were included incorrectly within the scope of license renewal and subject to an AMR. Therefore, the staff's concern described in RAI 2.3.3.B.13-13 is resolved.

In RAI 2.3.3.B.13-14 dated November 17, 2004, the staff stated that drawing LR-44A-0 depicts a piping section in the system as within the scope of license renewal and subject to an AMR but not another piece of piping within this section connecting the system to a valve. Therefore, the staff requested that the applicant explain the apparent exclusion of this portion of the fire detection and protection system from requiring an AMR.

In its response by letter dated December 17, 2004, the applicant stated that drawing LR-44A-0 is incorrect and does not properly show the components within the scope of license renewal and subject to an AMR. The components in question should have been highlighted on the LR drawing showing them within the scope of license renewal and subject to an AMR but inadvertently were not highlighted. The components in question have been included within the scope of license renewal and subject to an AMR.

Based on its review, the staff found the applicant's response to RAI 2.3.3.B.13-14 acceptable because it adequately explained that the components in question are within the scope of license renewal under 10 CFR 54.4(a) and subject to an AMR under 10 CFR 54.21(a) but inadvertently were not highlighted on the LR drawing. The staff concludes that there is reasonable assurance that the components were included correctly within the scope of license renewal and subject to an AMR. Therefore, the staff's concern described in RAI 2.3.3.B.13-14 is resolved.

In RAI 2.3.3.B.13-15 dated November 17, 2004, the staff stated that drawing LR-44A-0 depicts drain line piping in some portions of the system as within the scope of license renewal and subject to an AMR but not drain line piping in other portions of the system. Therefore, the staff requested that the applicant explain the apparent exclusion of this portion of the fire detection and protection system from requiring an AMR.

In its response by letter dated December 17, 2004, the applicant stated that drawing LR-44A-0 is incorrect and does not properly show the components within the scope of license renewal and subject to an AMR. The components in question should have been highlighted on the LR drawing showing that they are within the scope of license renewal and subject to an AMR but inadvertently were not highlighted. The components in question have been included within the scope of license renewal and subject to an AMR.

Based on its review, the staff found the applicant's response to RAI 2.3.3.B.13-15 acceptable because it adequately explained that the components in question are within the scope of license renewal under 10 CFR 54.4(a) and subject to an AMR under 10 CFR 54.21(a) but inadvertently were not highlighted on the LR drawing. The staff concludes that there is reasonable assurance that the components were included correctly within the scope of license renewal and subject to an AMR. Therefore, the staff's concern described in RAI 2.3.3.B.13-15 is resolved.

In RAI 2.3.3.B.13-16 dated November 17, 2004, the staff stated that drawing LR-44A-0 depicts piping in some portions of the system as within the scope of license renewal and subject to an AMR but not other sections of piping connecting the system to valves. The applicant was asked to explain the apparent exclusion of this portion of the fire detection and protection system from requiring an AMR.

In its response by letter dated December 17, 2004, the applicant stated that drawing LR-44A-0 is incorrect and does not properly show the components within the scope of license renewal and subject to an AMR. The components in question should have been highlighted on the LR drawing showing them within the scope of license renewal and are subject to an AMR but inadvertently were not highlighted. The components in question have been included within the scope of license renewal and are subject to an AMR.

Based on its review, the staff found the applicant's response to RAI 2.3.3.B.13-16 acceptable because it adequately explained that the components in question are within the scope of license renewal under 10 CFR 54.4(a) and subject to an AMR under 10 CFR 54.21(a) but inadvertently were not highlighted on the LR drawing. The staff concludes that there is reasonable assurance that the components were included correctly within the scope of license renewal and subject to an AMR. Therefore, the staff's concern described in RAI 2.3.3.B.13-16 is resolved.

In RAI 2.3.3.B.13-17 dated November 17, 2004, the staff stated that drawing LR-45C-0 depicts several solenoid valves in some portions of the system as within the scope of license renewal and subject to an AMR but not other similar solenoid valves in the system. The applicant was asked to explain the apparent exclusion of such valves in the fire detection and protection system from requiring an AMR.

In its response by letter dated December 17, 2004, the applicant stated that Drawing LR-45C-0 is incorrect and does not properly show the components within the scope of license renewal and subject to an AMR. The components in question should have been highlighted on the LR drawing showing them within the scope of license renewal and are subject to an AMR but inadvertently were not highlighted. The components in question have been included within the scope of license renewal and are subject to an AMR.

Based on its review, the staff found the applicant's response to RAI 2.3.3.B.13-17 acceptable because it adequately explained that the components in question are within the scope of license renewal under 10 CFR 54.4(a) and subject to an AMR under 10 CFR 54.21(a) but inadvertently were not highlighted on the LR drawing. The staff concludes that there is reasonable assurance that the components were included correctly within the scope of license renewal and subject to an AMR. Therefore, the staff's concern described in RAI 2.3.3.B.13-17 is resolved.

In RAI 2.3.3.B.13-18 dated November 17, 2004, the staff stated that drawing LR-45C-0 depicts piping sections associated with several solenoid valves in portions of the system as within the scope of license renewal and subject to an AMR but not other similar piping sections associated with similar solenoid valves in the system. Therefore, the staff requested that the applicant explain the apparent exclusion of these valves of the fire detection and protection system from requiring an AMR.

. In its response by letter dated December 17, 2004, the applicant stated that drawing LR-45C-0 is correct and properly shows some of the identified piping associated with SV153 excluded from the scope of license renewal and from being subject to an AMR. The applicant stated that although some of the components in question were highlighted incorrectly none of the identified piping downstream of the solenoid valves should have been highlighted on the LR drawing in question, LR-45C-0, showing them within the scope of license renewal and subject to an AMR, that the piping downstream of the solenoid valves is associated with cardox hose reels and provides venting and that the piping described is not required for the cardox system to perform its intended function and therefore is not subject to an AMR.

Based on its review, the staff found the applicant's response to RAI 2.3.3.B.13-18 acceptable because it adequately explained that all the similar components in question are not within the scope of license renewal under 10 CFR 54.4(a) and not subject to an AMR under 10 CFR 54.21(a) but that some were highlighted inadvertently on the LR drawing. The staff concludes that there is reasonable assurance that the components were excluded correctly from the scope of license renewal. Therefore, the staff's concern described in RAI 2.3.3.B.13-18 is resolved.

In RAI 2.3.3.B.13-19 dated November 17, 2004, the staff stated that drawing LR-45C-0 depicts piping sections associated with several solenoid valves in portions of the cardox system as within the scope of license renewal and subject to an AMR but not other similar piping sections associated with similar solenoid valves in the system. The applicant was asked to explain the apparent exclusion of these sections of the fire detection and protection system from requiring an AMR.

In its response by letter dated December 17, 2004, the applicant stated that drawing LR-45C-0 is correct and properly shows some of the components in question excluded from the scope of license renewal and requiring an AMR. Although some of the components in question were highlighted incorrectly none of the components in question should have been highlighted on the LR drawing showing them within the scope of license renewal and subject to an AMR. The components in question are associated with cardox zone discharge piping and provide venting. The piping described is not required for the cardox system to perform its intended function. Therefore, they are not subject to an AMR.

Based on its review, the staff found the applicant's response to RAI 2.3.3.B.13-19 acceptable because it adequately explained that all the similar components in question are not within the scope of license renewal under 10 CFR 54.4(a) and not subject to an AMR in under CFR 54.21(a) but that some were highlighted inadvertently on the LR drawing. The staff concludes that there is reasonable assurance that the components were correctly excluded from the scope of license renewal. Therefore, the staff's concern described in RAI 2.3.3.B.13-19 is resolved.

In RAI-2.3.3.B.13-20 dated November 17, 2004, the staff stated that drawing LR-46-0 depicts piping to a specific solenoid valve at a halon storage bottle in a portion of the halon system as within the scope of license renewal and subject to an AMR but not similar piping to other solenoid valves in the system, The applicant was asked to explain the apparent exclusion of this portion of the fire detection and protection system from requiring an AMR.

In its response by letter dated December 17, 2004, the applicant stated that drawing LR-46A-0 is incorrect and does not properly show the components within the scope of license renewal and subject to an AMR. The components in question should have been highlighted on the LR drawing showing them within the scope of license renewal and are subject to an AMR but inadvertently were not highlighted. The components in question have been included within the scope of license renewal and subject to an AMR.

Based on its review, the staff found the applicant's response to RAI 2.3.3.B.13-20 acceptable because it adequately explained that the components in question are within the scope of license renewal under 10 CFR 54.4(a) and subject to an AMR under 10 CFR 54.21(a) but inadvertently were not highlighted on the LR drawing. The staff concludes that there is reasonable assurance that the components were included correctly within the scope of license renewal and subject to an AMR. Therefore, the staff's concern described in RAI 2.3.3.B.13-20 is resolved.

In RAI-2.3.3.B.13-21 dated November 17, 2004, the staff stated that various LR drawings for the fire detection and protection system depict tamper switches associated with several water supply valves in the system as within the scope of license renewal and subject to an AMR but not other similar tamper switches in the foam and the cardox subsystems. Therefore, the staff requested that the applicant explain the apparent exclusion of these switches of the fire detection and protection system from requiring an AMR.

In its response by letter dated December 17, 2004, the applicant stated that the identified LR drawings are incorrect and do not properly show the components within the scope of license renewal and subject to an AMR. The applicant stated that the tamper switches in question should not have been highlighted on the LR drawing showing them within the scope of license renewal and subject to an AMR and that the components in question have been included within the scope of license renewal but are active components and therefore excluded from being subject to an AMR.

Based on its review, the staff found the applicant's response to RAI 2.3.3.B.13-21 acceptable because it adequately explained that the components in question are within the scope of license renewal under 10 CFR 54.4(a) but are active and therefore not subject to an AMR under CFR 54.21(a). The staff concludes that there is reasonable assurance that the components correctly were included within the scope of license renewal and excluded from being subject to an AMR. Therefore, the staff's concern described in RAI 2.3.3.B.13-21 is resolved.

In RAI-2.3.3.B.13-22 dated November 17, 2004, the staff stated USAR Section 9A.3.1.2.5, "Detailed Fire Hazard Analysis by Building," includes descriptions of drains and smoke removal for various buildings and also describes the floor drains provided to collect and remove fire water detection and protection system water discharge. This section states that drains are designed for sufficient capacity for this purpose. The applicant was asked whether drainage of such capacity was included within the scope of license renewal and subject to an AMR.

In its response by letter dated December 17, 2004, the applicant stated that floor drains are included in the floor and equipment drains system, which is within the scope of license renewal, for performing an NSR functional support intended function and are subject to an AMR. The applicant further stated that the scoping and screening LRA table for the floor and equipment drains system identifies these components and that the AMR table describes their environments.

Based on its review, the staff found the applicant's response to RAI 2.3.3.B.13-22 acceptable because it adequately explained that the floor and equipment drains of sufficient capacity to provide drainage from a discharge of the fire detection and protection system are within the scope of license renewal under 10 CFR 54.4(a) and subject to an AMR under 10 CFR 54.21(a). Therefore, the staff's concern described in RAI 2.3.3.B.13-22 is resolved.

In RAI 2.3.3.B.13-23 dated November 17, 2004, the staff stated that drawing LR-43C-0 depicts a portion of the dry pipe sprinkler system for the reactor building railroad access bay as being excluded from the scope of license renewal and requiring an AMR. The applicant was asked to explain the apparent exclusion of this portion of the fire detection and protection system from requiring an AMR.

In its response by letter dated December 17, 2004, the applicant stated that drawing LR-43C-0 is correct and properly shows the components in question excluded from the scope of license renewal and from requiring an AMR. The dry-pipe sprinkler system in the railroad access bay is not credited to meet the requirements of 10 CFR 50.48 and therefore not within the scope of license renewal because it has no license renewal intended functions. The dry-pipe sprinkler system in question is depicted correctly on its LR drawing and is not subject to an AMR.

In evaluating this response the staff found it incomplete and that review of LRA Section 2.3.3.B.13 could not be completed. Although it explains that the dry-pipe sprinkler system in question is not credited to meet the requirements of 10 CFR 50.48 the staff found this explanation contrary to the USAR, which includes the original NMP2 fire protection SER as CLB. The NMP2 USAR includes a description of this sprinkler system. The staff held a telephone conference with the applicant on January 25, 2005, to discuss information necessary to resolve the concern in RAI 2.3.3.B.13-23. The product of the telephone conference was an agreement by the applicant to transmit the required information by a follow-up letter.

By letter dated February 11, 2005, the applicant provided references in the USAR that describe the railroad access bay as within the standby gas treatment building. Further, USAR Table 9A.3-3 notes that there is no SR equipment in the railroad access bay. The USAR states that the only SR equipment in the standby gas treatment building is the standby gas treatment system units. These units are located in the standby gas treatment system rooms, which are separated from the railroad access bay by 3-hour fire walls and floors. Additionally, the USAR states that the safe shutdown analysis summarized in USAR Section 9B.8 does not identify any equipment located in the railroad access bay (Fire Area 4, Zone 242 NW) required for safe shutdown of the plant. Thus, the applicant stated that the dry-pipe sprinkler system in the railroad access bay does not perform or support any fire protection intended functions.

Based on its review, the staff found the applicant's response to RAI 2.3.3.B.13-23 including the information in the teleconference and letter dated February 11, 2005, acceptable because it adequately provided the CLB references that eliminate the dry-pipe sprinkler system in the railroad access bay from compliance with 10 CFR 50.48. Further, this elimination conforms to LRA Section 2.1.4.3.1, "Fire Protection (FP)," which describes the scoping criteria to determine whether systems and structures require inclusion within scope of license renewal under 10 CFR 54.4(a)(3). Therefore, the staff's concern described in RAI 2.3.3.B.13-23 is resolved.

In RAI 2.3.3.B.13-24 dated November 17, 2004, the staff stated that USAR 9.5.1.2.14 describes structural steel fire protection coating but that it was not clear from review of the LRA that the fire protection coatings for structural steel and steel embedded in fire barriers are included within the scope of license renewal and subject to an AMR. The applicant was asked to identify where in the LRA the fire protection coating for structural steel is evaluated or to explain its exclusion.

In its response by letter dated December 17, 2004, the applicant stated that the structural steel fire protection coatings are within the scope of license renewal and subject to an AMR. They are included in the scoping and screening table for the fire stops and seals commodity system of the LRA represented by the component type "Fire Wrap in Air." Further, the applicant explained that their aging management is described in the AMR table for this commodity system.

Based on its review, the staff found the applicant's response to RAI 2.3.3.B.13-24 acceptable because it adequately explained that fire protection coatings for structural steel are within the scope of license renewal under 10 CFR 54.4(a) and subject to an AMR under 10 CFR 54.21(a). Therefore, the staff's concern described in RAI 2.3.3.B.13-24 is resolved.

In RAI 2.3.3.B.13-25 dated November 17, 2005, the staff stated that USAR 9.5.1.2.16 describes criteria for fire resistance of interior finishes but that it was not clear from review of the LRA that interior finishes are included within the scope of license renewal. The applicant was asked to confirm that interior finishes are within the scope of license renewal and subject to an AMR or to explain their exclusion.

In its response by letter dated December 17, 2004, the applicant explained that the USAR states that noncombustible and fire-resistive building and interior finish materials are used wherever practical throughout the plant, particularly in structures containing safety-related systems and components. The applicant stated that the interior finishes, which consist of paint and floor coverings, serve no intended function and are not in scope for license renewal, and that the materials used to seal structural gaps and joints that have an intended function for fire protection can be found in the LRA Section 2.4.C.2, "Fire Stops and Seals."

In evaluating this response the staff found it incomplete and that review of original LRA Section 2.3.3.B.13 could not be completed. Although it explains that the interior finishes in question are not credited with meeting the requirements of 10 CFR 50.48 the staff found this explanation contrary to the USAR, which includes the original NMP2 fire protection SE as CLB. The USAR includes a description of the fire resistance of interior finishes. The staff held a telephone conference with the applicant on January 25, 2005, to discuss information necessary to resolve the concern in RAI 2.3.3.B.13-25. The product of the teleconference was an agreement by the applicant to transmit the required information by a follow-up letter.

By letter dated February 11, 2005, the applicant provided references in the USAR that describe the use and characteristics of interior finishes required for building construction. USAR Section 9.5.1.2.16 provides the required flame spread and smoke and fuel contribution rating applicable to structures containing SR systems and components. Further, USAR Section 9.5.1.4 satisfactorily evaluates the requirements of interior finishes against the basis for the original NMP2 fire protection SE and determines that they are in compliance. Additionally, the USAR and the SE do not describe finish materials as performing a fire barrier or fire proofing function. Thus, interior finish materials do not perform or support any fire protection intended functions.

Based on its review, the staff found the applicant's response to RAI 2.3.3.B.13-25, including the information in the teleconference and letter dated February 11, 2005, acceptable because it adequately provided the CLB references stating the use and characteristics of interior finishes required for building construction containing SR systems and components. The USAR and the staff SE do not describe finish materials as performing a fire barrier or fire proofing function. Thus, interior finish materials do not perform or support any fire protection intended functions. Further, this finding conforms to the original LRA Section 2.1.4.3.1, which describes the scoping criteria to determine whether systems and structures require inclusion within scope of license renewal under 10 CFR 54.4(a)(3). Therefore, the staff's concern described in RAI 2.3.3.B.13-25 is resolved.

In RAI 2.3.3.B.13-26 dated November 17, 2004, the staff stated that the original LRA Section 2.4.B does not include the condensate storage tank structure and the normal switchgear building as structures within the scope of license renewal. These structures are found in the USAR Appendix 9A as the fire protection licensing basis and thus should be considered within the scope of license renewal. Because these structures support fire protection intended functions, the applicant was asked to explain their apparent exclusion from requiring an AMR.

In its response by letter dated December 17, 2004, the applicant explained that the normal switchgear building and the condensate storage building are located in the protected area and considered to be nonessential yard structures. The applicant explained further that these structures do not meet any of the three criteria of 10 CFR 54.4; that they contain no SR equipment, equipment required for safe plant shutdown, or radioactive material; that fire protection equipment in these structures is for asset protection only; and that neither of these buildings is credited for 10 CFR 50.48 fire protection.

In evaluating this response the staff found it incomplete and that review of LRA Section 2.3.3.B.13 could not be completed. Although it explained that the normal switchgear building and the condensate storage building are not credited to meet the requirements of 10 CFR 50.48, the staff found this explanation contrary to the USAR, which includes the original NMP2 fire protection SER as CLB. The staff held a teleconference with the applicant on January 25, 2005, to discuss information necessary to resolve the concern in RAI 2.3.3.B.13-26. The result of the telephone conference was an agreement by the applicant to transmit the required information by a follow-up letter.

By letter dated February 11, 2005, the applicant provided references in the USAR that describe the attributes of both the condensate storage facility and the normal switchgear building. The condensate storage facility is shown in USAR Section 9.2.6.1 to include the condensate storage tanks, condensate storage tank building, and access way to the turbine building. The USAR states that condensate storage system is not required to effect or support safe shutdown of the reactor. USAR Section 9A.3.1.2.5.8 describes the normal switchgear building NSR switchgear and ventilation systems as not required to effect safe shutdown of the reactor. Further, the applicant explained that neither the condensate storage facility nor the normal switchgear building has any intended functions and have been excluded from the scope of license renewal under 10 CFR 54.4.

Based on its review, the staff found the applicant's response to RAI 2.3.3.B.13-26, including the information in the teleconference and letter dated February 11, 2005, acceptable because it adequately provided the CLB references stating that the condensate storage facility and the normal switchgear building do not have intended functions under 10 CFR 54.4 and have been correctly excluded from the scope of license renewal. Further, this exclusion conforms to LRA Section 2.1.4.3.1, which describes the scoping criteria to determine whether systems and structures require inclusion within scope of license renewal under 10 CFR 54.4(a)(3). Therefore, the staff's concern described in RAI 2.3.3.B.13-26 is resolved.

In RAI 2.3.3.B.13-27, dated November 17, 2004, the staff stated that USAR Section 9A.3.6.2.6 requires at least 350 gallons of fuel in the fire pump diesel fuel oil storage tank. Drawing LR-43A-0 shows air tubing and other components supplying the level indicating instrumentation for the fuel oil storage tank excluded from the scope of license renewal and requiring an AMR. Therefore, the staff requested that the applicant explain the apparent exclusion of these components from the scope of license renewal and from requiring an AMR.

In its response by letter dated December 17, 2004, the applicant explained that the LR drawing identifies the diesel fire pump fuel storage tank, which has a capacity of 650 gallons, as subject to an AMR and that lines running from the tank to the fuel storage tank level transmitter are also within the scope of license renewal and subject to an AMR. However, the fuel storage tank level indicator and the level switch in question are not subject to an AMR because they are active components.

Based on its review, the staff found the applicant's response to RAI 2.3.3.B.13-27 acceptable because it adequately explained that level instrumentation in question for the diesel fire pump fuel storage tank is within the scope of license renewal under 10 CFR 54.4(a) but are active components and are therefore not subject to an AMR under 10 CFR 54.21(a). Therefore, the staff's concern described in RAI 2.3.3.B.13-27 is resolved.

2.3B.3.13.3 Conclusion

The staff reviewed the ALRA, RAI responses, and accompanying scoping boundary drawings to determine whether any SSCs that should be within the scope of license renewal had not been identified by the applicant. No omissions were identified. In addition, the staff performed a review to determine whether any components that should be subject to an AMR had not been identified by the applicant. No omissions were identified. On the basis of its review, the staff concludes that there is reasonable assurance that the applicant had adequately identified the fire detection and protection system components that are within the scope of license renewal, as required by 10 CFR 54.4(a), and the fire detection and protection system components that are subject to an AMR, as required by 10 CFR 54.21(a)(1).

2.3B.3.14 NMP2 Floor and Equipment Drains System

2.3B.3.14.1 Summary of Technical Information in the Amended Application

In ALRA Section 2.3.3.B.14, the applicant described the floor and equipment drains system. The floor and equipment drains system collects, holds, monitors, and discharges drainage from floor and equipment drain subsystems from various buildings/areas and provides for the proper handling and disposal of radioactive and nonradioactive effluents. The floor and equipment drains system consists of the drywell and reactor building equipment drains/floor drains, the standby diesel generator building floor and equipment drains, the miscellaneous floor and equipment drains, the radwaste building floor and equipment drains, the auxiliary service building floor and equipment drains, the turbine building equipment and floor drains, and the turbine plant miscellaneous drains subsystems. Floor and equipment drain systems are designed to prevent contamination of the storm drain system with effluent from sumps containing radioactive or potentially radioactive drainage. The floor and equipment drain systems serving buildings that house SR equipment have sufficient capacity to prevent excessive drain buildup that could affect the operability of the equipment. Each equipment and floor drain sump receiving radioactive influent is lined with either stainless steel or fiberglass to prevent migration of its contents. Flow from floor and equipment drains that has no potential for radioactive contamination is discharged to the storm drainage system.

The floor and equipment drains system contains SR components that are relied upon to remain functional during and following DBEs. The failure of NSR SSCs in the floor and equipment drains system could potentially prevent the satisfactory accomplishment of an SR function. In addition, the floor and equipment drains system performs functions that support fire protection, EQ, and SBO.

The intended functions within the scope of license renewal include the following:

- maintains mechanical and structural integrity of NSR components to prevent spatial interactions with SR components

- provides removal and/or holdup of fission products

- provides pressure retaining boundary

- converts liquid into spray

- maintains mechanical and structural integrity of NSR components that provide structural support to attached SR components

- provides flow restriction

In ALRA Table 2.3.3.B.14-1, the applicant identified the following floor and equipment drains system component types that are within the scope of license renewal and subject to an AMR:

- bolting
- drain tank
- floor drains
- flow elements
- piping and fittings
- pumps
- orifices
- spray nozzle
- strainers
- valves

2-188

2.3B.3.14.2 Staff Evaluation

The staff reviewed ALRA Section 2.3.3.B.14 and USAR Section 9.3.3 using the evaluation methodology described in SER Section 2.3. The staff conducted its review in accordance with the guidance described in SRP-LR Section 2.3.

In conducting its review, the staff evaluated the system functions described in the ALRA and USAR in accordance with the requirements of 10 CFR 54.4(a) to verify that the applicant had not omitted from the scope of license renewal any components with intended functions delineated under 10 CFR 54.4(a). The staff then reviewed those components that the applicant had identified as being within the scope of license renewal to verify that the applicant had not omitted any passive and long-lived components that should be subject to an AMR in accordance with the requirements of 10 CFR 54.21(a)(1).

The staff's review of LRA Section 2.3.3.B.14 identified areas in which additional information was necessary to complete the review of the applicant's scoping and screening results. The applicant responded to the staff's RAIs as discussed below.

In RAI 2.3.3.B.14-1 dated November 19, 2004, the staff stated that LR drawings LR-63C, LR-63D, LR-63E, and LR-66B do not show in-scope flagging as depicted in the typical boundary flagging legend on each drawing. Both red-colored and black-colored piping and fittings are shown beyond the license renewal floor and equipment drains' blue flagging on the drawings. Therefore, the staff requested that the applicant explain why the black-colored piping and fittings are shown beyond the license renewal blue flagging. The staff also requested that the applicant discuss if the black-colored piping and fittings are within scope of license renewal under 10 CFR 54.4(a)(2), and if not, justify how their failure would not affect the pressure boundary function of the in-scope piping with which this piping connects.

In its response by letter dated December 22, 2004, the applicant stated:

> The AMR boundary flags were purposely drawn this way so as not to obscure the depiction of the floor on the drawings. In all cases, the AMR boundary is correctly shown by the components highlighted in red. NSR components of this system containing liquid in the Auxiliary Service Building, Control Room Building, Diesel Generator Building, Main Stack, Primary Containment Structure, Radwaste Building, Reactor Building (secondary containment), Screenwell Building, and Turbine Building are in-scope and subject to an AMR per criterion 10 CFR 54.4(a)(2). Per the convention adopted for the LR drawings, components in-scope for LR and subject to an AMR for criterion (a)(2) only are not identified in red on the LR drawings.

Based on its review, the staff found the applicant's response to RAI 2.3.3.B.14-1 acceptable because it adequately explained that the AMR boundary flags were purposely drawn as shown on the subject LR drawings so as not to obscure the depiction of the individual floor levels on the drawings. In addition the applicant stated that NSR components of the floor and equipment drains system containing liquid in the auxiliary service building, control room building, diesel generator building, main stack, primary containment structure, radwaste building, reactor building (secondary containment), screenwell building, and turbine building are within the scope of license renewal, subject to an AMR, and are shown therefore only in black on the drawings instead of in red. Therefore, the staff's concern described in RAI 2.3.3.B.14-1 is resolved.

In RAI 2.3.3.B.14-2 dated November 19, 2004, the staff requested that the applicant clarify information given on drawing LR-67A concerning SSCs in scope of license renewal. The staff stated in the RAI that drawing LR-67A shows the drywell equipment drain tank 1, associated discharge piping and fittings, and downstream valves and downstream equipment drain pumps in red and within blue flagging boundaries, indicating that these components are in scope for license renewal per 10 CFR 54.4(a)(3). However, the inlet piping and fittings to the drywell equipment drain tank 1 upstream valves and upstream drywell equipment drain cooler are shown in black, indicating these components are functionally outside the scope of license renewal. The staff also requested that the applicant identify the intended function of the portion of the system beyond the drywell equipment drain tank 1 that satisfies 10 CFR 54.4(a)(3) and explain how the function is performed without reliance on the inlet piping to the tank to be functional and within the scope of license renewal.

In its response by letter dated December 22, 2004, the applicant stated that, consistent with the description of equipment subject to an AMR in LRA Section 2.3.3.B.14, the identified liquid-filled inlet piping and components are within the scope of licence renewal and subject to an AMR per 10 CFR 54.4(a)(2) because they are located in the reactor building in the vicinity of SR components and, per the convention adopted for the LR drawings components within the scope of license renewal and subject to an AMR for 10 CFR 54.4 (a)(2) only, are not to be identified in red on the LR drawings.

In its ALRA dated July 14, 2005, the applicant provided the staff with a revised LR drawing which identifies all SSCs in scope of license renewal and subject to an AMR including those subject under 10 CFR 54.4(a)(2).

Based on review of the information submitted in the ALRA the staff's concern described in RAI 2.3.3.B.14-2 is resolved.

2.3B.3.14.3 Conclusion

The staff reviewed the ALRA, RAI responses, and accompanying scoping boundary drawings to determine whether any SSCs that should be within the scope of license renewal had not been identified by the applicant. No omissions were identified. In addition, the staff performed a review to determine whether any components that should be subject to an AMR had not been identified by the applicant. No omissions were identified. On the basis of its review, the staff concludes that there is reasonable assurance that the applicant had adequately identified the floor and equipment drains system components that are within the scope of license renewal, as required by 10 CFR 54.4(a), and the floor and equipment drains system components that are subject to an AMR, as required by 10 CFR 54.21(a)(1).

2.3B.3.15 NMP2 Generator Standby Lube Oil System

2.3B.3.15.1 Summary of Technical Information in the Amended Application

In ALRA Section 2.3.3.B.15, the applicant described the generator standby lube oil system. The generator standby lube oil system is designed to lubricate the engine bearings, turbocharger, and other moving parts of the emergency diesel generators. Additionally, this system preheats the oil, prelubricates the engine, warms the jacket water, cools the pistons, and keeps the inside of the engine clean by preventing rust and corrosion. The generator standby lube oil system also features a generator standby temperature system that preheats the lubricating oil and jacket water to enhance long-term engine reliability and first-try starting of the diesel engine. When the engine starts the circulating oil pump stops and the main engine-driven oil pump takes over. A thermostatic valve controls the oil temperature to the engine by regulating the flow to the oil cooler.

The generator standby lube oil system contains SR components that are relied upon to remain functional during and following DBEs. In addition, the generator standby lube oil system performs functions that support fire protection.

The intended functions within the scope of license renewal include the following:

- provides heat transfer
- provides pressure retaining boundary

In ALRA Table 2.3.3.B.15-1 the applicant identified the following generator standby lube oil system component types that are within the scope of license renewal and subject to an AMR:

- bolting
- filters/strainers
- heat exchangers
- piping and fittings
- pumps
- orifices
- sight glasses
- valves

2.3B.3.15.2 Staff Evaluation

The staff reviewed ALRA Section 2.3.3.B.15 and USAR Section 9.5.7 using the evaluation methodology described in SER Section 2.3. The staff conducted its review in accordance with the guidance described in SRP-LR Section 2.3.

In conducting its review, the staff evaluated the system functions described in the ALRA and USAR in accordance with the requirements of 10 CFR 54.4(a) to verify that the applicant had not omitted from the scope of license renewal any components with intended functions delineated under 10 CFR 54.4(a). The staff then reviewed those components that the applicant had identified as being within the scope of license renewal to verify that the applicant had not

omitted any passive and long-lived components that should be subject to an AMR in accordance with the requirements of 10 CFR 54.21(a)(1).

The staff's review of LRA Section 2.3.3.B.15 identified areas in which additional information was necessary to complete the review of the applicant's scoping and screening results. The applicant responded to the staff's RAIs as discussed below.

In RAI 2.3.3.B.15-1 dated November 19, 2004, the staff stated that LR drawing 104E-0 shows several sight glasses not highlighted, indicating that they are not subject to an AMR. Sight glasses are passive and long-lived components. Therefore, the staff requested that the applicant clarify or justify the exclusion of this component from requiring an AMR under 10 CFR 54.21(a)(1).

In its response by letter dated December 22, 2004, the applicant stated that the sight glasses in question were excluded inadvertently from within the scope of license renewal. They should have been identified as within the scope of license renewal and subject to an AMR, and highlighted in red on the drawing LR-104E-0. In its response, the applicant provided revisions to LRA Tables 2.3.3.B.15-1 and 3.3.3.B-15, and LRA Section 3.3.2.B.15 that included sight glasses with the intended function of pressure boundary as within the scope of license renewal. The applicant stated that the components connecting the sight glasses to the system piping are included with the "Piping and Fittings" component type.

Based on its review, the staff found the applicant's response to RAI 2.3.3.B.15-1 acceptable because the it clarified that the sight glasses had been excluded inadvertently from the scope of license renewal, stated that sight glasses on the LR drawing should be highlighted in red, and provided the LRA revisions to include the components in question. Therefore, the staff's concern described in RAI 2.3.3.B.15-1 is resolved.

In RAI 2.3.3.B.15-2 dated November 19, 2004, the staff stated that LR drawing 104E-0 shows the turbo lube oil pressure trip valves, PEV-18, for the Division I and II diesels and connecting tubing as not subject to an AMR. It appears that failure of this component and its connecting tubing could prevent its associated standby diesel generator from performing its intended function. Therefore, the staff requested that the applicant describe the function of this component and its effects on the intended function for the standby diesel generator. Further, the staff requested that the applicant include PEV-18 as a component requiring an AMR if found to have an intended function.

In its response by letter dated December 22, 2004, the applicant stated that AMR boundary depiction shown on drawing LR-104E-0 for valve PEV-18 and its associated tubing components is incorrect and does not show the components in question properly as within the scope of license renewal and subject to an AMR. The applicant further explained that valve PEV-18 is the tubocharger lube oil pressure trip valve and its function is to trip the diesel engine on low turbo oil pressure; as such PEV-18 (similar to valve PEV-14) and connecting tubing as well as 2EGS*HV118A, -B, -C, and -D are within the scope of license renewal and subject to an AMR. The applicant further stated that the LRA table already represents these components under the component types "Valves" and "Piping and Fittings."

Based on its review the staff found the applicant's response to RAI 2.3.3.B.15-2 acceptable because it adequately explained that (1) the components in question are within the scope of license renewal under 10 CFR 54.4(a) and subject to an AMR under 10 CFR 54.21(a) but inadvertently were not highlighted on the LR drawing and (2) the valves and tubing are already covered in the LRA tables. Therefore, the staff's concern described in RAI 2.3.3.B.15-2 is resolved.

In RAI 2.3.3.B.15-3 dated November 19, 2004, the staff stated that LR drawing 104E-0 shows three restricting orifices for the Division III diesel as subject to an AMR. LRA Table 2.3.3.B.15-1 includes the "Orifices" component type with a pressure boundary intended function. Restricting orifices also have a flow restriction function (as defined in LRA Table 2.0-1) that has not been identified in the LRA table. Therefore, the staff requested that the applicant confirm that the loss of flow restriction is not an intended function for restricting orifices in the generator lube oil system requiring an AMR.

In its response by letter dated December 22, 2004, the applicant stated that failure of the flow restriction function of the orifices in question would not prevent the system from performing its intended function. The applicant further stated that "pressure boundary" is the only intended function credited for these components as identified in the LRA Table 2.3.3.B.15-1.

Based on its review, the staff found the applicant's response to RAI 2.3.3.B.15-3 acceptable because it adequately explained that the components in question perform only a pressure boundary function and that the loss of the flow restriction function would not prevent the system from performing its intended function. Therefore, the staff's concern described in RAI 2.3.3.B.15-3 is resolved.

In RAI 2.3.3.B.15-4 dated November 19, 2004, the staff stated that drawing LR-104E-0 shows Y-strainers for the Division III diesel generator as subject to an AMR. LRA Table 2.3.3.B.15-1 includes the "Filters/Strainers" component type with an intended function of pressure boundary. Y-strainers also have a filtration function that has not been identified in the LRA table. Therefore, the staff requested that the applicant confirm that the loss of filtration is not an intended function for Y-strainers in the generator lube oil system requiring an AMR.

In its response by letter dated December 22, 2004, the applicant stated that the filtration function of the Y-strainers is to establish an "initial condition" for the generator lube oil system and that the failure of this function would not prevent the system from performing its intended function. The applicant further stated that pressure boundary is the only intended function credited for these components as identified in the LRA Table 2.3.3.B.15-1.

Based on its review, the staff found the applicant's response to RAI 2.3.3.B.15-4 acceptable because it adequately explained that the components in question perform only a pressure boundary function and that the loss of the filtration function would not prevent the system from performing its intended function. Therefore, the staff's concern described in RAI 2.3.3.B.15-4 is resolved.

In RAI 2.3.3.B.15-5 dated November 19, 2004, the staff stated that USAR Section 9.5.7 states that each standby diesel generator has an independent lubrication system to lubricate engine bearings and other moving parts. LR drawing 104E-0 shows a line labeled "To Engine Bearings" at location C8 for the Division I and II diesel generators as not requiring an AMR. This line appears to support the intended function for the generator standby lube oil system. Therefore, the staff requested that the applicant explain why this line is not subject to an AMR.

In its response by letter dated December 22, 2004, the applicant stated that drawing LR-104E-0 is incorrect. The lubrication to the engine bearing is within the scope of license renewal, subject to an AMR, and included in LRA Table 2.3.3.B.15-1 under the "Piping and Fittings" component type and as such should be highlighted in red on the drawing.

Based on its review, the staff found the applicant's response to RAI 2.3.3.B.15-5 acceptable because it adequately explained that the components in question are within the scope of license renewal under 10 CFR 54.4(a) and subject to an AMR under 10 CFR 54.21(a) but inadvertently were not highlighted on the LR drawing. The staff concludes that there is reasonable assurance that the components were included correctly within the scope of license renewal and subject to an AMR, as represented by the "Piping and Fittings" component type in LRA Table 2.3.3.B.15-1. Therefore, the staff's concern described in RAI 2.3.3.B.15-5 is resolved.

2.3B.3.15.3 Conclusion

The staff reviewed the ALRA, RAI responses, and accompanying scoping boundary drawings to determine whether any SSCs that should be within the scope of license renewal had not been identified by the applicant. No omissions were identified. In addition, the staff performed a review to determine whether any components that should be subject to an AMR had not been identified by the applicant. No omissions were identified. On the basis of its review, the staff concludes that there is reasonable assurance that the applicant had adequately identified the generator standby lube oil system components that are within the scope of license renewal, as required by 10 CFR 54.4(a), and the generator standby lube oil system components that are subject to an AMR, as required by 10 CFR 54.21(a)(1).

2.3B.3.16 NMP2 Glycol Heating System (Removed)

In ALRA Section 2.3.3.B.16, the applicant stated that the glycol heating system has been removed from the scope of license renewal since it has been determined that it does not meet any of the criteria of 10 CFR 54.4. The original LRA included this system within the scope of license renewal in accordance with 10 CFR 54.4(a)(2). However, based upon the applicant's detailed explanation that this system is not credited for mitigation of any CLB event, this system does not contain any SR/NSR interfaces, nor does this system introduce any spatial interactions with SR SSCs, the staff agrees with the applicant that the glycol heating system is not within the scope of license renewal.

2.3B.3.17 NMP2 Hot Water Heating System

2.3B.3.17.1 Summary of Technical Information in the Amended Application

In ALRA Section 2.3.3.B.17, the applicant described the hot water heating system. The hot water heating system functions with the glycol heating system to heat outdoor makeup air used for ventilation. Hot water is generated from steam and circulated through glycol heat exchangers. The hot water heating system is equipped with piping connections to allow this system to be connected to a temporary hot water heating plant. This is only used if the hot water heating system is not available and glycol heating in the RB is needed. The system is closed loop. Depending upon the operating mode, electric boiler, auxiliary steam, or extraction steam is supplied to the shell side of both the building heating auxiliary heat exchangers and the intermediate heat exchangers.

The failure of NSR SSCs in the hot water heating system could potentially prevent the satisfactory accomplishment of an SR function.

The intended function within the scope of license renewal is to maintain mechanical and structural integrity to prevent spatial interactions.

In ALRA Table 2.3.3.B.17-1 the applicant identified the following hot water heating system component types that are within the scope of license renewal and subject to an AMR:

- bolting
- piping and fittings
- valves

2.3B.3.17.2 Staff Evaluation

The staff reviewed ALRA Section 2.3.3.B.17 and USAR Section 9.4.12 using the evaluation methodology described in SER Section 2.3. The staff conducted its review in accordance with the guidance described in SRP-LR Section 2.3.

In conducting its review, the staff evaluated the system functions described in the ALRA and USAR in accordance with the requirements of 10 CFR 54.4(a) to verify that the applicant had not omitted from the scope of license renewal any components with intended functions delineated under 10 CFR 54.4(a). The staff then reviewed those components that the applicant had identified as being within the scope of license renewal to verify that the applicant had not omitted any passive and long-lived components that should be subject to an AMR in accordance with the requirements of 10 CFR 54.21(a)(1).

The staff's review of LRA Section 2.3.3.B.17 identified areas in which additional information was necessary to complete the review of the applicant's scoping and screening results. The applicant responded to the staff's RAIs as discussed below.

In RAI 2.3.3.B.17-1 dated November 19, 2004, the staff requested that the applicant provide information to describe the interface between the reactor water and the hot water system that precludes the interface from being a pressure boundary. The staff stated that the system description section of LRA Section 2.3.3.B.17 indicates that reactor water may be supplied to the shell side of the building heating auxiliary heat exchangers and the intermediate heat exchangers. No LR drawings were provided for this system. LRA Table 2.3.3.B.17 indicates that NSR piping, fittings, and equipment with the intended function of preventing failure from affecting safety-related equipment are subject to an AMR.

The applicant stated in the response by letter dated December 22, 2004, that the description in LRA Section 2.3.3.B.17 is incorrect, that reactor water does not interface with the hot water heating system, and that LRA Section 2.3.3.B.17 will be revised to remove "reactor water" and replace it with "auxiliary steam."

In its ALRA dated July 14, 2005, the applicant provided the staff with a revised LRA system description and LR drawing which identify SSCs within the scope of license renewal and subject to an AMR under 10 CFR 54.4. The applicant also provided clarification on the component groups included in LRA Table 2.3.3.B.17-1 to address the concern in RAI 2.3.3.B.17-1. Therefore, the staff's concern described in RAI 2.3.3.B17-1 is resolved.

In RAI 2.3.3.B.17-2 dated November 19, 2004, the staff stated that the system description section of LRA Section 2.3.3.B.17 states that components subject to an AMR include the NSR piping, fittings, and equipment containing liquid in the control room building, reactor building (secondary containment), radwaste building, screenwell building, standby gas treatment building, and turbine building. No LR drawings were provided for this system. Therefore, the staff requested that the applicant provide information that describes the boundaries of this system and confirms that there are no other components subject to an AMR.

In its response by letter dated December 22, 2004, the applicant stated, "Consistent with LR drawing convention, marked-up LR drawings were not provided for systems where the only system intended function was to meet the 10 CFR 54.4(a)(2) criterion." The applicant further stated that the components subject to an AMR for this system include the NSR piping, fittings, and equipment containing liquid or steam physically located in the control room building, radwaste building, reactor containment (secondary containment), screenwell building, standby gas treatment building, and turbine building. The applicant also referred to the diagram provided with its response to RAI 2.3.3.B.16-1 and stated that the system description from LRA Section 2.3.3.B.17 in conjunction with the diagram provides an adequate description of the components that are subject to an AMR.

In evaluating the applicant's response the staff found it incomplete and that review of LRA Section 2.3.3.B.17 could not be completed because the applicant did not describe the SR components in the radwaste building with which hot water heating system components can interact. The staff held a teleconference with the applicant on January 27, 2005, to discuss information necessary to resolve the concern in RAI 2.3.3.B.17-2. The result of this teleconference was an agreement by the applicant to transmit the required information in a follow-up letter.

In its follow-up response dated January 31, 2005, the applicant stated that the methodology employed for determining those NSR systems and components in the radwaste building within the scope of license renewal was based on the building in which the system/component was located, that this approach is conservative in that it brings into scope many more components than would be required if detailed walkdowns were performed, and that as such NMPNS did not specifically identify the SR components with which the hot water heating system could interact because the entire system within the radwaste building is subject to an AMR under 10 CFR 54.21(a).

Prior to the issuing of its ALRA, on July 14, 2005, the applicant performed a new scoping of the system, using the methdology from ALRA Section 2.1. In its ALRA, the applicant stated that the components subject to AMR for this system include the Hot water heating system NSR piping, fittings, and valves in the reactor building that are in the vicinity of the liquid poison tanks. The remaining portions of this system are not credited for the mitigation of CLB events, do not contain any SR/NSR interfaces, and are not located in the vicinity of SR SSCs. Therefore, those portions are not within the scope of license renewal.

Based on its review, the staff found the applicant's response to RAI 2.3.3.B.17-2, acceptable because it adequately explained what NSR piping, fittings, and equipment is subject to an AMR because of the possibility of spatial interaction with SR components. Therefore, the staff's concern described in RAI 2.3.3.B.17-2 is resolved.

2.3B.3.17.3 Conclusion

The staff reviewed the ALRA, RAI responses, and accompanying scoping boundary drawings to determine whether any SSCs that should be within the scope of license renewal had not been identified by the applicant. No omissions were identified. In addition, the staff performed a review to determine whether any components that should be subject to an AMR had not been identified by the applicant. No omissions were identified. On the basis of its review, the staff concludes that there is reasonable assurance that the applicant had adequately identified the hot water heating system components that are within the scope of license renewal, as required by 10 CFR 54.4(a), and the hot water heating system components that are subject to an AMR, as required by 10 CFR 54.21(a)(1).

2.3B.3.18 NMP2 Makeup Water System

2.3B.3.18.1 Summary of Technical Information in the Amended Application

In ALRA Section 2.3.3.B.18, the applicant described the makeup water system. The makeup water system is designed to provide demineralized makeup water for the TB closed loop cooling water system and the RB closed loop cooling water system. The makeup water system produces demineralized water by removing dissolved and suspended solids from city water using a portable demineralizer. The makeup water system also stores and distributes demineralized water from the makeup water treatment system. The system consists of the makeup water treatment system and the makeup water storage and transfer system. Additionally, the makeup water system meets plant requirements for demineralized water, including the suppression pool and the spent fuel pool.

The failure of NSR SSCs in the makeup water system could potentially prevent the satisfactory accomplishment of an SR function.

The intended functions within the scope of license renewal include the following:

- maintains mechanical and structural integrity of NSR components to prevent spatial interactions with SR components
- maintains mechanical and structural integrity of NSR components that provide structural support to attached SR components

In ALRA Table 2.3.3.B.18-1 the applicant identified the following makeup water system component types that are within the scope of license renewal and subject to an AMR:

- bolting
- piping and fittings
- valves

2.3B.3.18.2 Staff Evaluation

The staff reviewed ALRA Section 2.3.3.B.18 and the USAR Section 1.2.10.9 using the evaluation methodology described in SER Section 2.3. The staff conducted its review in accordance with the guidance described in SRP-LR Section 2.3.

In conducting its review, the staff evaluated the system functions described in the ALRA and USAR in accordance with the requirements of 10 CFR 54.4(a) to verify that the applicant had not omitted from the scope of license renewal any components with intended functions delineated under 10 CFR 54.4(a). The staff then reviewed those components that the applicant had identified as being within the scope of license renewal to verify that the applicant had not omitted any passive and long-lived components that should be subject to an AMR in accordance with the requirements of 10 CFR 54.21(a)(1).

2.3B.3.18.3 Conclusion

The staff reviewed the ALRA to determine whether any SSCs that should be within the scope of license renewal had not been identified by the applicant. No omissions were identified. In addition, the staff performed a review to determine whether any components that should be subject to an AMR had not been identified by the applicant. No omissions were identified. On the basis of its review, the staff concludes that there is reasonable assurance that the applicant had adequately identified the makeup water system components that are within the scope of license renewal, as required by 10 CFR 54.4(a), and the makeup water system components that are subject to an AMR, as required by 10 CFR 54.21(a)(1).

2.3B.3.19 NMP2 Neutron Monitoring System

2.3B.3.19.1 Summary of Technical Information in the Amended Application

In ALRA Section 2.3.3.B.19, the applicant described the neutron monitoring system. The neutron monitoring system is designed to provide neutron flux level monitoring of the reactor in three separate ranges, source range monitoring, intermediate range monitoring, and power range monitoring. It is used to monitor and aid the operator in controlling the reactor from startup through full power, inputs to the reactor manual control system (not within scope of license renewal) to initiate rod blocks if preset flux limits are exceeded, and it inputs signals to the reactor protection system to initiate a scram if limits are exceeded.

The neutron monitoring system has five subsystems. The source range monitoring subsystem measures the flux from startup through criticality. The intermediate range monitoring subsystem overlaps the source range monitoring subsystem and extends well into the power range. The power range is monitored by detectors that make up the local power range monitor subsystem. The average power range monitor subsystem is composed of core-wide sets of local power range monitor detectors that are averaged to provide a core average neutron flux. The traversing in-core probe subsystem provides a means for calibrating the local power range monitor subsystem.

The neutron monitoring system contains SR components that are relied upon to remain functional during and following DBEs. In addition, the neutron monitoring system performs functions that support fire protection, EQ, ATWS, and SBO.

The intended function within the scope of license renewal is to provide pressure retaining boundary.

In ALRA Table 2.3.3.B.19-1, the applicant identified the following neutron monitoring system component types that are within the scope of license renewal and subject to an AMR:

- bellows
- bolting
- piping and fittings
- valves

2.3B.3.19.2 Staff Evaluation

The staff reviewed ALRA Section 2.3.3.B.19 and the USAR using the evaluation methodology described in SER Section 2.3. The staff conducted its review in accordance with the guidance described in SRP-LR Section 2.3.

In conducting its review, the staff evaluated the system functions described in the ALRA and USAR in accordance with the requirements of 10 CFR 54.4(a) to verify that the applicant had not omitted from the scope of license renewal any components with intended functions delineated under 10 CFR 54.4(a). The staff then reviewed those components that the applicant had identified as being within the scope of license renewal to verify that the applicant had not omitted any passive and long-lived components that should be subject to an AMR in accordance with 10 CFR 54.21(a)(1).

2.3B.3.19.3 Conclusion

The staff reviewed the ALRA to determine whether any SSCs that should be within the scope of license renewal had not been identified by the applicant. No omissions were identified. In addition, the staff performed a review to determine whether any components that should be subject to an AMR had not been identified by the applicant. No omissions were identified. On the basis of its review, the staff concludes that there is reasonable assurance that the applicant had adequately identified the neutron monitoring system components that are within the scope of license renewal, as required by 10 CFR 54.4(a), and the neutron monitoring system components that are subject to an AMR, as required by 10 CFR 54.21(a)(1).

2.3B.3.20 NMP2 Primary Containment Purge System

2.3B.3.20.1 Summary of Technical Information in the Amended Application

In ALRA Section 2.3.3.B.20, the applicant described the primary containment purge system. The primary containment purge system is designed to inert the primary containment with nitrogen, and to limit oxygen and hydrogen concentrations in the primary containment and ensure a combustible atmosphere does not occur following a LOCA. The primary containment purge system is also designed to de-inert and ventilate the primary containment during plant shutdown for the purpose of drywell entry. The primary containment purge system operates as a subsystem of the reactor building HVAC system. Inerting the primary containment is accomplished by feed and bleed. To inert, nitrogen gas from the nitrogen system is fed into the drywell or suppression chamber. Air is exhausted into and processed by the standby gas treatment system before it is discharged through the main stack.

The primary containment purge system contains SR components that are relied upon to remain functional during and following DBEs. The failure of NSR SSCs in the primary containment purge system could potentially prevent the satisfactory accomplishment of an SR function. In addition, the primary containment purge system performs functions that support EQ.

The intended functions within the scope of license renewal include the following:

- provides filtration

- provides removal and/or holdup of fission products

- provides pressure retaining boundary

- maintains mechanical and structural integrity of NSR components that provide structural support to attached SR components

In ALRA Table 2.3.3.B.20-1, the applicant identified the following primary containment purge system component types that are within the scope of license renewal and subject to an AMR:

- bolting
- debris screens
- flow element
- piping and fittings
- valves

2.3B.3.20.2 Staff Evaluation

The staff reviewed ALRA Section 2.3.3.B.20 and the USAR Section 9.4.2.2.2 using the evaluation methodology described in SER Section 2.3. The staff conducted its review in accordance with the guidance described in SRP-LR Section 2.3.

In conducting its review, the staff evaluated the system functions described in the ALRA and USAR in accordance with the requirements of 10 CFR 54.4(a) to verify that the applicant had not omitted from the scope of license renewal any components with intended functions delineated under 10 CFR 54.4(a). The staff then reviewed those components that the applicant had identified as being within the scope of license renewal to verify that the applicant had not omitted any passive and long-lived components that should be subject to an AMR in accordance with the requirements of 10 CFR 54.21(a)(1).

2.3B.3.20.3 Conclusion

The staff reviewed the ALRA to determine whether any SSCs that should be within the scope of license renewal had not been identified by the applicant. No omissions were identified. In addition, the staff performed a review to determine whether any components that should be subject to an AMR had not been identified by the applicant. No omissions were identified. On

the basis of its review, the staff concludes that there is reasonable assurance that the applicant had adequately identified the primary containment purge system components that are within the scope of license renewal, as required by 10 CFR 54.4(a), and the primary containment purge system components that are subject to an AMR, as required by 10 CFR 54.21(a)(1).

2.3B.3.21 NMP2 Process Sampling System

2.3B.3.21.1 Summary of Technical Information in the Amended Application

In ALRA Section 2.3.3.B.21, the applicant described the process sampling system. The process sampling system is designed to monitor selected plant process streams and provide grab sample points to back-up the continuous analyzers and allow laboratory analysis of other process streams. The process sampling system is a water chemistry analysis system involving multipoint sample panels and grab sample sinks in the RB, TB, and radwaste building (RWB). Miscellaneous sample points are provided on individual process systems where needed.

The process sampling system consists of the following subsystems: post-accident sampling system, RWB sampling system, reactor plant sampling system, and turbine plant sampling system. The post-accident sampling system is designed to obtain representative liquid and gas samples from within the primary containment for radiological analysis in association with the possible consequences of a LOCA. The radwaste building sampling system is used for obtaining grab samples for monitoring the radioactive liquid waste management and radwaste auxiliary steam system drain coolers. The reactor plant sampling system monitors the quality of reactor coolant and various reactor plant fluids. The turbine plant sampling system monitors the quality of reactor grade water flowing in the TB.

The failure of NSR SSCs in the process sampling system could potentially prevent the satisfactory accomplishment of an SR function.

The intended functions within the scope of license renewal include the following:

* maintains mechanical and structural integrity of NSR components to prevent spatial interactions with SR components
* provides removal and/or holdup of fission products

In ALRA Table 2.3.3.B.21-1, the applicant identified the following process sampling system component types that are within the scope of license renewal and subject to an AMR:

* bolting
* flow indicators
* heat exchangers
* piping and fittings
* valves

2-201

2.3B.3.21.2 Staff Evaluation

The staff reviewed ALRA Section 2.3.3.B.21 and USAR Sections 1.2.10.7, 1.10.II.B.3, and 9.3.2 using the evaluation methodology described in SER Section 2.3. The staff conducted its review in accordance with the guidance described in SRP-LR Section 2.3.

In conducting its review, the staff evaluated the system functions described in the ALRA and USAR in accordance with the requirements of 10 CFR 54.4(a) to verify that the applicant had not omitted from the scope of license renewal any components with intended functions delineated under 10 CFR 54.4(a). The staff then reviewed those components that the applicant had identified as being within the scope of license renewal to verify that the applicant had not omitted any passive and long-lived components that should be subject to an AMR in accordance with the requirements of 10 CFR 54.21(a)(1).

The staff's review of LRA Section 2.3.3.B.21 identified an area in which additional information was necessary to complete the review of the applicant's scoping and screening results. The applicant responded to the staff's RAI as discussed below.

In RAI 2.3.3.B.21-1 dated November 19, 2004, the staff requested that the applicant clarify drawing inconsistencies that the staff encountered in its review. The staff stated that LR drawing LRA-31 sheet D (location G7) and sheet E (location D8) show valves 2-RHS-SOV-35A and 2-RHS-SOV-35B highlighted in red and within the residual heat removal system AMR boundary flags. However, LR drawing LR-17 sheet G (locations I1, K1) does not show these valves within the AMR boundary of the residual heat removal system. The staff requested that the applicant confirm that valves 2-RHS-SOV-35A and 2-RHS-SOV-35B are within the scope of license renewal and subject to an AMR or explain the discrepancy between the LR drawings.

In its response by letter dated December 22, 2004, the applicant stated that valves 2RHS*SOV35A and 2RHS*SOV35B are SR within the scope of license renewal and subject to an AMR, that drawing LR-17G for these valves and their associated piping should be highlighted in red consistent with the depiction of the respective valves on drawings LR-31D and LR-31E, and that at continuation locations I1 and K1 on drawing LR-17G the continuation flags should be marked as "LR-31D" and "LR-31E" as appropriate and include LR continuation flags showing "LR-RHS" in both directions.

In its ALRA dated July 14, 2005, the applicant provided the staff with revised LR drawings correcting the inconsistencies and accurately depicting all the components subject to an AMR, including those subject under 10 CFR 54.4(a)(2). Therefore, the staff's concern described in RAI 2.3.3.B.21-1 is resolved.

2.3B.3.21.3 Conclusion

The staff reviewed the ALRA, RAI responses, and accompanying scoping boundary drawings to determine whether any SSCs that should be within the scope of license renewal had not been identified by the applicant. No omissions were identified. In addition, the staff performed a review to determine whether any components that should be subject to an AMR had not been identified by the applicant. No omissions were identified. On the basis of its review, the staff concludes that there is reasonable assurance that the applicant had adequately identified the process sampling system components that are within the scope of license renewal, as required by 10 CFR 54.4(a), and the process sampling system components that are subject to an AMR, as required by 10 CFR 54.21(a)(1).

2.3B.3.22 NMP2 Radiation Monitoring System

2.3B.3.22.1 Summary of Technical Information in the Amended Application

In ALRA Section 2.3.3.B.22, the applicant described the radiation monitoring system. The radiation monitoring system is designed to initiate appropriate manual or automatic protective action to limit the potential release of radioactive materials from the reactor vessel, primary and secondary containment, and fuel storage areas if predetermined radiation levels are exceeded in major process/effluent streams and to provide main control room personnel with radiation level indication throughout the course of an accident.

The radiation monitoring system consists of a computer-based digital radiation monitoring system, a computer-based gaseous effluent monitoring system, and the main steam line radiation monitors. The digital radiation monitoring system measures, evaluates, and reports radioactivity in process streams and liquid effluents and annunciates and/or initiates an automatic control function for abnormal system or plant operating conditions. The gaseous effluent monitoring system measures, evaluates, and reports radioactivity in gaseous effluents. It also provides annunciation if release levels approach limits specified in the offsite dose calculation manual. The gaseous effluent monitoring system also provides real time noble gas isotopic analysis and continuous iodine and particulate sample collection for main stack, RWB, and RB vent releases. The main steam line radiation monitoring system monitors the gamma radiation level exterior to the main steam lines. In the event of a gross release of fission products from the core this monitoring system provides annunciation in the control room.

The radiation monitoring system contains SR components that are relied upon to remain functional during and following DBEs. The failure of NSR SSCs in the radiation monitoring system could potentially prevent the satisfactory accomplishment of an SR function.

The intended functions within the scope of license renewal include the following:

- provides filtration
- provides pressure retaining boundary
- provides flow restriction

In ALRA Table 2.3.3.B.22-1, the applicant identified the following radiation monitoring system component types that are within the scope of license renewal and subject to an AMR:

- bolting
- filters
- flow elements
- piping and fittings
- pumps
- valves

2.3B.3.22.2 Staff Evaluation

The staff reviewed ALRA Section 2.3.3.B.22 and USAR Sections 11.5.2 and 12.3.4 using the evaluation methodology described in SER Section 2.3. The staff conducted its review in accordance with the guidance described in SRP-LR Section 2.3.

In conducting its review, the staff evaluated the system functions described in the ALRA and USAR in accordance with the requirements of 10 CFR 54.4(a) to verify that the applicant had not omitted from the scope of license renewal any components with intended functions delineated under 10 CFR 54.4(a). The staff then reviewed those components that the applicant had identified as being within the scope of license renewal to verify that the applicant had not omitted any passive and long-lived components that should be subject to an AMR in accordance with the requirements of 10 CFR 54.21(a)(1).

The staff's review of LRA Section 2.3.3.B.22 identified an area in which additional information was necessary to complete the evaluation of the applicant's scoping and screening results. The applicant responded to the staff's RAI as discussed below.

In RAI 2.3.3.B.22-1 dated November 19, 2004, the staff stated that the applicant did not identify the radiation monitoring system components within the scope of license renewal under 10 CFR 54.4(a)(1) and (2). Furthermore, an LR drawing for the NMP2 radiation monitoring system was not provided to show the portions of this system containing components within the scope of license renewal. Therefore, the staff requested that the applicant identify the components of the radiation monitoring system within the scope of license renewal under 10 CFR 54.4(a)(1) and (2) and justify the exclusion of these components from being subject to an AMR under 10 CFR 54.21(a)(1).

In its response by letter dated December 22, 2004, the applicant stated that SR radiation monitors and their inclusive mechanical components are in-scope for LR and subject to an AMR for a "Pressure Boundary" intended function that the subject components that perform the LR intended function will be identified, and that AMR of those components will be performed. The applicant stated that LRA revisions to incorporate the AMR results and any other associated LRA changes would be submitted to the staff.

In its ALRA dated July 14, 2005, the applicant provided the staff with revised LR drawings accurately depicting all the components subject to an AMR, including those subject under 10 CFR 54.4(a)(2). The applicant also added ALRA Table 2.3.3.B.22-1 to provide a list of component types requiring an AMR and their intended functions. Therefore, the staff's concern described in RAI 2.3.3.B.22-1 is resolved.

2.3B.3.22.3 Conclusion

The staff reviewed the ALRA, RAI response, and accompanying scoping boundary drawings to determine whether any SSCs that should be within the scope of license renewal had not been identified by the applicant. No omissions were identified. In addition, the staff performed a review to determine whether any components that should be subject to an AMR had not been identified by the applicant. No omissions were identified. On the basis of its review, the staff concludes that there is reasonable assurance that the applicant had adequately identified the radiation monitoring system components that are within the scope of license renewal, as required by 10 CFR 54.4(a), and the radiation monitoring system components that are subject to an AMR, as required by 10 CFR 54.21(a)(1).

2.3B.3.23 NMP2 Reactor Building Closed Loop Cooling Water System

2.3B.3.23.1 Summary of Technical Information in the Amended Application

In ALRA Section 2.3.3.B.23, the applicant described the reactor building closed loop cooling water system. The RBCLC water system is designed to remove heat from various reactor auxiliary equipment located in the primary containment, RB, and TB. The RBCLC water system is cooled by the service water system and makeup water is supplied from the makeup water system. During normal plant operation the system provides an intermediate barrier between systems containing radioactive products and the service water system, which precludes a direct release of radioactive products into the environment.

The RBCLC water system contains SR components that are relied upon to remain functional during and following DBEs. The failure of NSR SSCs in the RBCLC water system could potentially prevent the satisfactory accomplishment of an SR function. In addition, the RBCLC water system performs functions that support fire protection and EQ.

The intended functions within the scope of license renewal include the following:

- maintains mechanical and structural integrity of NSR components to prevent spatial interactions with SR components

- provides structural support to NSR components whose failure could prevent accomplishment of SR function(s)

- provides pressure retaining boundary

- maintains mechanical and structural integrity of NSR components that provide structural support to attached SR components

In ALRA Table 2.3.3.B.23-1, the applicant identified the following RBCLC water system component types that are within the scope of license renewal and subject to an AMR:

- bolting
- flow element
- heat exchangers
- piping and fittings
- unit coolers
- valves

2.3B.3.23.2 Staff Evaluation

The staff reviewed ALRA Section 2.3.3.B.23 and USAR Section 9.2.2 using the evaluation methodology described in SER Section 2.3. The staff conducted its review in accordance with the guidance described in SRP-LR Section 2.3.

In conducting its review, the staff evaluated the system functions described in the ALRA and USAR in accordance with the requirements of 10 CFR 54.4(a) to verify that the applicant had not omitted from the scope of license renewal any components with intended functions delineated under 10 CFR 54.4(a). The staff then reviewed those components that the applicant had identified as being within the scope of license renewal to verify that the applicant had not omitted any passive and long-lived components that should be subject to an AMR in accordance with the requirements of 10 CFR 54.21(a)(1).

2.3B.3.23.3 Conclusion

The staff reviewed the ALRA to determine whether any SSCs that should be within the scope of license renewal had not been identified by the applicant. No omissions were identified. In addition, the staff performed a review to determine whether any components that should be subject to an AMR had not been identified by the applicant. No omissions were identified. On the basis of its review, the staff concludes that there is reasonable assurance that the applicant had adequately identified the RBCLC water system components that are within the scope of license renewal, as required by 10 CFR 54.4(a), and the RBCLC water system components that are subject to an AMR, as required by 10 CFR 54.21(a)(1).

2.3B.3.24 NMP2 Reactor Building HVAC System

2.3B.3.24.1 Summary of Technical Information in the Amended Application

In ALRA Section 2.3.3.B.24, the applicant described the reactor building HVAC system. The reactor building HVAC system is designed to remove heat generated within the drywell and maintain ambient temperature within design limits, thus providing an environment that ensures optimum performance of equipment. Additionally, the reactor building HVAC system is an alternative system for venting the primary containment to the atmosphere if necessary.

The reactor building HVAC system consists of the following subsystems: drywell cooling, primary containment purge, and all other RB areas. The drywell cooling system conditions the air inside the drywell where unit coolers control drywell temperature and pressure for all other reactor building areas subsystems. The supply ventilation air handling unit assembly consists of an air intake, prefilter, filter, heating coil, cooling coil, dampers, controls, and supply fans. The system operates in both a normal operation mode and an emergency operation mode.

The reactor building HVAC system contains SR components that are relied upon to remain functional during and following DBEs. In addition, the reactor building HVAC system performs functions that support fire protection and EQ.

The intended functions within the scope of license renewal include the following:

- provides rated fire barrier
- provides heat transfer
- provides pressure retaining boundary

In ALRA Table 2.3.3.B.24-1, the applicant identified the following reactor building HVAC system component types that are within the scope of license renewal and subject to an AMR:

- bolting
- ducting
- filters/strainers
- flow elements
- piping and fittings
- pumps
- radiation sample points
- unit coolers
- valves and dampers (includes fire dampers)

2.3B.3.24.2 Staff Evaluation

The staff reviewed ALRA Section 2.3.3.B.24 and USAR Section 9.4.2 using the evaluation methodology described in SER Section 2.3. The staff conducted its review in accordance with the guidance described in SRP-LR Section 2.3.

In conducting its review, the staff evaluated the system functions described in the ALRA and USAR in accordance with the requirements of 10 CFR 54.4(a) to verify that the applicant had not omitted from the scope of license renewal any components with intended functions delineated under 10 CFR 54.4(a). The staff then reviewed those components that the applicant had identified as being within the scope of license renewal to verify that the applicant had not omitted any passive and long-lived components that should be subject to an AMR in accordance with the requirements of 10 CFR 54.21(a)(1).

2.3B.3.24.3 Conclusion

The staff reviewed the ALRA to determine whether any SSCs that should be within the scope of license renewal had not been identified by the applicant. No omissions were identified. In addition, the staff performed a review to determine whether any components that should be subject to an AMR had not been identified by the applicant. No omissions were identified. On the basis of its review, the staff concludes that there is reasonable assurance that the applicant had adequately identified the reactor building HVAC system components that are within the scope of license renewal, as required by 10 CFR 54.4(a), and the reactor building HVAC system components that are subject to an AMR, as required by 10 CFR 54.21(a)(1).

2.3B.3.25 NMP2 Reactor Water Cleanup System

2.3B.3.25.1 Summary of Technical Information in the Amended Application

In ALRA Section 2.3.3.B.25, the applicant described the RWCU system. The purpose of the RWCU system is to maintain high reactor water quality and remove excess reactor coolant from the RPV during all modes of plant operation. High water quality is maintained to minimize the fouling of heat transfer surfaces and limit impurities available for neutron activation. The RWCU system provides the means to maintain water chemistry within the limits outlined in Regulatory Guide (RG) 1.56, Revision 1. The RWCU system recirculates a portion of reactor coolant through a filter demineralizer to remove particulate and dissolved impurities from the reactor coolant. It also removes excess coolant from the reactor system under controlled conditions. The major components of the RWCU system are located outside the drywell.

The RWCU system contains SR components that are relied upon to remain functional during and following DBEs. The failure of NSR SSCs in the RWCU system could potentially prevent the satisfactory accomplishment of an SR function. In addition, the RWCU system performs functions that support fire protection, ATWS, and SBO.

The intended functions within the scope of license renewal include the following:

- maintains mechanical and structural integrity of NSR components to prevent spatial interactions with SR components

- provides removal and/or holdup of fission products

- provides pressure retaining boundary

- maintains mechanical and structural integrity of NSR components that provide structural support to attached SR components

- provides flow restriction

In ALRA Table 2.3.3.B.25-1, the applicant identified the following RWCU system component types that are within the scope of license renewal and subject to an AMR:

- bolting
- filter/strainer
- flow elements
- heat exchanger
- piping and fittings
- orifices
- valves
- pumps

2.3B.3.25.2 Staff Evaluation

The staff reviewed ALRA Section 2.3.3.B.25 and the USAR Section 5.4.8 using the evaluation methodology described in SER Section 2.3. The staff conducted its review in accordance with the guidance described in SRP-LR Section 2.3.

In conducting its review, the staff evaluated the system functions described in the ALRA and USAR in accordance with the requirements of 10 CFR 54.4(a) to verify that the applicant had not omitted from the scope of license renewal any components with intended functions delineated under 10 CFR 54.4(a). The staff then reviewed those components that the applicant had identified as being within the scope of license renewal to verify that the applicant had not omitted any passive and long-lived components that should be subject to an AMR in accordance with the requirements of 10 CFR 54.21(a)(1).

The staff's review of LRA Section 2.3.3.B.25 identified areas in which additional information was necessary to complete the evaluation of the applicant's scoping and screening results. The applicant responded to the staff's RAIs as discussed below.

In RAI 2.3.3.B.25-1 dated November 19, 2004, the staff stated that LR drawings LR-37A-1 and LR-37B-2 show components not highlighted indicating that they require no AMR. Therefore, the staff requested that the applicant justify the exclusion of these components from within the scope of license renewal and not subject to an AMR. The applicant's response by letter dated December 22, 2004, has been incorporated in the ALRA as discussed below.

In its ALRA dated July 14, 2005, the applicant provided the staff with a revised LRA Table 2.3.3.B.25-1 identifying components subject to an AMR and revised LR drawings which identify SSCs within the scope of license renewal and subject to an AMR under 10 CFR 54.4. Therefore, the staff's concern described in RAI 2.3.3.B.25-1 is resolved.

In RAI 2.3.3.B.25-2 dated November 19, 2004, the staff stated that the introduction to NMP2 USAR Table 3.9B-2 states that this table lists the major SR components in the plant. Item W and X identify the RWCU system pumps and the RWCU heat exchangers respectively as parts of this table. However, neither of these components is highlighted on drawing LR-37B-0 as within the scope of license renewal and subject to an AMR. Also the LRA Table 2.3.3.B.25-1 does not include the component type "Pumps or Heat Exchangers." The staff believes that these components meet 10 CFR 54.4(a)(1) criteria and should require a 10 CFR 54.21(a)(1) AMR. Therefore, the staff requested that the applicant justify the exclusion of these components from within the scope of license renewal and AMR.

In its response by letter dated December 22, 2004, the applicant stated that NMP2 USAR Table 3.2-1 describes in additional detail the portions of the RWCU system that perform a safety function and are therefore within the scope of license renewal for 10 CFR 54.4(a)(1) or (3). According to the applicant, the RWCU system pumps and heat exchangers are not SR per NMP2 USAR Table 3.2-1 or the NMP2 MEL and are not required for safe shutdown of the reactor. The applicant indicated that consistent with 10 CFR 54.4(b) these components do not support any system intended functions pursuant to 10 CFR 54.4(a)(1) or (3).

The applicant further clarified that the NSR liquid filled piping and components shown in black on drawing LR-37B-0 are in-scope for license renewal and subject to 10 CFR 54.4(a)(2) AMR because all of these components are located in the reactor building in the vicinity of SR components. However, these components are not highlighted on the LR drawing because they are within the scope of license renewal and subject to an AMR for 10 CFR 54.4(a)(2) only.

In evaluating this response the staff found it incomplete and that review of LRA Section 2.3.3.B.25 could not be completed because (1) it does not explain why RWCU heat exchangers and pumps are listed in NMP2 USAR Table 3.9B-2 as the major safety-related component in the plant if they are not safety-related and (2) USAR Table 3.2-1 classifies RWCU heat exchangers and pumps as the American Society of Mechanical Engineers (ASME) Safety Class 3 and differs from the applicant's response. The staff held a teleconference with the applicant on January 27, 2005, to discuss information necessary to resolve the concern in RAI 2.3.3.B.25-2. The result of the teleconference was an agreement by the applicant to transmit the required information by a follow-up letter.

By letter dated February 11, 2005, the applicant acknowledged a discrepancy between USAR sections, in that the applicant has categorized the RWCU pump and heat exchangers as nonsafety-related components for the purpose of license renewal, but these components also are listed in NMP2 UFSAR Table as a major safety-related component and in the UFSAR Table 3.2-1 as ASME safety class 3, which differs from the applicant's previous response to the RAI. In the February 11, 2005, the applicant stated that NMPNS applied design criteria that exceed those required for the RWCU component safety classifications. The USAR Section 3.9B.3.1.14 states:

> The RWCU pump and regenerative and non-regenerative heat exchangers are not part of a safety system and are not designed to Category I requirements.

> The requirements of ASME Boiler and Pressure Vessel Code, Section III, Safety Class 3 components are used as guidelines in evaluating the RWCU system pump and heat exchanger components. The loading conditions, stress criteria, and calculated and allowable stresses are summarized in Tables 3.9B-2w and 3.9B-2x.

The applicant further stated in the February 11, 2005, letter, that (1) USAR Section 5.4.8 provides the description of the RWCU system and states that the portion of the system from the RPV to the outboard isolation valves is SR and the remainder of the system is NSR, and that (2) the discrepancy between the USAR sections has been placed into the site CAP for resolution. The applicant stated that it did not consider the resolution of this issue to be an impact to license renewal since the components are in-scope under 10 CFR 54.4(a)(2) and subject to an AMR.

Based on its review of the above applicant's explanation that the discrepancy between the USAR sections has been placed into the site CAP for resolution, and that the RWCU pump and heat exchangers were evaluated under ASME Boiler and Pressure Vessel Code, Section III, Safety Class 3 guide lines, the staff found the applicant's response to RAI 2.3.3.B.25-2, including the information in the letter dated February 11, 2005, acceptable, because it adequately explained that the USAR had a discrepancy that will be resolved and that the components in question are within the scope of license renewal. Therefore, the staff's concern described in RAI 2.3.3.B.25-2 is resolved.

2.3B.3.25.3 Conclusion

The staff reviewed the ALRA, RAI responses, and accompanying scoping boundary drawings to determine whether any SSCs that should be within the scope of license renewal had not been identified by the applicant. No omissions were identified. In addition the staff performed a review to determine whether any components that should be subject to an AMR had not been identified by the applicant. No omissions were identified. On the basis of its review, the staff concludes that there is reasonable assurance that the applicant had adequately identified the RWCU system components that are within the scope of license renewal, as required by 10 CFR 54.4(a), and the RWCU system components that are subject to an AMR, as required by 10 CFR 54.21(a)(1).

2.3B.3.26 NMP2 Seal Water System (Removed)

The NMP2 seal water system has been removed from the scope of license renewal since it has been determined that it does not meet any of the criteria of 10 CFR 54.4. The original LRA included this system within the scope of license renewal under 10 CFR 54.4(a)(2). However, based upon detailed evaluations, this system is not credited for mitigation of any CLB event, does not contain any SR/NSR interfaces, nor introduce any spatial interactions with SR SSCs. Therefore, the seal water system is not within the scope of license renewal.

2.3B.3.27 NMP2 Service Water System

2.3B.3.27.1 Summary of Technical Information in the Amended Application

In ALRA Section 2.3.3.B.27, the applicant described the service water system. The service water system is designed to provide a reliable supply of cooling water for essential components and systems. The service water system provides cooling water to the secondary sides of the RBCLC water system and TBCLC water system heat exchangers during normal plant operation and planned outages. Service water is also supplied to the secondary side of the RHR system heat exchangers during planned unit outages. In addition, the system is designed to provide makeup water to the circulating water system and cooling water to miscellaneous nonessential TB and RB components during normal plant operation. The service water system at is a once-through system which utilizes raw lake water from Lake Ontario.

The service water system contains SR components that are relied upon to remain functional during and following DBEs. The failure of NSR SSCs in the service water system could potentially prevent the satisfactory accomplishment of an SR function. In addition, the service water system performs functions that support fire protection and EQ.

The intended functions within the scope of license renewal include the following:

- provides filtration
- maintains mechanical and structural integrity of NSR components to prevent spatial interactions with SR components
- provides pressure retaining boundary
- maintains mechanical and structural integrity of NSR components that provide structural support to attached SR components
- provides flow restriction

In ALRA Table 2.3.3.B.27-1, the applicant identified the following service water system component types that are within the scope of license renewal and subject to an AMR:

- bolting
- filters/strainers
- flow elements
- orifices
- piping and fittings

- pumps
- temperature elements
- valves

2.3B.3.27.2 Staff Evaluation

The staff reviewed ALRA Section 2.3.3.B.27 and USAR Section 9.2.1 using the evaluation methodology described in SER Section 2.3. The staff conducted its review in accordance with the guidance described in SRP-LR Section 2.3.

In conducting its review, the staff evaluated the system functions described in the ALRA and USAR in accordance with the requirements of 10 CFR 54.4(a) to verify that the applicant had not omitted from the scope of license renewal any components with intended functions delineated under 10 CFR 54.4(a). The staff then reviewed those components that the applicant had identified as being within the scope of license renewal to verify that the applicant had not omitted any passive and long-lived components that should be subject to an AMR in accordance with the requirements of 10 CFR 54.21(a)(1).

The staff's review of LRA Section 2.3.3.B.27 identified areas in which additional information was necessary to complete the evaluation of the applicant's scoping and screening results. The applicant responded to the staff's RAIs as discussed below.

In RAI 2.3.3.B.27-1 dated November 19, 2004, the staff stated that drawing LR-11 for the service water system shows a certain component within the AMR boundary of the service water system, an expansion joint or bellows. Similarly, the same LR drawing shows two components that appear to be expansion joints or bellows connecting to the high pressure core spray diesel generator cooler. The LRA table associated with the service water system does not list expansion joints or bellows as a component subject to an AMR within the NMP2 service water system. In addition a review of the AMR table associated with the service water system did not identify expansion joints or bellows as a component type. The applicant was asked to confirm that this component is included in the AMR of piping and fittings or explain the basis for omitting these component types from LRA Table 2.3.3.B.27-1 and from requiring an AMR.

In its response by letter dated December 22, 2004, the applicant stated that the components addressed in question are expansion joints in the suction lines for the service water pumps P1D, B, F, and C respectively. These expansion joints are within the scope of license renewal and are shown as components subject to an AMR under the component type "Piping and Fittings" in the LRA tables for the NMP2 service water system.

Based on its review, the staff found the applicants response acceptable because it identifies the expansion joints in question and confirmed their inclusion within the scope of license renewal and subject to an AMR. Therefore, the staff's concerns described in RAI 2.3.3.B.27-1 are resolved.

In RAIs 2.3.3.B.27-2 through 2.3.3.B.27-4 dated November 19, 2004, the staff requested that the applicant clarify information given on such LR drawings as LR-11 sheet B, LR-11J-0, LR-43 sheet G, LR-11 sheet E concerning SSCs for NMP2 service water systems that are within the scope of license renewal per 10 CFR 54.4(a). The applicant's response by letter dated

December 22, 2004, provided adequate information to the staff and the information has been incorporated in the ALRA as discussed below.

In its ALRA dated July 14, 2005, the applicant provided the staff with revised LR drawings which identify all SSCs within the scope of license renewal and subject to an AMR including those under 10 CFR 54.4(a)(2). Based on review of the information submitted in the ALRA the staff's concerns described in RAI 2.3.3.A.27-2 through 2.3.3.A.27-4 are resolved.

In RAI 2.3.3.B.27-5 dated November 19, 2004, the staff stated that drawing LR-11 shows two flow element root valves, V53B and V54B, and associated piping outside the scope of license renewal and not subject to an AMR. Failure of these pipes could affect the integrity of the service water system. Therefore, the staff requested that the applicant explain the basis for excluding these components from being subject to an AMR.

In its response by letter dated December 22, 2004, the applicant stated that the root valves V53B and V54B and their associated piping highlighted in black are SR globe root valves. Therefore, these valves and their associated piping back to flow element FE-161B should be highlighted in red as within the scope of license renewal and subject to an AMR.

Based on its review, the staff found the applicant's response acceptable because the applicant concurred that the flow element root valves and associated piping are within the scope of license renewal and subject to an AMR. Therefore, the staff's concern described in RAI 2.3.3.B.27-5 is resolved.

2.3B.3.27.3 Conclusion

The staff reviewed the ALRA, RAI responses, and accompanying scoping boundary drawings to determine whether any SSCs that should be within the scope of license renewal had not been identified by the applicant. No omissions were identified. In addition, the staff performed a review to determine whether any components that should be subject to an AMR had not been identified by the applicant. No omissions were identified. On the basis of its review, the staff concludes that there is reasonable assurance that the applicant had adequately identified the service water system components that are within the scope of license renewal, as required by 10 CFR 54.4(a), and the service water system components that are subject to an AMR, as required by 10 CFR 54.21(a)(1).

2.3B.3.28 NMP2 Spent Fuel Pool Cooling and Cleanup System

2.3B.3.28.1 Summary of Technical Information in the Amended Application

In ALRA Section 2.3.3.B.28, the applicant described the spent fuel pool cooling and cleanup system. The spent fuel pool cooling and cleanup system is designed to remove the decay heat released from the spent fuel elements and maintain a specified fuel pool water temperature, water clarity, and water level. The spent fuel pool cooling and cleanup system is also designed to provide cooling to the spent fuel pool, reactor cavity pool, and reactor internals during plant refueling outages. The cooling section can operate independently from the cleanup section.

The spent fuel pool cooling and cleanup system contains SR components that are relied upon to remain functional during and following DBEs. The failure of NSR SSCs in the spent fuel pool cooling and cleanup system could potentially prevent the satisfactory accomplishment of an SR function. In addition, the spent fuel pool cooling and cleanup system performs functions that support fire protection.

The intended functions within the scope of license renewal include the following:

- provides heat transfer

- maintains mechanical and structural integrity of NSR components to prevent spatial interactions with SR components

- provides structural support to NSR components whose failure could prevent accomplishment of SR function(s)

- provides pressure retaining boundary

- maintains mechanical and structural integrity of NSR components that provide structural support to attached SR components

In ALRA Table 2.3.3.B.28-1, the applicant identified the following spent fuel pool cooling and cleanup system component types that are within the scope of license renewal and subject to an AMR:

- bolting
- filters/strainers
- flow elements
- heat exchangers
- orifices
- piping and fittings
- pumps
- tanks
- valves

2.3B.3.28.2 Staff Evaluation

The staff reviewed ALRA Section 2.3.3.B.28 and the USAR Section 9.1.3 using the evaluation methodology described in SER Section 2.3. The staff conducted its review in accordance with the guidance described in SRP-LR Section 2.3.

In conducting its review, the staff evaluated the system functions described in the ALRA and USAR in accordance with the requirements of 10 CFR 54.4(a) to verify that the applicant had not omitted from the scope of license renewal any components with intended functions delineated under 10 CFR 54.4(a). The staff then reviewed those components that the applicant had identified as being within the scope of license renewal to verify that the applicant had not omitted any passive and long-lived components that should be subject to an AMR in accordance with the requirements of 10 CFR 54.21(a)(1).

The staff's review of LRA Section 2.3.3.B.28 identified areas in which additional information was necessary to complete the evaluation of the applicant's scoping and screening results. The applicant responded to the staff's RAIs as discussed below.

In RAI 2.3.3.B.28-1 dated November 19, 2004, the staff stated that the spent fuel pool cooling and cleanup system is shown primarily on LR drawings LR-38A, -B, and -C. Because the cooling and cleanup systems operate independently of one another and the majority of the cleanup system does not contain components subject to an AMR the drawings supplied by the applicant do not contain all of the detail the staff needs for complete review to understand the configuration of the components requiring an AMR. This information is not depicted clearly in the LR drawings supplied. The staff requested that the applicant supply figures from the NMP2 USAR for the spent fuel pool cooling and cleanup system and for drawing LR-38D-0.

In its response by letter dated December 22, 2004, the applicant stated that NMP2 USAR Figures 9.1-5a through 9.1-5d were enclosed. The applicant also clarified that the drawing LR-38D-0 does not exist since there are on that drawing no components in-scope to meet the requirements of 10 CFR 54.4(a)(1) or (a)(3).

Based on its review, the staff found the applicant's response acceptable because the applicant provided the requested USAR figures and clarified that there is no drawing LR-38D-0. Therefore, the staff's concern described in RAI 2.3.3.B.28-1 is resolved.

In RAI 2.3.3.B.28-2 dated November 19, 2004, the staff stated that there are spargers noted on LR drawings at the bottom of the spent fuel pool (LR-38B-0), the reactor refueling cavity (LR-38A-0), and the reactor internals storage pit (LR-38A-0) as subject to an AMR. The staff asked the applicant to clarify whether these spargers are included in the component type "Piping and Fittings" in the LRA Table 2.3.3.B.28-1 or indicate if they are included in the table under any another component type.

In its response by letter dated December 22, 2004, the applicant stated that these spargers, also known as spray nozzles, are included with the component type "Piping and Fittings" in Table 2.3.3.B.28-1.

Based on its review, the staff found the applicant's response acceptable because it stated that the spargers are included with the component type "Piping and Fittings" in the AMR tables for this system. Therefore, the staff's concern described in RAI 2.3.3.B.28-2 is resolved.

In RAI 2.3.3.B.28-3 dated November 19, 2004, the staff stated that LRA Tables 2.3.3.B.28-1 and 3.3.2.B-27 list the component "Filter/Strainer" as subject to an AMR. In LR drawings for the spent fuel pool cooling and cleanup system no filters or strainers were found. Therefore, the staff requested that the applicant clarify whether there are any filters or strainers in the spent fuel cooling and cleanup system subject to an AMR and, if not, that the applicant remove the reference from the LRA tables.

In its response by letter dated December 22, 2004, the applicant stated that drawing LR-038A-0 identifies one of the strainers, STRT1B, in question, and drawing LR-038B-0 identifies the other, STRT1A. Both components are within the scope of license renewal and subject to an AMR. Both components are included with the component type "Filters/Strainers" in LRA Tables 2.3.3.B.28-1 and 3.3.2.B-27.

Based on its review, the staff found the applicant's response to RAI 2.3.3.B.28-3 acceptable because it adequately identified the strainers on the LR drawings. Therefore, the staff's concern described in RAI 2.3.3.B.28-3 is resolved.

2.3B.3.28.3 Conclusion

The staff reviewed the ALRA, RAI responses, and accompanying scoping boundary drawings to determine whether any SSCs that should be within the scope of license renewal had not been identified by the applicant. No omissions were identified. In addition, the staff performed a review to determine whether any components that should be subject to an AMR had not been identified by the applicant. No omissions were identified. On the basis of its review, the staff concludes that there is reasonable assurance that the applicant had adequately identified the spent fuel pool cooling and cleanup system components that are within the scope of license renewal, as required by 10 CFR 54.4(a), and the spent fuel pool cooling and cleanup system components that are subject to an AMR, as required by 10 CFR 54.21(a)(1).

2.3B.3.29 NMP2 Standby Diesel Generator Fuel Oil System

2.3B.3.29.1 Summary of Technical Information in the Amended Application

In ALRA Section 2.3.3.B.29, the applicant described the standby diesel generator fuel oil system. The standby diesel generator fuel oil system is designed to deliver sufficient fuel oil flow to the EDGs and provide fuel oil storage capacity for each diesel generator for seven days of continuous diesel generator operation without interconnection to any other onsite fuel oil system. The EDGs are equipped with a fuel oil day tank which has enough fuel for approximately one hour of running time plus a margin of 10 percent at the highest allowed gravity. The day tank is elevated above the EDG and kept full of fuel oil from the fuel oil storage tank by the fuel oil transfer pumps. The elevated location of the tank provides adequate net positive suction head to the engine-driven fuel pump of the diesel engine. Each storage tank is filled from its own tank truck fill station located in the yard. Electric oil transfer pumps mounted on top of each tank permit the transfer of fuel oil to the day tanks. One fuel oil transfer pump is capable of supplying the maximum fuel demand of a standby diesel generator. Each pump discharges through a strainer with an automatic shutoff in case of high differential pressure. After passing through the strainer the fuel oil discharges into the day tank.

The standby diesel generator fuel oil system contains SR components that are relied upon to remain functional during and following DBEs. The failure of NSR SSCs in the standby diesel generator fuel oil system could potentially prevent the satisfactory accomplishment of an SR function. In addition, the standby diesel generator fuel oil system performs functions that support fire protection.

The intended functions within the scope of license renewal include the following:

- provides filtration
- provides heat transfer
- provides pressure retaining boundary

- maintains mechanical and structural integrity of NSR components that provide structural support to attached SR components

In ALRA Table 2.3.3.B.29-1, the applicant identified the following standby diesel generator fuel oil system component types that are within the scope of license renewal and subject to an AMR:

- bolting
- filters/strainers
- flow elements
- heat exchangers
- piping and fittings
- pumps
- tanks
- valves

2.3B.3.29.2 Staff Evaluation

The staff reviewed ALRA Section 2.3.3.B.29 and USAR Section 9.5.4 using the evaluation methodology described in SER Section 2.3. The staff conducted its review in accordance with the guidance described in SRP-LR Section 2.3.

In conducting its review, the staff evaluated the system functions described in the ALRA and USAR in accordance with the requirements of 10 CFR 54.4(a) to verify that the applicant had not omitted from the scope of license renewal any components with intended functions delineated under 10 CFR 54.4(a). The staff then reviewed those components that the applicant had identified as being within the scope of license renewal to verify that the applicant had not omitted any passive and long-lived components that should be subject to an AMR in accordance with the requirements of 10 CFR 54.21(a)(1).

The staff's review of LRA Section 2.3.3.B.29 identified areas in which additional information was necessary to complete the evaluation of the applicant's scoping and screening results. The applicant responded to the staff's RAIs as discussed below.

In RAI 2.3.3.B.29-1 dated November 19, 2004, the staff stated that on LR drawing LR-104F-0 orifices are shown to be within the scope of license renewal and subject to an AMR. However, "Orifice" is not included as a component type in LRA Table 2.3.3.B.29-1. LRA Table 2.0-1 identifies "Flow Restriction" as a component intended function applicable to an orifice. The staff asked the applicant to clarify whether this component is included with another component type within the scope of license renewal and subject to an AMR and, if not, to justify its exclusion from the scope of license renewal and from being subject to an AMR or update the corresponding table to include this component.

In its response by letter dated December 22, 2004, the applicant stated that flow orifices are included under the component type "Flow Element" in LRA Table 2.3.3.B.29-1. The applicant also explained that not all orifices have an intended function of "Flow Restriction," which is an intended function only if required for the system in which it is installed to meet an intended functions for safe shutdown of the plant. Most orifices have a "Pressure Boundary" function only.

Based on its review, the staff found the applicant's response acceptable because it adequately explained that the flow orifices in question are within the scope of license renewal under 10 CFR 54.4(a) and subject to an AMR under 10 CFR 54.21(a). Further, the applicant stated that the flow orifices are represented in the LRA table by the component type flow element. Therefore, the staff's concern described in RAI 2.3.3.B.29-1 is resolved.

In RAI 2.3.3.B.29-2 dated November 19, 2004, the staff stated that on LR drawing LR-103B-0, flexible hoses are shown within the scope of license renewal and subject to an AMR. However, flexible hose is not included as a component type in LRA Table 2.3.3.B.29-1 associated with the standby diesel generator fuel oil system. Therefore, the staff requested that the applicant clarify whether this component is included with another component type within the scope of license renewal and subject to an AMR and, if not, to justify its exclusion from the scope of license renewal and from being subject to an AMR or update the corresponding table to include this component.

In its response by letter dated December 22, 2004, the applicant stated that the flexible hoses in question are stainless steel and included under the component type "Piping and Fittings" in the LRA Table 2.3.3.B.29-1.

Based on its review, the staff found the applicant's response to RAI 2.3.3.B.29-2 acceptable because it adequately explained that the flexible hoses in question are within the scope of license renewal under 10 CFR 54.4(a) and subject to an AMR under 10 CFR 54.21(a). Further, the applicant stated that the flexible hoses are represented in the LRA Table by the component type "Piping and Fittings." Therefore, the staff's concern described in RAI 2.3.3.B.29-2 is resolved.

In RAI 2.3.3.B.29-3 dated November 19, 2004, the staff stated that NMP2 USAR Section 9.5.4 states that the standby diesel generator fuel oil storage and transfer system consists of six electric motor-driven, vertical, turbine-type fuel oil transfer pumps. The pumps are mounted in duplex sets on top of each fuel oil storage tank and each duplex set is connected in parallel to its respective day tank to permit the transfer of fuel oil by the pumps. Drawing LR-104C-0 shows two of these pumps and drawing LR-104B-0 shows four within the scope of license renewal and subject to an AMR. However, the LR drawings do not show the piping within the standby diesel generator storage tanks connecting to these pumps as subject to an AMR. The staff believed that this piece of piping should be within the scope of license renewal and subject to an AMR for the standby diesel generator fuel oil storage tanks to perform their intended function to ensure the transfer of fuel oil to the day tanks. Therefore, the staff requested that the applicant justify the exclusion of this piece of piping from the scope of license renewal and from being subject to an AMR.

In its response by letter dated December 22, 2004, the applicant agreed with the staff that the suction piping inside the diesel fuel oil storage tanks is in-scope for license renewal and subject to an AMR, that the piping is included with the component type "Piping and Fittings" in LRA Table 2.3.3.B.29-1, and that LR drawings LR-104B and LR-104C should show these piping segments highlighted in red.

Based on its review, the staff found the applicant's response acceptable because it adequately explained that the piping in question is within the scope of license renewal and subject to an AMR and included with the component type "Piping and Fittings." The staff concludes that there is reasonable assurance that the components were included correctly within the scope of license renewal and subject to an AMR. Therefore, the staff's concern described in RAI 2.3.3.B.29-3 is resolved.

In RAI 2.3.3.B.29-4 dated November 19, 2004, the staff stated that drawing LR-104B-0 shows piping within the standby diesel generator fuel oil storage and transfer system connecting to nine level switches and drawing LR-104C-0 shows such piping connecting to four level switches as not within the scope of license renewal or subject to an AMR. The staff believed that the level switches are used to monitor the oil level in their associated day tanks and that the piping connecting to these level switches should be within the scope of license renewal and subject to an AMR. Therefore, the staff requested that the applicant justify the exclusion of this piping from the scope of license renewal and from being subject to an AMR.

In its response by letter dated December 22, 2004, the applicant stated that the sensing lines up to the referenced level switches are within the scope of license renewal and subject to an AMR. The components in question should have been highlighted on drawing LR-104B-0 showing them within the scope of license renewal and are subject to an AMR but inadvertently were not highlighted.

Based on its review, the staff found the applicant's response acceptable because it adequately explained that the components in question are within the scope of license renewal under 10 CFR 54.4(a) and subject to an AMR under 10 CFR 54.21(a) but inadvertently were not highlighted on the LR drawing. The staff concludes that there is reasonable assurance that the components were included correctly within the scope of license renewal and subject to an AMR. Therefore, the staff's concern described in RAI 2.3.3.B.29-4 is resolved.

In RAI 2.3.3.B.29-5 dated November 19, 2004, the staff stated that on LR drawings LR-104B-0, LR-104C-0, and LR-104F-0 vents are shown to be within the scope of license renewal and subject to an AMR. However, vents are not included as a component type in the LRA Table 2.3.3.B.29-1. Therefore, the staff requested that the applicant clarify whether this component is included with another component type within the scope of license renewal and subject to an AMR and, if not, to justify its exclusion from the scope of license renewal and from being subject to an AMR or update the corresponding table to include this component.

In its response by letter dated December 22, 2004, the applicant stated that vents were included under the component type "Piping and Fittings" in LRA Table 2.3.3.B.29-1.

Based on its review, the staff found the applicant's response acceptable because it adequately explained that the vents are represented in the LRA table by the component type "Piping and Fittings." Therefore, the staff's concern described in RAI 2.3.3.B.29-5 is resolved.

2.3B.3.29.3 Conclusion

The staff reviewed the ALRA, RAI responses, and accompanying scoping boundary drawings to determine whether any SSCs that should be within the scope of license renewal had not been identified by the applicant. No omissions were identified. In addition, the staff performed a review to determine whether any components that should be subject to an AMR had not been identified by the applicant. No omissions were identified. On the basis of its review, the staff concludes that there is reasonable assurance that the applicant had adequately identified the standby diesel generator fuel oil system components that are within the scope of license renewal, as required by 10 CFR 54.4(a), and the standby diesel generator fuel oil system components that are subject to an AMR, as required by 10 CFR 54.21(a)(1).

2.3B.3.30 NMP2 Standby Diesel Generator Protection (Generator) System

2.3B.3.30.1 Summary of Technical Information in the Amended Application

In ALRA Section 2.3.3.B.30, the applicant described the standby diesel generator protection (generator) system. The standby diesel generator protection (generator) system is designed to provide for the operation of emergency systems and ESFs during and following the shutdown of the reactor when the preferred power supply is not available. The standby power supply system consists of three standby diesel generators. One generator is dedicated to each of the three divisions of the SR electric power distribution system feeding each Class 1E load group. Any two of the three standby diesel generators have sufficient capacity to start and supply all needed ESFs and emergency shutdown loads in case of a LOCA or LOOP. The standby diesel generators are normally maintained in a standby status. In case of a LOOP or degraded offsite voltage condition the standby diesel generators automatically start, accelerate to rated speed and voltage, and start picking up loads sequentially. In case of a LOCA the standby diesel generators automatically start, accelerate to rated speed, voltage, and frequency, and run unloaded. Should any subsequent LOOP occur, the standby diesel generators would then energize their respective busses. The standby diesel generator protection (generator) system also includes the generator support systems for cooling water and lube oil discussed in the generator standby lube oil system.

The standby diesel generator protection (generator) system contains SR components that are relied upon to remain functional during and following DBEs. In addition, the standby diesel generator protection (generator) system performs functions that support fire protection.

The intended functions within the scope of license renewal include the following:

- provides heat transfer
- provides pressure retaining boundary

In ALRA Table 2.3.3.B.30-1, the applicant identified the following standby diesel generator protection (generator) system component types that are within the scope of license renewal and subject to an AMR:

- bolting
- heat exchangers
- piping and fittings
- pumps
- sight glass
- tank
- valves

2.3B.3.30.2 Staff Evaluation

The staff reviewed ALRA Section 2.3.3.B.30 and USAR Sections 1.2.9.17 and 8.3.1 using the evaluation methodology described in SER Section 2.3. The staff conducted its review in accordance with the guidance described in SRP-LR Section 2.3.

In conducting its review, the staff evaluated the system functions described in the ALRA and USAR in accordance with the requirements of 10 CFR 54.4(a) to verify that the applicant had not omitted from the scope of license renewal any components with intended functions delineated under 10 CFR 54.4(a). The staff then reviewed those components that the applicant had identified as being within the scope of license renewal to verify that the applicant had not omitted any passive and long-lived components that should be subject to an AMR in accordance with the requirements of 10 CFR 54.21(a)(1).

The staff's review of LRA Section 2.3.3.B.30 identified areas in which additional information was necessary to complete the evaluation of the applicant's scoping and screening results. The applicant responded to the staff's RAIs as discussed below.

In RAI 2.3.3.B.30-1 dated November 19, 2004, the staff stated that drawing LR-104D-0 (location C-3) shows the turbocharger as not subject to an AMR. The turbocharger is required for the proper operation of the diesel and has a passive pressure boundary function. This component meets 10 CFR 54.4(a)(1) criteria. Therefore, the staff requested that the applicant provide the basis for excluding the turbocharger from the scope of license renewal and from being subject to an AMR.

In its response by letter dated December 22, 2004, the applicant stated that the emergency diesel generators and its turbochargers, are within the scope of license renewal; however, since they are active components, they are not subject to an AMR. This statement is consistent with the guidance provided in NEI 95-10, Revision 3, Appendix B.

Based on its review, the staff found the applicant's response acceptable because it adequately explained that the component in question is within the scope of license renewal under 10 CFR 54.4(a) but not subject to an AMR under 10 CFR 54.21(a) because it is an active component. Therefore, the staff's concern described in RAI 2.3.3.B.30-1 is resolved.

In RAI 2.3.3.B.30-2 dated November 19, 2004, the staff stated that drawing LR-104D-0 (locations H4 and H5) shows that lube oil coolers, fuel oil coolers, and inter coolers were not highlighted as subject to an AMR. These components have a passive pressure boundary function and meet 10 CFR 54.4(a)(1) criteria. Therefore, the staff requested that the applicant provide the basis for omitting these components from the scope of license renewal and from being subject to an AMR.

In its response by letter dated December 22, 2004, the applicant stated that drawing LR-104D-0 (locations H4 and H5) identifies intercoolers, fuel oil coolers, and lube oil coolers. These heat exchangers are passive and long-lived. They are within the scope of license renewal and subject to an AMR. As such they should be shown in red on this drawing. They are included in LRA Sections 2.3.3.B.29, 2.3.3.B.30, 3.3.2.B.28, and 3.3.2.B.29, and in Tables 3.3.2.B-29 and 3.3.2.B-30.

Based on its review, the staff found the applicant's response acceptable because it adequately explained that the components in question are within the scope of license renewal under 10 CFR 54.4(a) and subject to an AMR under 10 CFR 54.21(a) but inadvertently were not highlighted on the LR drawing. The staff concludes that there is reasonable assurance that the components were included correctly within the scope of license renewal and subject to an AMR. Therefore, the staff's concern described in RAI 2.3.3.B.30-2 is resolved.

In RAI 2.3.3.B.30-3 dated November 19, 2004, the staff stated that drawing LR-104D-0 shows jacket water circulation heaters as not subject to an AMR. The USAR does not describe clearly how the heater functions. If the immersion heater works by immersing the heating element in the cooling fluid, then the heater containing the cooling fluid has a passive pressure boundary function and is subject.to an AMR per 10 CFR 54.4(a)(1). Because drawing LR-104E-0 shows a similar heater as subject to an AMR the applicant was asked to provide the basis for not including these heaters on drawing LR-104D-0 in the license renewal boundary.

In its response by letter dated December 22, 2004, the applicant stated that drawing LR-104D-0 is incorrect, that the jacket water heaters in question 2EGT*CH4/5 and immersion heater 2EGT*CH1 have a pressure boundary function similar to that of lube oil heaters 2EGT*CH2 and 2EGT*CH3 on drawing LR-104E-0, that the chamber for 2EGT*CH4/5 is within the scope of license renewal and subject to an AMR. The applicant explained that these chambers are treated as part of the "Piping and Fittings" component type. The applicant stated that, however, these heaters themselves are also in-scope but are active components per NEI 95-10, Revision 3, Appendix B, and as such not subject to an AMR.

Based on its review, the staff found the applicant's response acceptable because it adequately explained that the components in question are within the scope of license renewal under 10 CFR 54.4(a) and subject to an AMR under 10 CFR 54.21(a) but inadvertently were not highlighted on the LR drawing. The staff concludes that there is reasonable assurance that the components were included correctly within the scope of license renewal and subject to an AMR as part of the "Piping and Fittings" component type. The heaters themselves are also in-scope but are active components not subject to an AMR. Therefore, the staff's concern described in RAI 2.3.3.B.30-3 is resolved.

In RAI 2.3.3.B.30-4 dated November 19, 2004, the staff stated that drawing LR-104D-0 shows the water expansion tank and overflow line as subject to an AMR. However, the tubing leading to the level switches and the sight glass on the expansion tank is specifically excluded from AMR. LRA Table 2.3.3.B.30-1 does not identify tanks as a component type requiring an AMR. Therefore, the staff requested that the applicant provide the basis for excluding the sight glass and tubing from the scope of license renewal and omitting the component type "Tank" from the referenced table. The applicant's response by letter dated December 22, 2004, has been incorporated in the ALRA as discussed below.

In its ALRA dated July 14, 2005, the applicant provided the staff with a revised LR drawing which identifies SSCs in scope of license renewal and subject to an AMR under 10 CFR 54.4(a)(2). The applicant also provided clarification on the component groups included in LRA Table 2.3.3.B.30 -1. Therefore, the staff's concern described in RAI 2.3.3.B.30-4 is resolved.

In RAI 2.3.3.B.30-5 dated November 19, 2004, the staff stated that on drawing LR-104D-0 the tube sides of jacket water coolers are not highlighted as subject to an AMR. Therefore, the staff requested that the applicant provide the basis for not including this portion of the component within the scope of license renewal.

In its response by letter dated December 22, 2004, the applicant stated that drawing LR-104D-0 identifies the jacket water coolers in question, 2EGS*E1A, 2EGS*E1B, 2EGS*E2A, and 2EGS*E2B. The applicant stated that, these coolers and their subcomponents are safety-related, in-scope for license renewal, and subject to an AMR. The LR drawing therefore should show the tube side of the jacket water coolers in red.

Based on its review, the staff found the applicant's response acceptable because it adequately explained that the components in question are within the scope of license renewal under 10 CFR 54.4(a) and subject to an AMR under 10 CFR 54.21(a) but inadvertently were not highlighted on the LR drawing. The staff concludes that there is reasonable assurance that the component type was included correctly within the scope of license renewal and subject to an AMR as part of the "Piping and Fittings" component type. Therefore, the staff's concern described in RAI 2.3.3.B.30-5 is resolved.

In RAI 2.3.3.B.30-6 dated November 19, 2004, the staff stated that on drawing LR-104E-0 the LR boundary stops at an open valve HV31J. The tubing beyond this valve has a pressure boundary function subject to an AMR. The same concern exists at location B-7 for piping downstream of valve HV18C. Therefore, the staff requested that the applicant provide the basis for not subjecting the piping/tubing down stream of an open valve to AMR.

In its response by letter dated December 22, 2004, the applicant stated that drawing LR-104E-0 has several errors such as at location D10, hand valves 2EGS*HV131J, 2EGS*HV131K, 2EGS*HV231J, and 2EGS*HV231K are shown in black. These components are safety-related, in-scope for license renewal, and should be shown in red as subject to an AMR. In addition the associated piping (from 2EGS*V165 and 2EGS*V265 up to but not including 2EGS*PS4002A and 2EGS*PS4002B) also should be shown in red as in-scope and subject to an AMR.

With respect to hand valve 2EGS*HV118C the same issue was also addressed by the applicant in its response to RAI 2.3.3.B.15-2 (discussed in SER Section 2.3.3.B.15).

Based on its review, the staff found the applicant's response acceptable because it adequately explained that the components in question are within the scope of license renewal under 10 CFR 54.4(a) and subject to an AMR under 10 CFR 54.21(a) but inadvertently were not highlighted on the LR drawing. The staff concludes that there is reasonable assurance that the component types were included correctly within the scope of license renewal and subject to an AMR as part of the "Piping and Fittings" and "Valves" component types. Therefore, the staff's concern described in RAI 2.3.3.B.30-6 is resolved.

In RAI 2.3.3.B.30-7 dated November 19, 2004, the staff noted a discrepancy between drawings LR-104B-0 and LR-104F-0 in that LR-104B-0 showed the interconnecting piping on LR-104F-0 as not subject to an AMR while LR-104F-0 showed the piping as subject to an AMR. Therefore, the staff requested that the applicant explain the apparent discrepancy.

In its response by letter dated December 22, 2004, the applicant assumed that the location referenced for drawing LR-104B-0 is in error and that the correct location reference in the RAI for drawing LR-104B-0 appears to be M8, and that the LR continuation flag at location M8 should indicate "LR-EGF" in both directions as opposed to "LR-EGF|solid blue." The drawing continuation flag directs the reader to "LR-104F (G4)." The LR continuation flag at location G4 on drawing LR-104F-0, showing "LR-EGF" in both directions is correct and the lower of the two drawing continuation flags at this location correctly sends the reader to "LR-104B (M8)."

Based on its review, the staff found the applicant's response to RAI 2.3.3.B.30-7 acceptable because it adequately explained that the LR renewal continuation flags at location M8 should have indicated "LR-EGF" in both directions as opposed to "LR-EGF" in one direction and "solid blue" in the other. The components on both sides of the continuation flags in question are within the scope of license renewal under 10 CFR 54.4(a) and subject to an AMR under 10 CFR 54.21(a). The staff concludes that there is reasonable assurance that the components were included correctly within the scope of license renewal and subject to an AMR. Therefore, the staff's concern described in RAI 2.3.3.B.30-7 is resolved.

2.3B.3.30.3 Conclusion

The staff reviewed the ALRA, RAI responses, and accompanying scoping boundary drawings to determine whether any SSCs that should be within the scope of license renewal had not been identified by the applicant. No omissions were identified. In addition, the staff performed a review to determine whether any components that should be subject to an AMR had not been identified by the applicant. No omissions were identified. On the basis of its review, the staff concludes that there is reasonable assurance that the applicant had adequately identified the standby diesel generator protection (generator) system components that are within the scope of license renewal, as required by 10 CFR 54.4(a), and the standby diesel generator protection (generator) system components that are subject to an AMR, as required by 10 CFR 54.21(a)(1).

2.3B.3.31 NMP2 Standby Liquid Control System

2.3B.3.31.1 Summary of Technical Information in the Amended Application

In ALRA Section 2.3.3.B.31, the applicant described the standby liquid control system. The standby liquid control system is designed to inject a boron solution into the reactor when needed to bring the core to a subcritical condition, providing an alternate method to shut down the reactor in the event that sufficient control rods cannot be inserted in the reactor core to accomplish shutdown and cooldown in the normal manner. This system is designed to provide sufficient negative reactivity to shut down the reactor and keep the reactor from going critical as it cools by mixing a neutron absorber with the primary reactor coolant. The neutron absorber is injected within the core zone via the HPCS system injection line. The standby liquid control system can be initiated manually or automatically by the redundant reactivity control system. The sodium pentaborate solution is discharged radially over the top of the core through the high pressure core spray sparger.

The standby liquid control system contains SR components that are relied upon to remain functional during and following DBEs. The failure of NSR SSCs in the standby liquid control system could potentially prevent the satisfactory accomplishment of an SR function. In addition, the standby liquid control system performs functions that support ATWS.

The intended functions within the scope of license renewal include the following:

- maintains mechanical and structural integrity of NSR components to prevent spatial interactions with SR components
- provides structural support to NSR components whose failure could prevent accomplishment of SR function(s)
- provides pressure retaining boundary
- maintains mechanical and structural integrity of NSR components that provide structural support to attached SR components
- provides flow restriction

In ALRA Table 2.3.3.B.31-1, the applicant identified the following standby liquid control system component types that are within the scope of license renewal and subject to an AMR:

- bolting
- expansion joint
- filters/strainers
- flow elements
- piping and fittings
- pumps
- orifices
- tanks
- temperature elements
- valves

2.3B.3.31.2 Staff Evaluation

The staff reviewed ALRA Section 2.3.3.B.31 and USAR Section 9.3.5 using the evaluation methodology described in SER Section 2.3. The staff conducted its review in accordance with the guidance described in SRP-LR Section 2.3.

In conducting its review, the staff evaluated the system functions described in the ALRA and USAR in accordance with the requirements of 10 CFR 54.4(a) to verify that the applicant had not omitted from the scope of license renewal any components with intended functions delineated under 10 CFR 54.4(a). The staff then reviewed those components that the applicant had identified as being within the scope of license renewal to verify that the applicant had not omitted any passive and long-lived components that should be subject to an AMR in accordance with the requirements of 10 CFR 54.21(a)(1).

In reviewing LRA Section 2.3.3.B.31, the staff identified areas in which additional information was necessary to complete the evaluation of the applicant's scoping and screening results. The applicant responded to the staff's RAIs as discussed below.

In RAI 2.3.3.B.31-1 dated November 19, 2004, the staff stated that drawing LR-36A-0 shows Y-strainers at locations G5 and G9 and a strainer element plate at location B9 as subject to an AMR. LRA Table 2.3.3.B.31-1 includes the component type "Filters/Strainers" with an intended function of "Pressure Boundary." Strainers also have a filtration function (as defined in LRA Table 2.0-1) not identified in LRA Table 2.3.3.B.31-1. Therefore, the staff requested that the applicant confirm that filtration is not an intended function for the strainers in the standby liquid control system that requires AMR.

In its response by letter dated December 22, 2004, the applicant stated that a component function would be considered an intended function only if failure of that component function would cause the failure of a system intended function. The applicant also stated that failure of the "Filtration" function of the Y-strainers in question would not prevent the system from performing its intended function the only intended function credited for these components is "Pressure Boundary" as identified in the LRA Table.

Based on its review, the staff found the applicant's response to RAI 2.3.3.B.31-1 acceptable because it adequately explained that the components in question perform only a pressure boundary function and that the loss of the filtration function would not prevent the system from performing its intended function. Therefore, the staff's concern described in RAI 2.3.3.B.31-1 is resolved.

IN RAI 2.3.3.B.31-2 dated November 19, 2004, the staff stated that drawing LR-36A-0 shows a manhole as subject to an AMR. However, manhole is not listed in the LRA Table 2.3.3.B.31-1 as a component type subject to an AMR. Manholes serve a pressure boundary intended function and are passive, long-lived components. To clarify whether this component is a sub-component of a component type listed in LRA Table 2.3.3.B.31-1 the staff requested that the applicant justify the exclusion of the manhole component from being subject to an AMR under 10 CFR 54.21(a)(1).

In its response dated December 22, 2004, the applicant stated that manholes are within the scope of license renewal, subject to an AMR, and included under the component type "Tanks" in the LRA Table.

Based on its review, the staff found the applicant's response to RAI 2.3.3.B.31-2 acceptable because it adequately explained that the manhole component in question is within the scope of license renewal under 10 CFR 54.4(a) and subject to an AMR under 10 CFR 54.21(a). Further, the applicant stated that the manhole component is represented in the LRA Table by the component type "Tanks." Therefore, the staff's concern described in RAI 2.3.3.B.31-2 is resolved.

In RAI 2.3.3.B.31-3 dated November 19, 2004, the staff requested that the applicant clarify information on drawing LR-36A-0 concerning SSCs in the scope of license renewal. The applicant's response by letter dated December 22, 2004, has been incorporated in the ALRA as discussed below.

In its ALRA dated July 14, 2005, the applicant provided the staff with revised LR drawings which identify all SSCs within the scope of license renewal and subject to an AMR including those subject under 10 CFR 54.4(a)(2). Based on review of the information submitted in the amended LRA the staff's concern described in RAI 2.3.3.B.31-3 is resolved.

In RAI 2.3.3.B.31-4 dated November 19, 2004, the staff stated that drawing LR-36A-0 shows the pneumatic signals from the FIC103, LT-103, and LIX103 to the storage tank TK1 as subject to an AMR. However, the flow indicator controller FIC103 is shown as excluded from requiring an AMR. This instrument as shown on LR-36A-0 is installed in-line for isolation of the air supply to the level instruments. Therefore, FIC103 serves a pressure boundary intended function. The applicant was asked to explain why FIC103 had been excluded from the scope of license renewal and from requiring an AMR under 10 CFR 54.4(a) and 10 CFR 54.21(a)(1).

Furthermore, drawing LR-36A-0 does not show how this pneumatic signal (line) extends inside the storage tank. Therefore, the staff requested that the applicant clarify whether the pneumatic line portions inside the storage tank are within the scope of license renewal and subject to an AMR and, if not, justify their exclusion under 10 CFR 54.4(a) and 10 CFR 54.21(a)(1).

In its response by letter dated December 22, 2004, the applicant stated that the pneumatic lines are shown incorrectly on drawing LR-36A-0 as in-scope for license renewal and subject to an AMR. The applicant stated that these lines are NSR and should be shown in black. Additionally, the applicant expalined that the instruments are NSR and as such are not within the scope of license renewal. As active devices if they were within scope they would not be subject to an AMR.

Based on its review, the staff found that the applicant's NSR level instruments in question are not needed to perform safe shutdown of the reactor and do not have an intended function under 10 CFR 54.4(a). Therefore, the staff's concern described in RAI 2.3.3.B.31-4 is resolved.

In RAI 2.3.3.B.31-5 dated November 19, 2004, the staff stated that drawing LR-36A-0 shows pipeline 2-MWS-001-68-4 at location B1, to 2-SLC-001-28-4, at location C-1. The staff therefore sought the following additional information:

a. The acronym MWS, apparently the makeup water system that provides demineralized water, is not defined in the license renewal boundary drawing LR–000-2F-0. Define the MWS acronym.

b. The check valve V3 at location C1 function is to isolate the in-scope portion of the pipeline 2-SLC-001-28-4 from the out of scope pipeline 2-MWS-001-68-4. Check valves are passive and long-lived components. Justify the exclusion of V3 from the scope of license renewal and from being subject to an AMR under 10 CFR 54.4(a) and 10 CFR 54.21(a)(1).

In its response by letter dated December 22, 2004, the applicant stated:

a. "MWS" does stand for Make-up Water System. This acronym was inadvertently omitted from drawing LR-000-2F-0.

Based on its review, the staff found the applicant's response to RAI 2.3.3.B.31-5a acceptable because it adequately explained that the acronym MWS stands for make-up water system. Therefore, the staff's concern described in RAI 2.3.3.B.31-5a is resolved.

In its response to part b the applicant further stated:

b.   See response to RAI 2.3.3.B.31-3. The described correction in that RAI includes the incorrect depiction of valve 2SLS-V3 on drawing LR-36A-0.

Based on its review, the staff found the applicant's response to RAI 2.3.3.B.31-5b acceptable because it adequately explained that the components in question are within the scope of license renewal under 10 CFR 54.4(a) and subject to an AMR under 10 CFR 54.21(a) but inadvertently were not highlighted on the LR drawing. There is reasonable assurance that the components were included correctly within the scope of license renewal and subject to an AMR. Therefore, the staff's concern described in RAI 2.3.3.B.31-5b is resolved.

2.3B.3.31.3 Conclusion

The staff reviewed the ALRA, RAI responses, and accompanying scoping boundary drawings to determine whether any SSCs that should be within the scope of license renewal had not been identified by the applicant. No omissions were identified. In addition, the staff performed a review to determine whether any components that should be subject to an AMR had not been identified by the applicant. No omissions were identified. On the basis of its review, the staff concludes that there is reasonable assurance that the applicant had adequately identified the standby liquid control system components that are within the scope of license renewal, as required by 10 CFR 54.4(a), and the standby liquid control system components that are subject to an AMR, as required by 10 CFR 54.21(a)(1).

2.3B.3.32 NMP2 Yard Structures Ventilation System

2.3B.3.32.1 Summary of Technical Information in the Amended Application

In ALRA Section 2.3.3.B.32, the applicant described the yard structures ventilation system. The yard structures ventilation system is designed to provide heating and outside air ventilation for the service water pump bays, screenwell building, fire pump rooms, demineralizer water storage tank building, CST building, electrical bay, screenhouse, and chiller building. Each of the service water pump bays is also equipped with redundant unit coolers which maintain the space temperature within design limits by discharging heat to the service water system. The yard structures ventilation system also provides space cooling to the service water pump bays, ensuring that ambient temperature remains within the pump operating design limits.

The yard structures ventilation system contains SR components that are relied upon to remain functional during and following DBEs. In addition, the yard structures ventilation system performs functions that support fire protection and EQ.

The intended functions within the scope of license renewal include the following:

*   provides rated fire barrier
*   provides heat transfer
*   provides pressure retaining boundary

In ALRA Table 2.3.3.B.32-1, the applicant identified the following yard structures ventilation system component types that are within the scope of license renewal and subject to an AMR:

- blowers
- dampers (includes fire dampers)
- ducting
- unit coolers

2.3B.3.32.2 Staff Evaluation

The staff reviewed ALRA Section 2.3.3.B.32 and USAR Sections 9.4.7 and 9B.4.4.3.4 using the evaluation methodology described in SER Section 2.3. The staff conducted its review in accordance with the guidance described in SRP-LR Section 2.3.

In conducting its review, the staff evaluated the system functions described in the ALRA and USAR in accordance with the requirements of 10 CFR 54.4(a) to verify that the applicant had not omitted from the scope of license renewal any components with intended functions delineated under 10 CFR 54.4(a). The staff then reviewed those components that the applicant had identified as being within the scope of license renewal to verify that the applicant had not omitted any passive and long-lived components that should be subject to an AMR in accordance with the requirements of 10 CFR 54.21(a)(1).

2.3B.3.32.3 Conclusion

The staff reviewed the ALRA to determine whether any SSCs that should be within the scope of license renewal had not been identified by the applicant. No omissions were identified. In addition, the staff performed a review to determine whether any components that should be subject to an AMR had not been identified by the applicant. No omissions were identified. On the basis of its review, the staff concludes that there is reasonable assurance that the applicant had adequately identified the yard structures ventilation system components that are within the scope of license renewal, as required by 10 CFR 54.4(a), and the yard structures ventilation system components that are subject to an AMR, as required by 10 CFR 54.21(a)(1).

2.3B.3.33 NMP2 Auxiliary Boiler System

2.3B.3.33.1 Summary of Technical Information in the Amended Application

In ALRA Section 2.3.3.B.33, the applicant described the auxiliary boiler system. The auxiliary boiler system is designed to supply primary loads during plant shutdown including building heating, radwaste process reboiler system, and other auxiliary system heat exchangers. As the auxiliary boilers are not normally used to augment the auxiliary steam system, auxiliary boiler steam may be used to provide a heat source to the off-gas system and clean steam reboilers prior to start-up.

The failure of NSR SSCs in the auxiliary boiler system could potentially prevent the satisfactory accomplishment of an SR function.

The intended function within the scope of license renewal is to maintain mechanical and structural integrity to prevent spatial interactions.

In ALRA Table 2.3.3.B.33-1, the applicant identified the following auxiliary boiler system component types that are within the scope of license renewal and subject to an AMR:

- accumulator
- bolting
- filter housing
- heat exchanger
- piping and fittings
- pumps
- restricting orifices
- tanks
- valves

2.3B.3.33.2 Staff Evaluation

The staff reviewed ALRA Section 2.3.3.B.33 and USAR Section 9.5.10 using the evaluation methodology described in SER Section 2.3. The staff conducted its review in accordance with the guidance described in SRP-LR Section 2.3.

In conducting its review, the staff evaluated the system functions described in the ALRA and USAR in accordance with the requirements of 10 CFR 54.4(a) to verify that the applicant had not omitted from the scope of license renewal any components with intended functions delineated under 10 CFR 54.4(a). The staff then reviewed those components that the applicant had identified as being within the scope of license renewal to verify that the applicant had not omitted any passive and long-lived components that should be subject to an AMR in accordance with the requirements of 10 CFR 54.21(a)(1).

2.3B.3.33.3 Conclusion

The staff reviewed the ALRA to determine whether any SSCs that should be within the scope of license renewal had not been identified by the applicant. No omissions were identified. In addition, the staff performed a review to determine whether any components that should be subject to an AMR had not been identified by the applicant. No omissions were identified. On the basis of its review, the staff concludes that there is reasonable assurance that the applicant had adequately identified the auxiliary boiler system components that are within the scope of license renewal, as required by 10 CFR 54.4(a), and the auxiliary boiler system components that are subject to an AMR, as required by 10 CFR 54.21(a)(1).

2.3B.3.34 NMP2 Circulating Water System

2.3B.3.34.1 Summary of Technical Information in the Amended Application

In ALRA Section 2.3.3.B.34, the applicant described the circulating water system. The function of the circulating water system is to provide the main condenser with a continuous supply of cooling water. The water is used to remove the heat discharged from the turbine exhaust and turbine bypass steam as well as from other equipment over the full range of operating loads. Makeup water for the circulating water system is obtained from the service water system. During the winter months warm water from the circulating water system is used to temper the lake intake water.

The failure of NSR SSCs in the circulating water system could potentially prevent the satisfactory accomplishment of an SR function.

The intended functions within the scope of license renewal include the following:

- maintains mechanical and structural integrity of NSR components to prevent spatial interactions with SR components

- maintains mechanical and structural integrity of NSR components that provide structural support to attached SR components

In ALRA Table 2.3.3.B.34-1, the applicant identified the following circulating water system component types that are within the scope of license renewal and subject to an AMR:

- bolting
- piping and fittings
- valves

2.3B.3.34.2 Staff Evaluation

The staff reviewed ALRA Section 2.3.3.B.34 and USAR Section 10.4.5 using the evaluation methodology described in SER Section 2.3. The staff conducted its review in accordance with the guidance described in SRP-LR Section 2.3.

In conducting its review, the staff evaluated the system functions described in the ALRA and USAR in accordance with the requirements of 10 CFR 54.4(a) to verify that the applicant had not omitted from the scope of license renewal any components with intended functions delineated under 10 CFR 54.4(a). The staff then reviewed those components that the applicant had identified as being within the scope of license renewal to verify that the applicant had not omitted any passive and long-lived components that should be subject to an AMR in accordance with the requirements of 10 CFR 54.21(a)(1).

2.3B.3.34.3 Conclusion

The staff reviewed the ALRA to determine whether any SSCs that should be within the scope of license renewal had not been identified by the applicant. No omissions were identified. In addition, the staff performed a review to determine whether any components that should be subject to an AMR had not been identified by the applicant. No omissions were identified. On the basis of its review, the staff concludes that there is reasonable assurance that the applicant had adequately identified the circulating water system components that are within the scope of license renewal, as required by 10 CFR 54.4(a), and the circulating water system components that are subject to an AMR, as required by 10 CFR 54.21(a)(1).

2.3B.3.35 NMP2 Makeup Water Treatment System

2.3B.3.35.1 Summary of Technical Information in the Amended Application

In ALRA Section 2.3.3.B.35, the applicant described the makeup water treatment system. The makeup water treatment system processes domestic water to supply the makeup water storage and transfer system with demineralized water. The system will also provide domestic water from the filtered water storage tank for seal water to the circulating water system pumps. The domestic water is pumped by one of the filter pumps from either the filtered water storage tank or the waste water recovery tanks through the water treating filter to the Ecolochem trailer. After processing through the Ecolochem trailer, it is pumped to the demineralized water storage tanks.

The failure of NSR SSCs in the makeup water treatment system could potentially prevent the satisfactory accomplishment of an SR function.

The intended function within the scope of license renewal is to maintain mechanical and structural integrity to prevent spatial interactions.

In ALRA Table 2.3.3.B.35-1, the applicant identified the following makeup water treatment system component types that are within the scope of license renewal and subject to an AMR:

- bolting
- piping and fittings
- valves

2.3B.3.35.2 Staff Evaluation

The staff reviewed ALRA Section 2.3.3.B.35 and USAR Section 9.2.3.2 using the evaluation methodology described in SER Section 2.3. The staff conducted its review in accordance with the guidance described in SRP-LR Section 2.3.

In conducting its review, the staff evaluated the system functions described in the ALRA and USAR in accordance with the requirements of 10 CFR 54.4(a) to verify that the applicant had not omitted from the scope of license renewal any components with intended functions delineated under 10 CFR 54.4(a). The staff then reviewed those components that the applicant had identified as being within the scope of license renewal to verify that the applicant had not omitted any passive and long-lived components that should be subject to an AMR in accordance with the requirements of 10 CFR 54.21(a)(1).

2.3B.3.35.3 Conclusion

The staff reviewed the ALRA to determine whether any SSCs that should be within the scope of license renewal had not been identified by the applicant. No omissions were identified. In addition, the staff performed a review to determine whether any components that should be subject to an AMR had not been identified by the applicant. No omissions were identified. On the basis of its review, the staff concludes that there is reasonable assurance that the applicant had adequately identified the makeup water treatment system components that are within the scope of license renewal, as required by 10 CFR 54.4(a), and the makeup water treatment system components that are subject to an AMR, as required by 10 CFR 54.21(a)(1).

2.3B.3.36 NMP2 Radioactive Liquid Waste Management System

2.3B.3.36.1 Summary of Technical Information in the Amended Application

In ALRA Section 2.3.3.B.36, the applicant described the radioactive liquid waste management system. The radioactive liquid waste management system is conceptually divided into four subsystems: the waste collector subsystem, the floor drain collector subsystem, the regenerant waste subsystem, and the phase separator subsystem. The waste collection subsystem provides for collection, filtering, and demineralizing of generally low conductivity waste. The floor drain collector system pumps provide necessary head and flow for mixing, sampling, or processing. Floor drain system water is normally processed using vendor-supplied equipment (Thermex). The two cleanup phase separator tanks accept RWCU filter/demineralizer backwashes. The spent fuel pool phase separator tank accepts spent fuel pool filter/demineralizer backwashes. The two regenerant waste tanks receive waste transferred from the waste neutralizer tank at the demineralizer regeneration system or from the radwaste chemical sump. Regenerant waste pumps provide necessary head and flow for mixing, sampling, and processing to vendor-supplied equipment (Thermex) or through spent resin. The spent resin tank accepts transfers from the phase separator tanks, filter backwash tank, waste sludge tank, Thermex, and demineralizer regeneration system. The tank is decanted by gravity drain to the floor drain collector tanks. The remaining waste is transferred to the waste sludge tank for transfer to a liner using solid radwaste procedures.

The failure of NSR SSCs in the radioactive liquid waste management system could potentially prevent the satisfactory accomplishment of an SR function.

The intended function within the scope of license renewal is to maintain mechanical and structural integrity to prevent spatial interactions.

In ALRA Table 2.3.3.B.36-1, the applicant identified the following radioactive liquid waste management system component types within the scope of license renewal and subject to an AMR:

- bolting
- filters
- piping and fittings
- pumps
- restricting orifice
- tanks
- valves

2.3B.3.36.2 Staff Evaluation

The staff reviewed ALRA Section 2.3.3.B.36 and USAR Section 11.2.1 using the evaluation methodology described in SER Section 2.3. The staff conducted its review in accordance with the guidance described in SRP-LR Section 2.3.

In conducting its review, the staff evaluated the system functions described in the ALRA and USAR in accordance with the requirements of 10 CFR 54.4(a) to verify that the applicant had not omitted from the scope of license renewal any components with intended functions delineated under 10 CFR 54.4(a). The staff then reviewed those components that the applicant had identified as being within the scope of license renewal to verify that the applicant had not omitted any passive and long-lived components that should be subject to an AMR in accordance with the requirements of 10 CFR 54.21(a)(1).

2.3B.3.36.3 Conclusion

The staff reviewed the ALRA to determine whether any SSCs that should be within the scope of license renewal had not been identified by the applicant. No omissions were identified. In addition, the staff performed a review to determine whether any components that should be subject to an AMR had not been identified by the applicant. No omissions were identified. On the basis of its review, the staff concludes that there is reasonable assurance that the applicant had adequately identified the radioactive liquid waste management system components that are within the scope of license renewal, as required by 10 CFR 54.4(a), and the radioactive liquid waste management system components that are subject to an AMR, as required by 10 CFR 54.21(a)(1).

2.3B.3.37 NMP2 Roof Drainage System

2.3B.3.37.1 Summary of Technical Information in the Amended Application

In ALRA Section 2.3.3.B.37, the applicant described the roof drainage system. The roof drainage system is designed to collect water accumulation on building roofs and transport it to Lake Ontario.

The failure of NSR SSCs in the roof drainage system could potentially prevent the satisfactory accomplishment of an SR function.

The intended function within the scope of license renewal is to maintain mechanical and structural integrity to prevent spatial interactions.

In ALRA Table 2.3.3.B.37-1, the applicant identified that the "Piping and Fittings" component type of the roof drainage system is within the scope of license renewal and subject to an AMR.

2.3B.3.37.2 Staff Evaluation

The staff reviewed ALRA Section 2.3.3.B.37 and USAR Section 2.4.2.3 using the evaluation methodology described in SER Section 2.3. The staff conducted its review in accordance with the guidance described in SRP-LR Section 2.3.

In conducting its review, the staff evaluated the system functions described in the ALRA and USAR in accordance with the requirements of 10 CFR 54.4(a) to verify that the applicant had not omitted from the scope of license renewal any components with intended functions delineated under 10 CFR 54.4(a). The staff then reviewed those components that the applicant had identified as being within the scope of license renewal to verify that the applicant had not

omitted any passive and long-lived components that should be subject to an AMR in accordance with the requirements of 10 CFR 54.21(a)(1).

2.3B.3.37.3 Conclusion

The staff reviewed the ALRA to determine whether any SSCs that should be within the scope of license renewal had not been identified by the applicant. No omissions were identified. In addition, the staff performed a review to determine whether any components that should be subject to an AMR had not been identified by the applicant. No omissions were identified. On the basis of its review, the staff concludes that there is reasonable assurance that the applicant had adequately identified the roof drainage system components that are within the scope of license renewal, as required by 10 CFR 54.4(a), and the roof drainage system components that are subject to an AMR, as required by 10 CFR 54.21(a)(1).

2.3B.3.38 NMP2 Sanitary Drains and Disposal System

2.3B.3.38.1 Summary of Technical Information in the Amended Application

In ALRA Section 2.3.3.B.38, the applicant described the sanitary drains and disposal system. The sanitary drains and disposal system is designed to treat and dispose of the waste from all plumbing fixtures except lavatories, sinks, and drains containing waste that is contaminated or potentially contaminated with chemicals or radioactivity. Such contaminated or potentially contaminated waste is physically segregated from the sanitary drains and disposal system and is connected to the floor and equipment drains systems. Noncontaminated sanitary waste from NMP2 flows by gravity to an underground wetwell (11,500-gallon storage capacity). The wetwell is located adjacent to a sewage lift station equipped with two sewage pumps to transport the waste to an on-site sanitary waste treatment facility. All noncontaminated waste lines are vented to the atmosphere.

The failure of NSR SSCs in the sanitary drains and disposal system could potentially prevent the satisfactory accomplishment of an SR function.

The intended function within the scope of license renewal is to maintain mechanical and structural integrity to prevent spatial interactions.

In ALRA Table 2.3.3.B.38-1, the applicant identified that the "Piping and Fittings" component type of the sanitary drains and disposal system is within the scope of license renewal and subject to an AMR.

2.3B.3.38.2 Staff Evaluation

The staff reviewed ALRA Section 2.3.3.B.38 and USAR Section 9.2.4 using the evaluation methodology described in SER Section 2.3. The staff conducted its review in accordance with the guidance described in SRP-LR Section 2.3.

In conducting its review, the staff evaluated the system functions described in the ALRA and USAR in accordance with the requirements of 10 CFR 54.4(a) to verify that the applicant had not omitted from the scope of license renewal any components with intended functions delineated under 10 CFR 54.4(a). The staff then reviewed those components that the applicant had identified as being within the scope of license renewal to verify that the applicant had not omitted any passive and long-lived components that should be subject to an AMR in accordance with the requirements of 10 CFR 54.21(a)(1).

2.3B.3.38.3 Conclusion

The staff reviewed the ALRA to determine whether any SSCs that should be within the scope of license renewal had not been identified by the applicant. No omissions were identified. In addition, the staff performed a review to determine whether any components that should be subject to an AMR had not been identified by the applicant. No omissions were identified. On the basis of its review, the staff concludes that there is reasonable assurance that the applicant had adequately identified the sanitary drains and disposal system components that are within the scope of license renewal, as required by 10 CFR 54.4(a), and the sanitary drains and disposal system components that are subject to an AMR, as required by 10 CFR 54.21(a)(1).

2.3B.3.39 NMP2 Service Water Chemical Treatment System

2.3B.3.39.1 Summary of Technical Information in the Amended Application

In ALRA Section 2.3.3.B.39, the applicant described the service water chemical treatment system. The service water chemical treatment system provides biocides and detoxification to the service water system to control MIC. The biocides (sodium hypochlorite and sodium bromide) are dripped into the service water intake bay and the detoxification agent (sodium bisulfite) is introduced into the two 30-inch return lines. The chemicals are stored in the refurbished acid and hypochlorite tanks. The chemicals are delivered by six skid-mounted dosing pumps, two sodium bisulfite pumps, two sodium hypochlorite pumps, and two sodium bromide pumps. Demineralized carrier water is supplied from the makeup water system and is used to deliver chemicals from the dosing pumps to the appropriate delivery point.

The failure of NSR SSCs in the service water chemical treatment system could potentially prevent the satisfactory accomplishment of an SR function.

The intended functions within the scope of license renewal include the following:

- maintains mechanical and structural integrity of NSR components to prevent spatial interactions with SR components
- maintains mechanical and structural integrity of NSR components that provide structural support to attached SR components

In ALRA Table 2.3.3.B.39-1, the applicant identified the following service water chemical treatment system component types that are within the scope of license renewal and subject to an AMR:

- bolting
- piping and fittings
- valves

2.3B.3.39.2 Staff Evaluation

The staff reviewed ALRA Section 2.3.3.B.39 and USAR Section 9.2.4 using the evaluation methodology described in SER Section 2.3. The staff conducted its review in accordance with the guidance described in SRP-LR Section 2.3.

In conducting its review, the staff evaluated the system functions described in the ALRA and USAR in accordance with the requirements of 10 CFR 54.4(a) to verify that the applicant had not omitted from the scope of license renewal any components with intended functions delineated under 10 CFR 54.4(a). The staff then reviewed those components that the applicant had identified as being within the scope of license renewal to verify that the applicant had not omitted any passive and long-lived components that should be subject to an AMR in accordance with the requirements of 10 CFR 54.21(a)(1).

2.3B.3.39.3 Conclusion

The staff reviewed the ALRA to determine whether any SSCs that should be within the scope of license renewal had not been identified by the applicant. No omissions were identified. In addition, the staff performed a review to determine whether any components that should be subject to an AMR had not been identified by the applicant. No omissions were identified. On the basis of its review, the staff concludes that there is reasonable assurance that the applicant had adequately identified the service water chemical treatment system components that are within the scope of license renewal, as required by 10 CFR 54.4(a), and the service water chemical treatment system components that are subject to an AMR, as required by 10 CFR 54.21(a)(1).

2.3B.3.40 NMP2 Turbine Building Closed Loop Cooling Water System

2.3B.3.40.1 Summary of Technical Information in the Amended Application

In ALRA Section 2.3.3.B.40, the applicant described the turbine building closed loop cooling water (TBCLCW) system. The TBCLCW system is a demineralized water, closed-cycle, heat transfer system that consists of three 50-percent capacity pumps and heat exchangers along with appropriate controls and instrumentation to ensure adequate cooling capacity for the TB and RWB auxiliary systems and components during normal plant operation. Heat removed from components by the TBCLCW system is transferred to the service water system. A surge and makeup tank accommodates system volume changes due to temperature variations, maintains static head on the pumps, and allows detection of gross leaks in the system. It also provides for normal leakage in the system. Makeup water to the surge tank is provided by the makeup water system.

The failure of NSR SSCs in the TBCLCW system could potentially prevent the satisfactory accomplishment of an SR function.

The intended function within the scope of license renewal is to maintain mechanical and structural integrity to prevent spatial interactions.

In ALRA Table 2.3.3.B.40-1, the applicant identified the following TBCLCW system component types that are within the scope of license renewal and subject to an AMR:

- bolting
- piping and fittings
- sample cooler
- valves

2.3B.3.40.2 Staff Evaluation

The staff reviewed ALRA Section 2.3.3.B.40 and USAR Section 9.2.7 using the evaluation methodology described in SER Section 2.3. The staff conducted its review in accordance with the guidance described in SRP-LR Section 2.3.

In conducting its review, the staff evaluated the system functions described in the ALRA and USAR in accordance with the requirements of 10 CFR 54.4(a) to verify that the applicant had not omitted from the scope of license renewal any components with intended functions delineated under 10 CFR 54.4(a). The staff then reviewed those components that the applicant had identified as being within the scope of license renewal to verify that the applicant had not omitted any passive and long-lived components that should be subject to an AMR in accordance with the requirements of 10 CFR 54.21(a)(1).

2.3B.3.40.3 Conclusion

The staff reviewed the ALRA to determine whether any SSCs that should be within the scope of license renewal had not been identified by the applicant. No omissions were identified. In addition the staff performed a review to determine whether any components that should be subject to an AMR had not been identified by the applicant. No omissions were identified. On the basis of its review, the staff concludes that there is reasonable assurance that the applicant had adequately identified the TBCLCW system components that are within the scope of license renewal, as required by 10 CFR 54.4(a), and the TBCLCW system components that are subject to an AMR, as required by 10 CFR 54.21(a)(1).

2.3B.4 Steam and Power Conversion Systems

In ALRA Section 2.3.4.B, the applicant identified the structures and components of the NMP2 steam and power conversion systems that are subject to an AMR for license renewal.

The applicant described the supporting structures and components of the steam and power conversion systems in the following sections of the ALRA:

- 2.3.4.B.1   NMP2 main condenser air removal system
- 2.3.4.B.2   NMP2 condensate system
- 2.3.4.B.3   NMP2 feedwater system
- 2.3.4.B.4   NMP2 main steam system
- 2.3.4.B.5   NMP2 moisture separator and reheater system
- 2.3.4.B.6   NMP2 extraction steam and feedwater heater drain system
- 2.3.4.B.7   NMP2 turbine main system

The staff's review findings regarding ALRA Sections 2.3.4.B.1 through 2.3.4.B.7 are presented in SER Sections 2.3B.4.1 through 2.3B.4.7, respectively.

2.3B.4.1 NMP2 Main Condenser Air Removal System

2.3B.4.1.1 Summary of Technical Information in the Amended Application

In ALRA Section 2.3.4.B.1, the applicant described the main condenser air removal system. The purpose of the main condenser air removal system is to establish and maintain a main condenser vacuum by removing air and noncondensable gases from the main condenser. This system consists of two subsystems. The hogging subsystem is used to establish condenser vacuum during plant startup. The holding subsystem is used to maintain condenser vacuum during normal plant operations.

The failure of NSR SSCs in the main condenser air removal system could potentially prevent the satisfactory accomplishment of an SR function.

The intended function within the scope of license renewal is to maintain mechanical and structural integrity to prevent spatial interactions.

In ALRA Table 2.3.4.B.1-1, the applicant identified the following main condenser air removal system component types that are within the scope of license renewal and subject to an AMR:

- air ejectors
- bolting
- heat exchangers
- piping and fittings
- valves

2.3B.4.1.2 Staff Evaluation

The staff reviewed ALRA Section 2.3.4.B.1 and USAR Section 10.4.2 using the evaluation methodology described in SER Section 2.3. The staff conducted its review in accordance with the guidance described in SRP-LR Section 2.3.

In conducting its review, the staff evaluated the system functions described in the ALRA and USAR in accordance with the requirements of 10 CFR 54.4(a) to verify that the applicant had not omitted from the scope of license renewal any components with intended functions delineated under 10 CFR 54.4(a). The staff then reviewed those components that the applicant had identified as being within the scope of license renewal to verify that the applicant had not omitted any passive and long-lived components that should be subject to an AMR in accordance with the requirements of 10 CFR 54.21(a)(1).

2.3B.4.1.3 Conclusion

The staff reviewed the ALRA to determine whether any SSCs that should be within the scope of license renewal had not been identified by the applicant. No omissions were identified. In addition, the staff performed a review to determine whether any components that should be subject to an AMR had not been identified by the applicant. No omissions were identified. On the basis of its review, the staff concludes that there is reasonable assurance that the applicant had adequately identified the main condenser air removal system components that are within the scope of license renewal, as required by 10 CFR 54.4(a), and the main condenser air removal system components that are subject to an AMR, as required by 10 CFR 54.21(a)(1).

2.3B.4.2 NMP2 Condensate System

2.3B.4.2.1 Summary of Technical Information in the Amended Application

In ALRA Section 2.3.4.B.2, the applicant described the condensate system. The condensate system provides a reliable supply of condensate to the feedwater (FW) system. The condensate system consists of the main condenser, three condensate pumps, three condensate booster pumps, three trains of drain coolers and low pressure heaters, controls, instrumentation, piping, valves, and associated equipment to supply the FW system with heated, high quality condensate.

For license renewal purposes the condensate system also includes the following systems: condensate makeup and drawoff system, condensate demineralizer system, condensate demineralizer system – mixed bed system, condensate booster pump lube oil system, and auxiliary condensate system. The condensate makeup and drawoff system provides makeup water to various systems in the plant, serves as a source of water during refueling operations, serves as reserve for the RCIC system and the HPCS system, and provides for condenser hotwell level control. The condensate demineralizer system and condensate demineralizer system – mixed bed system are designed to maintain reactor feedwater purity by the removal of soluble and insoluble impurities from the condensate. They also provide a means of cleaning the condensate resins. The condensate booster pump lube oil system provides lubricating oil to the condensate booster pump seals. The auxiliary condensate system provides level controls and condensate removal functions for systems, structures, and components that are supplied with auxiliary steam.

The condensate system contains SR components that are relied upon to remain functional during and following DBEs. The failure of NSR SSCs in the condensate system could potentially prevent the satisfactory accomplishment of an SR function. In addition, the condensate system performs functions that support fire protection, EQ, ATWS, and SBO.

The intended functions within the scope of license renewal include the following:

- maintains mechanical and structural integrity of NSR components to prevent spatial interactions with SR components

- provides removal and/or holdup of fission products

- provides pressure retaining boundary

In ALRA Table 2.3.4.B.2-1, the applicant identified the following condensate system component types that are within the scope of license renewal and subject to an AMR:

- bolting
- flow element
- heat exchanger
- main condenser
- piping and fittings
- pump
- restriction orifice
- tanks
- valves

2.3B.4.2.2 Staff Evaluation

The staff reviewed ALRA Section 2.3.4.B.2 and USAR Sections 9.2.6, 10.4.1, 10.4.6, and 10.4.7 using the evaluation methodology described in SER Section 2.3. The staff conducted its review in accordance with the guidance described in SRP-LR Section 2.3.

In conducting its review, the staff evaluated the system functions described in the ALRA and USAR in accordance with the requirements of 10 CFR 54.4(a) to verify that the applicant had not omitted from the scope of license renewal any components with intended functions delineated under 10 CFR 54.4(a). The staff then reviewed those components that the applicant had identified as being within the scope of license renewal to verify that the applicant had not omitted any passive and long-lived components that should be subject to an AMR in accordance with the requirements of 10 CFR 54.21(a)(1).

The staff's review of LRA Section 2.3.4.B.2 identified areas in which additional information was necessary to complete the review of the applicant's scoping and screening results. The applicant responded to the staff's RAIs as discussed below.

In RAI 2.3.4.B.2-1 dated November 19, 2004, the staff stated that drawing LR-004B-0 shows that check valve *V298 is subject to an AMR. However, the line on which this valve is located is shown not requiring an AMR on both the upstream (line 2-CNS-006-44-4) and downstream (line 2-CNS-006-298-4) sides of the valve. The staff believes that a failure in these lines could affect structural support of the valve, cause a discontinuity in pressure boundary across the valve, and possibly prevent the valve from performing its intended function. The applicant was asked to describe the intended function of check valve *V298 and justify why the valve line would not require AMR.

In its response dated December 22, 2004, the applicant stated that the intended function for this check valve is "Pressure Boundary" to provide secondary containment integrity. The applicant stated that all support to the line in which the valve is located is seismic that and the associated piping upstream and downstream of the check valve is not required to be within the scope of license renewal for pressure boundary. The applicant stated further that the main supply line into the secondary containment contains this check valve at a low point which in case of a pipe break outside the containment is sealed by a 70-foot column of water. A line break within the reactor building would provide a preferential flow path for containment atmosphere leakage into the reactor building atmosphere and any gaseous leakage would be

2-241

collected by the standby gas treatment system and thus not be classified as bypass leakage. The applicant also stated that the associated piping upstream and downstream of 2CNS*V298 is not required to be within the scope of license renewal for the pressure boundary intended function. However, the piping, on either end of the valve that is contained within the reactor building and piping tunnel is within the scope of license renewal and subject to AMR pursuant to 10 CFR 54.4(a)(2) criterion.

Based on its review, the staff found the applicant's response to RAI 2.3.4.B.2-1 acceptable because it adequately explained that although the check valve in question provides a pressure boundary function to preserve secondary containment integrity and is therefore in the scope of license renewal under 10 CFR 54.4(a) and subject to an AMR under 10 CFR 54.21(a) its upstream and downstream piping is not necessary for secondary containment integrity. Therefore, the staff's concern described in RAI 2.3.4.B.2-1 is resolved.

In RAI 2.3.4.B.2-2 dated November 19, 2004, the staff stated that NMP2 USAR page 9.2-43 states that the "condensate storage facility's condensate makeup and drawoff (CNS) system" which contains both condensate storage tanks "is not required to effect or support safe shutdown of the reactor or to support the operation of any nuclear safety system." However, NMP2 USAR page 8.3-64 states that the CST inventory is monitored daily to assure the "availability, adequacy, and capability to achieve and maintain a safe plant shutdown and to recover from an SBO for the four-hour coping duration." Further, LRA Section 2.3.4.B.2 states that the CSTs are within the scope of license renewal and subject to an AMR. Therefore, the staff requested that the applicant explain the apparent discrepancy described above.

In its response by letter dated December 22, 2004, the applicant stated that NMP2 USAR Sections 9.2.6.1.1 and 9.2.6.3 properly state the safety design basis for the condensate storage facility CNS system. The applicant stated that the CNS system is NSR and is not required to prevent or mitigate DBEs but that, as discussed in USAR Section 8.3.1.5, the inventory in the condensate storage tanks is credited in the station blackout coping analysis. The applicant stated further that USAR Section 9.2.6.3 notes the condensate storage requirements related to station blackout. Therefore, there is no discrepancy and the condensate storage tanks and other components identified in the LRA section are included correctly within the scope of license renewal.

Based on its review, the staff found the applicant's response to RAI 2.3.4.B.2-2 acceptable because it adequately explained that although the nonsafety-related condensate storage tanks in question provide a pressure boundary function to recover from an SBO and are therefore in the scope of license renewal and subject to an AMR they are not required to prevent or mitigate DBEs under 10 CFR 54.4(a)(1). Therefore, the staff's concern described in RAI 2.3.4.B.2-2 is resolved.

In RAI 2.3.4.B.2-3 dated November 19, 2004, the staff requested that the applicant clarify inconsistencies in drawing LR-004A-0. The applicant's response by letter dated December 22, 2004, has been incorporated in the ALRA as discussed below.

In its ALRA dated July 14, 2005, the applicant provided the staff with a revised LR drawing correcting the inconsistencies and accurately depicting the components subject to an AMR under 10 CFR 54.4(a)(2). Therefore, the staff's concern described in RAI 2.3.3.B.2-3 is resolved.

In RAI 2.3.4.B.2-4 dated November 17, 2004, the staff stated that drawing LR-003A-0 shows lines from connections labeled 39 on condenser 1A and 1C to pressure transmitters 46A,B and 46C,D respectively. These instruments transmit condenser vacuum pressure. Upon loss of condenser vacuum the signal from these transmitters will effect a reactor scram and main turbine trip. However, only a segment of these lines from valves V2A and V2B to their respective transmitters is shown within the scope of license renewal. The segment from the condenser connection up to and including these valves is shown outside of scope. Since these transmitters perform a safety function the staff believed the entire line should have been within scope. Further, the drawing did not show corresponding lines and transmitters for condenser 1B. Therefore, the staff requested that the applicant justify exclusion of such line segments and valves from the scope of license renewal and to explain the absence of the corresponding lines and transmitters for condenser 1B.

In its response by letter dated December 22, 2004, the applicant stated that condenser 1B was not provided with the safety-related pressure transmitters per the original design and that the three condenser shells are connected by equalizing ducts in the condenser necks and by condensate pipes between the hotwells.

The applicant stated further that the SR pressure transmitters monitor main condenser vacuum and upon loss of vacuum will effect main steam line isolation and reactor trip, that the main condenser itself is NSR and not required for the safe shutdown of the reactor that the line segment from the main condenser up to and including valves addressed in the RAI form part of the NSR main condenser boundary, and that the SR pressure transmitters and the connecting tubing function to monitor loss of main condenser vacuum due to air in-leakage caused by failure of the NSR main condenser boundary.

Based on its review, the staff found the applicant's response to RAI 2.3.4.B.2-4 acceptable because it adequately explained that although the safety-related pressure transmitters effect a main steam line isolation and reactor trip on loss of condenser vacuum, the transmitters' piping and condensers they monitor are non-safety related, not required for safe shutdown of the reactor, and have no intended functions under 10 CFR 54.4(a). Therefore, the staff's concern described in RAI 2.3.4.B.2-4 is resolved.

In RAI 2.3.4.B.2-5 dated November 17, 2004, the staff stated that LR drawings LR-004A-0, LR-033B-0, and LR-035D-0 show that the vent on each condensate tank does not require AMR while the tank itself is within scope and subject to an AMR. The staff believed that failure of this vent could prevent the tank from performing its intended function by debris falling into the tank and blocking the supply lines to the RCIC and HPCS systems or tank collapse due to inadequate venting. Therefore, the staff requested that the applicant justify exclusion of the condensate tank vents from requiring an AMR.

In its response by letter dated December 22, 2004, the applicant stated that the NSR condensate storage tank vent line does not have an intended function for license renewal and therefore is not in-scope. The applicant stated that the condensate piping connects to the side of the tank and any potential negligible debris from general corrosion from the vent line would settle to the bottom of the tank with no possibility of condensate piping blockage. The applicant stated further that a sudden catastrophic failure of the vent line blocking the vent path is also implausible and that elimination of the vent piping would result in a hole in the tank which would act as a vent path.

Based on its review, the staff found the applicant's response to RAI 2.3.4.B.2-6 acceptable because it adequately explained that due to the configuration of the suction piping to the tank and the geometry of the vent line piping debris falling into the tank would not cause loss of the venting function. Therefore, the staff's concern described in RAI 2.3.4.B.2-5 is resolved.

In RAI 2.3.4.B.2-6 dated November 17, 2004, the staff stated that on drawing LR-033B-0, the acronyms "F1" and "GEX1" are shown encircled at various locations. However, they were not defined in the LRA, the USAR, the drawing legend, or on the drawing itself. The applicant was asked to define the acronyms F1 and GEX1.
In its response by letter dated December 22, 2004, the applicant stated that the "F1" and "GEX1" notations relate to the ASME in-service pressure test program.

Based on its review, the staff found the applicant's response to RAI 2.3.4.B.2-6 acceptable because it adequately explained that the acronyms F1 and GEX1 shown on LR drawings relate to the ASME in-service pressure test program and may be ignored for LRA purposes. Therefore, the staff's concern described in RAI 2.3.4.B.2-6 is resolved.

2.3B.4.2.3 Conclusion

The staff reviewed the ALRA, RAI responses, and accompanying scoping boundary drawings to determine whether any SSCs that should be within the scope of license renewal had not been identified by the applicant. No omissions were identified. In addition, the staff performed a review to determine whether any components that should be subject to an AMR had not been identified by the applicant. No omissions were identified. On the basis of its review, the staff concludes that there is reasonable assurance that the applicant had adequately identified the condensate system components that are within the scope of license renewal, as required by 10 CFR 54.4(a), and the condensate system components that are subject to an AMR, as required by 10 CFR 54.21(a)(1).

2.3B.4.3 NMP2 Feedwater System

2.3B.4.3.1 Summary of Technical Information in the Amended Application

In ALRA Section 2.3.4.B.3, the applicant described the FW system. The FW system provides a reliable supply of feedwater to the reactor at the temperature, pressure, quality, and flow rate required by the reactor. Connections from the zinc injection passivation system are provided on both the suction and discharge to the feedwater pumps. The RWCU system also connects to the FW system between the feedwater heaters and system isolation valves.

For license renewal purposes the FW system also includes the following systems: feedwater pump seals and leakoff system, feedwater pump recirculation balance drum leakoff system, and feedwater pump drive lube oil system. The feedwater pump seals and leakoff system provides seal water to the pump mechanical seals from the condensate booster pump discharge. The seal water minimizes pump mechanical seal leakage and cools the pump seals to minimize seal degradation. The feedwater pump recirculation balance drum leakoff system provides minimum flow protection for each feedwater pump via a recirculation line to the main condenser. The feedwater pump drive lube oil system provides lube oil to the reactor feed pumps.

The FW system contains SR components that are relied upon to remain functional during and following DBEs. The failure of NSR SSCs in the FW system could potentially prevent the satisfactory accomplishment of an SR function. In addition, the FW system performs functions that support EQ and ATWS.

The intended functions within the scope of license renewal include the following:

- maintains mechanical and structural integrity of NSR components to prevent spatial interactions with SR components

- provides removal and/or holdup of fission products

- provides pressure retaining boundary

- maintains mechanical and structural integrity of NSR components that provide structural support to attached SR components

In ALRA Table 2.3.4.B.3-1, the applicant identified the following FW system component types that are within the scope of license renewal and subject to an AMR:

- bolting
- flow element
- heat exchanger
- piping and fittings
- pump
- restriction orifice
- strainer
- valves

2.3B.4.3.2 Staff Evaluation

The staff reviewed ALRA Section 2.3.4.B.3 and USAR Section 10.4.7 using the evaluation methodology described in SER Section 2.3. The staff conducted its review in accordance with the guidance described in SRP-LR Section 2.3.

In conducting its review, the staff evaluated the system functions described in the ALRA and USAR in accordance with the requirements of 10 CFR 54.4(a) to verify that the applicant had not omitted from the scope of license renewal any components with intended functions delineated under 10 CFR 54.4(a). The staff then reviewed those components that the applicant had identified as being within the scope of license renewal to verify that the applicant had not omitted any passive and long-lived components that should be subject to an AMR in accordance with the requirements of 10 CFR 54.21(a)(1).

The staff's review of LRA Section 2.3.4.B.3 identified areas in which additional information was necessary to complete the review of the applicant's scoping and screening results. The applicant responded to the staff's RAIs as discussed below.

In RAI 2.3.4.B.3-1 dated November 19, 2004, the staff stated that drawing LR-037B shows segments of piping labeled 2-WCS-008-89-1 and 2-WCS-008-250-1 within the scope of license renewal and subject to an AMR and the branch lines that connect these segments to temperature elements TE79A and TE79B not subject to an AMR. In the LR drawing it appeared that these branch lines are exposed to the same fluid and not isolated from the lines they connect. Therefore, the staff believed they should be subject to an AMR. The applicant was asked to justify exclusion of the abovementioned branch lines from requiring an AMR.

In its response by letter dated December 22, 2004, the applicant stated that the LR drawing is incorrect and does not properly show the components within the scope of license renewal and subject to an AMR, that the components should have been highlighted on the LR drawing showing them within the scope of license renewal under 10 CFR 54.4(a) and are subject to an AMR under 10 CFR 54.21(a) but inadvertently were not highlighted. The components have been included within the scope of license renewal and subject to an AMR.

Based on its review, the staff found the applicant's response to RAI 2.3.4.B.3-1 acceptable because it adequately explained that the components in question are within the scope of license renewal under 10 CFR 54.4(a) and subject to an AMR under 10 CFR 54.21(a) but inadvertently were not highlighted on the LR drawing. The staff concludes that there is reasonable assurance that the components were included correctly within the scope of license renewal and subject to an AMR. Therefore, the staff's concern described in RAI 2.3.4.B.3-1 is resolved.

In RAI 2.3.4.B.3-2 dated November 19, 2004, the staff stated that on drawing LR-006A the only components shown within the scope of license renewal are eight valves located on the discharge side of the three reactor feed pumps (LV10A, -B, -C, FV2A, -B, -C, and LV55A, -B). However, the piping on both the upstream and downstream sides of these valves is shown outside of scope and not subject to an AMR. Further, on LRA page 2.3-203 the paragraph which describes the portions of the system containing components subject to an AMR does not reference these valves. Therefore, the staff requested that the applicant describe the intended function of the eight valves per 10 CFR 50.54(a)(1) or (a)(3) and discuss the effect of a pressure boundary breach in the lines housing these valves on the ability of the valves to perform their intended function.

In its response by letter dated December 22, 2004, the applicant stated that the valves located on the discharge side of the reactor feed pumps and the feedwater bypass valves to the condenser work in conjunction to mitigate the consequences of an anticipated transient without scram (ATWS) event by isolating feed flow to the reactor and diverting flow to the main condenser. The applicant stated that although the valves are within the scope of license renewal and NSR they do not have the pressure boundary intended function to accomplish the system level ATWS function. Further, the applicant explained that a pressure boundary breach in the lines housing these valves will not affect the ability of the valves to prevent feedwater flow to the reactor.

Based on its review, the staff found the applicant's response to RAI 2.3.4.B.3-2 acceptable because it adequately explained that although the nonsafety-related valves in question work in conjunction to isolate feedwater flow to the reactor to mitigate an ATWS and are therefore in the scope of license renewal under 10 CFR 54.4(a)(3) and subject to an AMR under

10 CFR 54.21(a) a pressure boundary breach in their associated piping would not affect this intended function. Therefore, the staff's concern described in RAI 2.3.4.B.3-2 is resolved.

2.3B.4.3.3 Conclusion

The staff reviewed the ALRA, RAI responses, and accompanying scoping boundary drawings to determine whether any SSCs that should be within the scope of license renewal had not been identified by the applicant. No omissions were identified. In addition, the staff performed a review to determine whether any components that should be subject to an AMR had not been identified by the applicant. No omissions were identified. On the basis of its review, the staff concludes that there is reasonable assurance that the applicant had adequately identified the FW system components that are within the scope of license renewal, as required by 10 CFR 54.4(a), and the FW system components that are subject to an AMR, as required by 10 CFR 54.21(a)(1).

2.3B.4.4 NMP2 Main Steam System

2.3B.4.4.1 Summary of Technical Information in the Amended Application

In ALRA Section 2.3.4.B.4, the applicant described the main steam system. The main steam system provides high pressure steam from the RPV to the main turbine and the reheating side of the moisture separator/reheater. The main steam system also provides steam to the RCIC system for operation of its turbine-driven pump. The main steam system consists of four main steam lines, eight main steam isolation valves, eighteen safety relief valves, controls, instrumentation, piping, valves, and associated equipment. Seven of the safety relief valves are used by the automatic depressurization system to depressurize the RPV during accident conditions.

For license renewal purposes the main steam system also includes the auxiliary steam system and the main steam safety valves vents and drains system. The auxiliary steam system provides reduced pressure steam to the steam jet air ejectors, offgas preheaters, clean steam reboiler, building heating intermediate heat exchanger and is the backup steam supply for the turbine gland seal system. The main steam safety valves vents and drains system directs high pressure steam from the safety relief valves to the suppression pool.

The main steam system contains SR components that are relied upon to remain functional during and following DBEs. The failure of NSR SSCs in the main steam system could potentially prevent the satisfactory accomplishment of an SR function. In addition, the main steam system performs functions that support fire protection, EQ, and SBO.

The intended functions within the scope of license renewal include the following:

- maintains mechanical and structural integrity of NSR components to prevent spatial interactions with SR components

- provides removal and/or holdup of fission products

- provides pressure retaining boundary

- maintains mechanical and structural integrity of NSR components that provide structural support to attached SR components
- provides flow restriction

In ALRA Table 2.3.4.B.4-1, the applicant identified the following main steam system component types that are within the scope of license renewal and subject to an AMR:

- "T" quenchers
- bolting
- condensing chambers ·
- flexible hose
- flow elements
- piping and fittings
- restriction orifices
- strainers
- valves

2.3B.4.4.2 Staff Evaluation

The staff reviewed ALRA Section 2.3.4.B.4 and USAR Sections 5.2.2, 5.4, and 10.3 using the evaluation methodology described in SER Section 2.3. The staff conducted its review in accordance with the guidance described in SRP-LR Section 2.3.

In conducting its review, the staff evaluated the system functions described in the ALRA and USAR in accordance with the requirements of 10 CFR 54.4(a) to verify that the applicant had not omitted from the scope of license renewal any components with intended functions delineated under 10 CFR 54.4(a). The staff then reviewed those components that the applicant had identified as being within the scope of license renewal to verify that the applicant had not omitted any passive and long-lived components that should be subject to an AMR in accordance with the requirements of 10 CFR 54.21(a)(1).

The staff's review of LRA Section 2.3.4.B.4 identified areas in which additional information was necessary to complete the review of the applicant's scoping and screening results. The applicant responded to the staff's RAIs as discussed below.

In RAI 2.3.4.B.4-1 dated November 19, 2004, the staff requested that the applicant provide information concerning the auxiliary steam system, specifically if it had been included within the scope of license renewal and subject to an AMR. The applicant's response by letter dated December 22, 2004, has been incorporated in the ALRA as discussed below.

The applicant stated in its response letter that it had transferred all of the auxiliary steam system components to the main steam system and thus evaluated them as part of this system in the LRA but that because none of the auxiliary steam system components were within the scope of license renewal per 10 CFR 54.4(a)(1) or (a)(3) they were not included on any LR drawings. Consistent with the system description in LRA Section 2.3.4.B.4, the applicant explained that any of the fluid-filled main steam or auxiliary steam system components located in the main steam tunnel, the reactor building, or the turbine building are within the scope of license renewal and subject to an AMR per 10 CFR 54.4(a)(2). Consistent with LR drawing convention, components within the scope of license renewal and subject to an AMR per 10 CFR 54.4(a)(2) only, are not to be shown in red.

In its ALRA dated July 14, 2005, the applicant provided the requested information. Based on review of the information submitted in the ALRA the staff's concern described in RAI 2.3.3.B.4-1 is resolved.

In RAI 2.3.4.B.4-2 dated November 17, 2004, the staff stated that LR drawings LR-1E-0 and LR-1F-0 show the inboard and outboard main steam isolation valves (MSIVs) respectively for each of the four main steam lines. These valves perform an SR function (system isolation) and are shown as requiring an AMR on the drawings. However, the pneumatic actuators for these valves are not shown to require AMR. Because the actuators are required to effect operation of the MSIVs the staff believes they likewise should be subject to an AMR. The applicant was asked to justify exclusion of the MSIV actuators from requiring an AMR.

In its response by letter dated December 22, 2004, the applicant stated that the MSIV pneumatic actuators are within the scope of license renewal and subject to an AMR for a pressure boundary intended function. The applicant further stated that AMR of these actuators will be performed and LRA revisions to incorporate the AMR results and any other associated LRA changes will be submitted by February 28, 2005.

In evaluating this response the staff found it incomplete and that review of LRA Section 2.3.4.B.4 could not be completed. Although it adequately explained that the MSIV pneumatic actuators in question are within the scope of license renewal and subject to an AMR, however, the applicant did not provide LRA revisions to incorporate the AMR results and any other associated LRA changes. The staff held a teleconference with the applicant on January 27, 2005, to discuss information necessary to resolve the concern in RAI 2.3.4.B.4-2. The product of the teleconference was an agreement by the applicant to transmit the required information by a follow-up letter.

By letter dated January 31, 2005, the applicant stated that NMPNS has reviewed the function of the NMP2 MSIV actuators and has concluded that an AMR is not required for license renewal. The eight MSIVs are air-operated valves normally open with a fail-safe position of closed. The actuator is a double-acting cylinder and air is used to move the valve in both the open and closed directions. The valves are also equipped with closing springs that will close them upon loss of air pressure. Valve closure following loss of air is assisted by air directed from an air tank accumulator to the top of the actuator cylinder but the closing spring forces are sufficient to meet the accident analysis time limit (3 to 10 seconds) for MSIV closure without the air assist feature. Therefore, the air pressure boundary function of the actuators is not required for the MSIVs to travel to their fail-safe (closed) positions. The applicant stated that more information regarding the design and evaluation of the MSIVs is available in NMP2 USAR Section 5.4.5.

Based on its review, the staff found the applicant's response to RAI 2.3.4.B.4-2, including the information in the teleconference and letter dated January 31, 2005, acceptable, because it adequately explained that the MSIV air cylinder actuators possess no pressure boundary intended function and therefore do not require an AMR. Additionally, the response described the USAR section applicable to these valve actuators. Therefore, the staff's concern described in RAI 2.3.4.B.4-2 is resolved.

2.3B.4.4.3 Conclusion

The staff reviewed the ALRA, RAI responses, and accompanying scoping boundary drawings to determine whether any SSCs that should be within the scope of license renewal had not been identified by the applicant. No omissions were identified. In addition, the staff performed a review to determine whether any components that should be subject to an AMR had not been identified by the applicant. No omissions were identified. On the basis of its review, the staff concludes that there is reasonable assurance that the applicant had adequately identified the main steam system components that are within the scope of license renewal, as required by 10 CFR 54.4(a), and the main steam system components that are subject to an AMR, as required by 10 CFR 54.21(a)(1).

2.3B.4.5 NMP2 Moisture Separator and Reheater System

2.3B.4.5.1 Summary of Technical Information in the Amended Application

In ALRA Section 2.3.4.B.5, the applicant described the moisture separator and reheater system. The moisture separator and reheater system removes entrained moisture from the high pressure turbine exhaust and reheats the dried steam to superheated conditions before it passes on to the low pressure turbine. The moisture separator and reheater system encompasses the cold reheat steam, hot reheat steam, moisture separator and reheater vents, and moisture separator vents and drains systems.

The failure of NSR SSCs in the moisture separator and reheater system could potentially prevent the satisfactory accomplishment of an SR function.

The intended function within the scope of license renewal is to maintain mechanical and structural integrity to prevent spatial interactions.

In ALRA Table 2.3.4.B.5-1, the applicant identified the following moisture separator and reheater system component types that are within the scope of license renewal and subject to an AMR:

- bolting
- heat exchanger
- piping and fittings
- restriction orifice
- strainer
- tank
- valve

2.3B.4.5.2 Staff Evaluation

The staff reviewed ALRA Section 2.3.4.B.5 and USAR Sections 10.1 and 10.2.2.1 using the evaluation methodology described in SER Section 2.3. The staff conducted its review in accordance with the guidance described in SRP-LR Section 2.3.

In conducting its review, the staff evaluated the system functions described in the ALRA and USAR in accordance with the requirements of 10 CFR 54.4(a) to verify that the applicant had not omitted from the scope of license renewal any components with intended functions delineated under 10 CFR 54.4(a). The staff then reviewed those components that the applicant had identified as being within the scope of license renewal to verify that the applicant had not omitted any passive and long-lived components that should be subject to an AMR in accordance with the requirements of 10 CFR 54.21(a)(1).

2.3B.4.5.3 Conclusion

The staff reviewed the ALRA to determine whether any SSCs that should be within the scope of license renewal had not been identified by the applicant. No omissions were identified. In addition, the staff performed a review to determine whether any components that should be subject to an AMR had not been identified by the applicant. No omissions were identified. On the basis of its review, the staff concludes that there is reasonable assurance that the applicant had adequately identified the moisture separator and reheater system components that are within the scope of license renewal, as required by 10 CFR 54.4(a), and the moisture separator and reheater system components that are subject to an AMR, as required by 10 CFR 54.21(a)(1).

2.3B.4.6 NMP2 Extraction Steam and Feedwater Heater Drain System

2.3B.4.6.1 Summary of Technical Information in the Amended Application

In ALRA Section 2.3.4.B.6, the applicant described the extraction steam and feedwater heater drain system. The extraction steam and feedwater heater drain system is designed to heat the reactor feedwater to meet reactor inlet requirements. The system also provides heating steam to the building heating intermediate heat exchangers and clean steam reboilers. The low-pressure section of the system consists of three independent strings of feedwater heaters, each containing two drain coolers and five closed feedwater heaters. The high-pressure section consists of three strings each with one closed feedwater heater.

The failure of NSR SSCs in the extraction steam and feedwater heater drain system could potentially prevent the satisfactory accomplishment of an SR function.

The intended function within the scope of license renewal is to maintain mechanical and structural integrity to prevent spatial interactions.

In ALRA Table 2.3.4.B.6-1, the applicant identified the following extraction steam and feedwater heater drain system component types that are within the scope of license renewal and subject to an AMR:

- bolting
- heat exchangers
- piping and fittings
- pumps
- tanks
- valves

2.3B.4.6.2 Staff Evaluation

The staff reviewed ALRA Section 2.3.4.B.6 and USAR Section 10.4.10 using the evaluation methodology described in SER Section 2.3. The staff conducted its review in accordance with the guidance described in SRP-LR Section 2.3.

In conducting its review, the staff evaluated the system functions described in the ALRA and USAR in accordance with the requirements of 10 CFR 54.4(a) to verify that the applicant had not omitted from the scope of license renewal any components with intended functions delineated under 10 CFR 54.4(a). The staff then reviewed those components that the applicant had identified as being within the scope of license renewal to verify that the applicant had not omitted any passive and long-lived components that should be subject to an AMR in accordance with the requirements of 10 CFR 54.21(a)(1).

2.3B.4.6.3 Conclusion

The staff reviewed the ALRA to determine whether any SSCs that should be within the scope of license renewal had not been identified by the applicant. No omissions were identified. In addition, the staff performed a review to determine whether any components that should be subject to an AMR had not been identified by the applicant. No omissions were identified. On the basis of its review, the staff concludes that there is reasonable assurance that the applicant had adequately identified the extraction steam and feedwater heater drain system components that are within the scope of license renewal, as required by 10 CFR 54.4(a), and the extraction steam and feedwater heater drain system components that are subject to an AMR, as required by 10 CFR 54.21(a)(1).

2.3B.4.7 NMP2 Turbine Main System

2.3B.4.7.1 Summary of Technical Information in the Amended Application

In ALRA Section 2.3.4.B.7, the applicant described the turbine main system. The turbine main system converts the thermal energy contained in the steam from the reactor into electrical energy. The turbine is a tandem-compound, single-stage reheat unit with 38-inch last-stage, low-pressure buckets. It consists of a double-flow, high-pressure turbine and three double-flow, low-pressure turbines.

The turbine main system consists of multiple subsystems including the main turbine system, turbine electric hydraulic oil and controls system, turbine generator gland seal and exhaust steam system, turbine generator lube oil system, turning gear and seal system, turbine generator oil conditioner and storage system, turbine main alarms and trips system, turbine main lube oil system, turbine main supervisory instrumentation system, and the turbine plant equipment vents system. Of these systems the turbine generator gland seal and exhaust steam system and the turbine electric hydraulic oil and controls system, specifically the turbine overspeed trip system, are in scope for license renewal. The turbine gland sealing system is designed to provide clean sealing steam for the turbine shaft and turbine steam control valves and to exhaust air drawn into the system to the stack. The sealing steam prevents steam leakage out through the high-pressure turbine shaft and turbine steam control valves (i.e., stop valves, control valves, bypass valves, and combined intermediate valves) and prevents air in-leakage through the low-pressure turbine shaft. The turbine generator has an emergency trip system which will close the main stop valves, control valves, and low-pressure turbine combined intermediate valves upon receipt of various protective signals, including a mechanical (110 percent) or electrical (112 percent) overspeed trip signal. These setpoints prevent the turbine rotor from exceeding the maximum transient speed of 120 percent (design overspeed) of rated turbine speed.

The failure of NSR SSCs in the turbine main system could potentially prevent the satisfactory accomplishment of an SR function.

The intended function within the scope of license renewal is to maintain mechanical and structural integrity to prevent spatial interactions.

In ALRA Table 2.3.4.B.7-1, the applicant identified the following turbine main system component types that are within the scope of license renewal and subject to an AMR:

- bolting
- heat exchanger
- piping and fittings
- restriction orifice
- tank
- valves

2.3B.4.7.2 Staff Evaluation

The staff reviewed ALRA Section 2.3.4.B.7 and USAR Section 10 using the evaluation methodology described in SER Section 2.3. The staff conducted its review in accordance with the guidance described in SRP-LR Section 2.3.

In conducting its review, the staff evaluated the system functions described in the ALRA and USAR in accordance with the requirements of 10 CFR 54.4(a) to verify that the applicant had not omitted from the scope of license renewal any components with intended functions delineated under 10 CFR 54.4(a). The staff then reviewed those components that the applicant had identified as being within the scope of license renewal to verify that the applicant had not omitted any passive and long-lived components that should be subject to an AMR in accordance with the requirements of 10 CFR 54.21(a)(1).

2.3B.4.7.3 Conclusion

The staff reviewed the ALRA to determine whether any SSCs that should be within the scope of license renewal had not been identified by the applicant. No omissions were identified. In addition, the staff performed a review to determine whether any components that should be subject to an AMR had not been identified by the applicant. No omissions were identified. On the basis of its review, the staff concludes that there is reasonable assurance that the applicant had adequately identified the turbine main system components that are within the scope of license renewal, as required by 10 CFR 54.4(a), and the turbine main system components that are subject to an AMR, as required by 10 CFR 54.21(a)(1).

2.4 Scoping and Screening Results: Structures and Component Supports

This section documents the staff's review of the applicant's scoping and screening results for structures and component supports. Specifically, this section discusses the following structures, component supports, and commodities:

- NMP1 structures
- NMP2 structures
- NMPNS structural commodities

In accordance with the requirements of 10 CFR 54.21(a)(1), the applicant identified and listed passive, long-lived SCs that are within the scope of license renewal and subject to an AMR. To verify that the applicant properly implemented its methodology, the staff focused its review on the implementation results. This approach allowed the staff to confirm that there were no omissions of structures and components that meet the scoping criteria and are subject to an AMR.

Staff Evaluation Methodology. The staff's evaluation of the information provided in the ALRA was performed in the same manner for all structures, components, and commodities. The objective of the review was to determine if the components and supporting structures for a specific structure, component, or commodity that appeared to meet the scoping criteria specified in 10 CFR Part 54 were identified by the applicant as within the scope of license renewal in accordance with 10 CFR 54.4. Similarly, the staff evaluated the applicant's screening results to verify that all long-lived, passive components were subject to an AMR in accordance with 10 CFR 54.21(a)(1).

Scoping. To perform its evaluation, the staff reviewed the applicable ALRA section and associated component drawings, focusing its review on components that had not been identified as within the scope of license renewal. The staff reviewed relevant licensing basis documents, including the NMP1 UFSAR and NMP2 USAR, for each structure, component, and commodity to determine if the applicant had omitted components with intended functions delineated under 10 CFR 54.4(a) from the scope of license renewal. The staff also reviewed the licensing basis documents to determine if all intended functions delineated under 10 CFR 54.4(a) were specified in the ALRA. If omissions were identified, the staff requested additional information to resolve the discrepancies.

Screening. Once the staff completed its review of the scoping results, the staff evaluated the applicant's screening results. For those structures, components, and commodities with intended functions, the staff sought to determine (1) if the functions are performed with moving parts or a change in configuration or properties or (2) if they are subject to replacement based on a qualified life or specified time period as described in 10 CFR 54.21(a)(1). For those that did not meet either of these criteria, the staff sought to confirm that these structures, components, and commodities were subject to an AMR as required by 10 CFR 54.21(a)(1). If discrepancies were identified, the staff requested additional information to resolve them.

2.4A NMP1 Structures

In ALRA Section 2.4.A, the applicant identified the NMP1 structures that are subject to an AMR for license renewal.

The applicant described the supporting systems and structures for the structures in the following sections of the ALRA:

- 2.4.A.1   NMP1 primary containment structure
- 2.4.A.2   NMP1 reactor building
- 2.4.A.3   NMP1 essential yard structures
- 2.4.A.4   NMP1 fuel handling system
- 2.4.A.5   NMP1 material handling system
- 2.4.A.6   NMP1 offgas building
- 2.4.A.7   NMP1 personnel/equipment access system
- 2.4.A.8   NMP1 radwaste solidification and storage building
- 2.4.A.9   NMP1 screen and pump house building
- 2.4.A.10   NMP1 turbine building
- 2.4.A.11   NMP1 vent stack
- 2.4.A.12   NMP1 waste disposal building

The staff's review findings regarding ALRA Sections 2.4.A.1 through 2.4.A.12 are presented in SER Sections 2.4A.1.1 through 2.4A.1.12 respectively.

2.4A.1 NMP1 Primary Containment Structure

2.4A.1.1 Summary of Technical Information in the Amended Application

In ALRA Section 2.4.A.1, the applicant described the primary containment structure (PCS). The PCS is a seismic Class I structure. The primary containment is a Mark I design that consists of a drywell, a suppression chamber in the shape of a torus, and a connecting vent system between the drywell and the suppression chamber. It also includes valves and piping associated with the vacuum breaker system and the structural portions of primary containment penetrations. The drywell is a steel pressure vessel in the shape of an inverted light bulb. The drywell is enclosed in reinforced concrete for shielding purposes. The stiffened pressure suppression chamber is a steel pressure vessel in the shape of a torus located below and encircling the drywell. The PCS is part of a multibarrier system with a primary barrier consisting of the primary containment with its pressure suppression system and a secondary barrier consisting of the RB. The PCS contains the released steam in the event of the design basis LOCA to limit the release to the RB of fission products associated with this accident. The PCS is an enclosure for the RPV, the reactor recirculation system, and other branch connections of the reactor coolant pressure boundary.

The PCS contains SR components that are relied upon to remain functional during and following DBEs. The failure of NSR SSCs in the PCS could potentially prevent the satisfactory accomplishment of an SR function.

The intended functions within the scope of license renewal include the following:

- provides spray shield or curbs for directing flow

- provides shielding against high energy line breaks

- provides missile barrier

- provides structural support to NSR components whose failure could prevent accomplishment of SR function(s)

- provides pressure retaining boundary

- provides shelter/protection to SR components

- provides shielding against radiation

- provides pressure boundary or essentially leaktight barrier

- provides structural and/or functional support to SR equipment

In ALRA Table 2.4.A.1-1, the applicant identified the following PCS component types that are within the scope of license renewal and subject to an AMR:

- beam seats
- bearing plates
- concrete & grout
- containment penetrations (electrical)
- containment penetrations (instrument)
- containment penetrations (mechanical)
- downcomer tie straps
- drywell
- drywell coating
- drywell equipment hatch
- drywell emergency airlock
- drywell floor
- drywell floor seal
- drywell head
- drywell head closure bolts
- drywell head manway
- drywell jet deflector
- drywell personnel airlock
- drywell ring girder
- drywell stabilizer hatches
- embedded structural plates
- expansion joints (mechanical)
- expansion/grouted anchors
- moisture barrier
- primary containment bellows
- primary containment sump
- reactor pedestal
- reactor pedestal anchor bolts
- reactor shield wall
- reactor stabilizers
- refueling seal platform

- refueling seal platform bellows
- refueling seal platform covers
- seals and gaskets
- structural beams
- structural columns
- structural fasteners
- structural steel: platforms, stairways, mezzanines
- torus
- torus access manhole fasteners
- torus access manholes
- vacuum breaker small bore piping
- vacuum relief piping
- vacuum relief valves
- vent header deflector
- vent header supports

2.4A.1.1.2 Staff Evaluation

The staff reviewed ALRA Section 2.4.A.1 and the UFSAR. The staff's review, using the evaluation methodology described in SER Section 2.4, was conducted in accordance with the guidance described in SRP-LR Section 2.4, "Scoping and Screening Results: Structures."

In conducting its review, the staff evaluated the structural component functions described in the ALRA and UFSAR in accordance with the requirements of 10 CFR 54.4(a) to verify that the applicant had not omitted from the scope of license renewal any components with intended functions delineated under 10 CFR 54.4(a). The staff then reviewed those components that the applicant had identified as being within the scope of license renewal to verify that the applicant had not omitted any passive and long-lived components that should be subject to an AMR in accordance with the requirements of 10 CFR 54.21(a)(1).

The staff's review of LRA Section 2.4.A.1 identified areas in which additional information was necessary to complete the review of the applicant's scoping and screening results. The applicant responded to the staff's RAIs as discussed below.

In RAI 2.4A-1, dated December 9, 2004, the staff requested that the applicant identify the location in the LRA where the following specific components that may perform SR functions per 10 CFR 50.54(a)(1) are addressed:

(a)   reactor vessel to biological shield stabilizers

(b)   biological shield to containment stabilizer

(c)   RPV male stabilizer attached to outside of drywell shell

(d)   RPV female stabilizer and anchor rods (also referred to as GIB) embedded in reactor building concrete wall

(e)   biological shield wall and anchor bolts

(f)   reactor vessel support skirt and anchor bolts

(g)   reactor vessel support ring girder and anchor bolts and reactor vessel support pedestal

(h)   drywell internal steel shear ring

(i)   drywell steel support skirt and anchor bolts

(j)   drywell head closure bolts and double gasket, tongue-and-groove seal arrangement.

In its response by letter dated January 10, 2005, the applicant provided the specific information requested. The staff reviewed the response and found the information acceptable because the applicant had included the structures and components and their intended functions adequately in LRA Table 2.4.A.1-1.

In follow-up to RAI 2.4.A-1 dated October 11, 2005, the staff noted that leakage through the refueling seals located at the top of the drywell potentially exposes the carbon steel drywell shell inner and outer surfaces to loss of material due to corrosion. This concern was particularly applicable to the embedded portion of the drywell shell. The staff pointed out that the corrosion detected on the outer shell surface in the sand pocket region in a number of Mark I steel containments has been attributed to leakage past the drywell-to-reactor building refueling seal coupled with clogging of the sand pocket drains. Leakage into the drywell past the reactor vessel-to-drywell refueling seal creates the potential for corrosion of the inaccessible portion of the inner surface of the drywell shell embedded in the concrete floor. Therefore, the staff requested that the applicant clarify whether all components of the drywell to the refueling cavity seal are within the scope of license renewal.

The staff further noted that from the information in ALRA Table 2.4.A.1-1 it was not clear (1) whether the refueling seals have been included in the license renewal scope or (2) if included how aging management is addressed. Therefore, in RAI 2.4.A-1 dated October 11, 2005, the staff also requested that the applicant provide the following information:

*   Verification that the refueling seals are included in the license renewal scope or a detailed explanation for their exclusion.

*   A detailed description of the plant-specific operating experience for the refueling seals including incidences of degradation, method of detection, root cause, corrective actions, and current inspection procedures.

*   A detailed description of the scoping, screening, and aging management review for the refueling seals.

*   The AMPs credited to manage aging of the refueling seals.

In its response by letter dated October 28, 2005, the applicant stated that the NMP1 refueling seals are within the scope of license renewal and subject to an AMR. The applicant described in detail the six components of the refueling seal. Two of the components are stainless steel bellows, one between the liner and the drywell shell and the other between the refueling seal platform and the reactor vessel flange. The third component is the carbon steel refueling seal platform, the fourth is the aluminum refueling seal platform cover, the fifth is the gaskets between the covers and the platform, and the sixth is the stainless steel bolting that fastens the platform covers to the platform proper. The applicant stated that all of these components are within the scope of license renewal except the gaskets. The gaskets between the cover pieces and the platform are within scope of license renewal but not subject to an AMR because new gaskets are used during each refueling outage.

The applicant also explained that the components within the scope of license renewal and subject to an AMR are addressed in the ALRA Table 2.4.A.1-1 as follows: (1) the bellows are included as the component "Refueling Seal Platform Bellows," (2) the refueling seal platform is included with the component "Refueling Seal Platform" as the component type "Structural Steel (Carbon Low Alloy Steel) in Air," (3) the refueling seal platform covers are included as the component "Refueling Seal Platform Covers," and (4) the bolting is included with the Component "Refueling Seal Platform" as the component type "Fasteners (Wrought Austenitic Stainless Steel) in Air." The indication in the ALRA that the bolting component type is "Fasteners (Carbon or Low Alloy Steel) in Air" is an error. The bolting is stainless steel. The environment is air because these components are in an air environment during normal operation. They are wetted only during refueling operations.

As requested in the RAI, the applicant stated that there had been no plant operating experience indicating leakage from the refueling seals at NMPNS. Furthermore, any corrosion of the drywell in visible areas would be detected and mitigated at each refueling outage when the refueling cavity is filled. Any leakage would be observed prior to settling in an inaccessible area of the drywell.

Additionally, the applicant noted that the NRC staff had requested that utilities mitigate and/or identify potential degradation of Mark I containments in Information Notice IN 86-99 and Generic Letter GL 87-05. Such degradation occurred at Oyster Creek Generating Station as a result of water intrusion in the air gap from leakage past the refueling seal and subsequent wetting of the sand cushion at the bottom of the air gap. NMPNS conducted several investigations and inspections which determined that water intrusion into the NMP1 sand cushion had not occurred and that periodic examination of the sand cushion area drain lines is not warranted.

Historically, the applicant's management of this area has been vigilant in detecting water leakage. The applicant has found no leakage from this area and stated in the response that NMPNS plans to manage the components of the refueling seal during the period of extended operation. The staff found the applicant's approach of managing the reactor cavity to drywell refueling seal acceptable. Therefore, the staff's concern described in RAI 2.4.A-1 is resolved.

2.4A.1.1.3 Conclusion

The staff reviewed the ALRA, RAI responses, and related structural components to determine whether any SSCs that should be within the scope of license renewal had not been identified by the applicant. No omissions were identified. In addition, the staff performed a review to determine whether any components that should be subject to an AMR had not been identified by the applicant. No omissions were identified. On the basis of its review, the staff concludes that there is reasonable assurance that the applicant had adequately identified the PCS components that are within the scope of license renewal, as required by 10 CFR 54.4(a), and the PCS components that are subject to an AMR, as required by 10 CFR 54.21(a)(1).

2.4A.1.2 NMP1 Reactor Building

2.4A.1.2.1 Summary of Technical Information in the Amended Application

In ALRA Section 2.4.A.2, the applicant described the RB. The RB is a seismic Class I structure which encloses the PCS pressure suppression system. The RB is a multi-floored structure, comprising a substantial reinforced concrete substructure with reinforced concrete walls extending up to the operating floor level and a steel framed superstructure above the operating floor level. Airlocks are provided on the areas of the building where access doors are provided. The reinforced concrete building substructure is founded on bedrock. Precast concrete panels and uninsulated metal wall panels are applied to the exterior of the reinforced concrete walls of the reactor building, except around the airlocks. However, these panels do not form a part of the building support. Metal wall panels and roofing above the operating floor are leak tight. This structure provides secondary containment when the pressure suppression system is in service, and primary containment during refueling, maintenance, or testing, when the PCS is open or not required. The RB houses the refueling and reactor servicing equipment, fresh and spent fuel storage facilities, and other reactor auxiliary or service equipment.

The RB contains SR components that are relied upon to remain functional during and following DBEs. The failure of NSR SSCs in the RB could potentially prevent the satisfactory accomplishment of an SR function. In addition, the RB performs functions that support fire protection.

The intended functions within the scope of license renewal include the following:

- provides spray shield or curbs for directing flow

- provides for thermal expansion and/or seismic separation

- provides rated fire barrier.

- provides flood protection barrier

- provides structural support to NSR components whose failure could prevent accomplishment of SR function(s)

- provides over-pressure protection

- provides shelter/protection to SR components

- provides shielding against radiation

- provides pressure boundary or essentially leaktight barrier

- provides structural and/or functional support to SR equipment

In ALRA Table 2.4.A.2-1, the applicant identified the following RB component types that are within the scope of license renewal and subject to an AMR:

- blowout panels
- compressible joints
- concrete & grout
- concrete columns

2-260

- concrete curbs
- concrete floors
- concrete slabs
- concrete walls
- doors and framing/hardware
- drywell shield wall
- drywell shield wall sleeves
- electrical and air duct sleeves
- embedded structural plates
- expansion/grouted anchors
- fuel pool gate gaskets
- fuel pool gates
- fuel transfer canal
- fuel transfer canal liner
- hatch cover seals
- hatch covers
- main steam tunnel
- masonry walls
- penetration seal clamps
- penetration seals
- penetration sleeves
- precast concrete panels
- RB foundation fill concrete
- RB foundation mat
- RB metal siding
- RB overhead crane rail clips and fasteners
- RB overhead crane rail crane girder
- RB sumps
- reactor head cavity
- reactor head cavity liner
- reactor internal storage pit
- reactor internal storage pit liner
- reactor shield plug liners
- reactor shield plugs
- refueling platform rubber seal
- refueling platform track anchor bolts
- refueling platform track and embedded plate
- removable masonry wall anchors
- removable masonry wall framing
- rock anchors
- sealing compounds
- seals and gaskets
- spent fuel storage pool
- spent fuel storage pool liner
- structural beams
- structural columns
- structural fasteners
- structural steel: platforms, stairways, mezzanines
- sump liner fasteners (RB and auxiliary bay)

- torus ring girder
- torus saddle anchors
- torus support column sway rod turnbuckles
- torus support column sway rods
- torus support columns
- torus support foundation

2.4A.1.2.2 Staff Evaluation

The staff reviewed ALRA Section 2.4.A.2 and the UFSAR using the evaluation methodology described in SER Section 2.4. The staff conducted its review in accordance with the guidance described in SRP-LR Section 2.4.

In conducting its review, the staff evaluated the structural component functions described in the ALRA and UFSAR in accordance with the requirements of 10 CFR 54.4(a) to verify that the applicant had not omitted from the scope of license renewal any components with intended functions delineated under 10 CFR 54.4(a). The staff then reviewed those components that the applicant had identified as being within the scope of license renewal to verify that the applicant had not omitted any passive and long-lived components that should be subject to an AMR in accordance with the requirements of 10 CFR 54.21(a)(1).

The staff's review of LRA Section 2.4.A.2 identified an area in which additional information was necessary to complete the review of the applicant's scoping and screening results. The applicant responded to the staff's RAI as discussed below.

In RAI 2.4.A-4 dated December 9, 2004, the staff stated that from LRA Table 2.4.A.2-1 it was not clear whether the entire enclosure building of the BWR reactor building with steel superstructure (including metal structure, metal panels) is within the scope of license renewal. RAI 2.4A.2-4 requested that the applicant clarify the extent to which items of the enclosure building are within the scope of license renewal and indicate the locations where its components are included in AMR in Table 3.5.2.A-2.

In its response by letter dated January 10, 2005, the applicant stated:

> The reactor building (RB) is a concrete structure up to the refueling floor elevation. Above this elevation, it is a steel-framed structure with metal wall and panels. The concrete structure of the RB is included under the component type "Concrete in Air" in LRA Tables 2.4.A.2-1 and 3.5.2.A-2. The steel structural members above the refueling floor are included under the component type "Structural Steel (Carbon and Low Alloy Steel) in Air" in LRA Tables 2.4.A.2-1 and 3.5.2.A-2. The metal panels are included under the component type Siding in Air" in LRA Table 2.4.A.2-1 and under the component type "Metal Siding in Air" in LRA Table 3.5.2.A-2. The concrete, steel, and metal siding are all within the scope of LR. The last sentence on LRA page 2.4-5 states: "The entire RB is made up of components that require AMR." This was meant to indicate that all of the components that comprise the RB are within the scope of LR and subject to AMR.

The staff found the applicant's explanation adequate. Therefore, the staff's concern described in RAI 2.4.A-4 is resolved.

2.4A.1.2.3 Conclusion

The staff reviewed the ALRA, RAI response, and related structural components to determine whether any SSCs that should be within the scope of license renewal had not been identified by the applicant. No omissions were identified. In addition, the staff performed a review to determine whether any components that should be subject to an AMR had not been identified by the applicant. No omissions were identified. On the basis of its review, the staff concludes that there is reasonable assurance that the applicant had adequately identified the RB components that are within the scope of license renewal, as required by 10 CFR 54.4(a), and the RB components that are subject to an AMR, as required by 10 CFR 54.21(a)(1).

2.4A.1.3 NMP1 Essential Yard Structures

2.4A.1.3.1 Summary of Technical Information in the Amended Application

In ALRA Section 2.4.A.3, the applicant described the essential yard structures (EYS). The EYS include the seismic Class I and Class II essential yard buildings, plus structures and civil foundation supports for SR electrical or mechanical equipment items located within the yard. The yard is defined as the owner controlled outside areas surrounding the major NMP1 plant buildings, both inside and outside the NMPNS protected area. The earthen structures, which provide flood protection to the site, are included in the EYS. Included in the EYS are the administration building and the administration building extension and the radwaste pipe tunnel extension. The administration building extension is a Class II structure and is seismically designed due to its proximity to the NMP1 diesel generator rooms. Also included are SR tank foundations. There are no class 1E ductlines or manholes in the yard at NMP1. The EYS also include the structures that support the equipment and high voltage lines in the 115KV switchyard for SBO. The SBO components are addressed in the 115KV AC electrical distribution system.

The EYS contain SR components that are relied upon to remain functional during and following DBEs. The failure of NSR SSCs in the EYS could potentially prevent the satisfactory accomplishment of an SR function. In addition, the EYS perform functions that support fire protection and SBO.

The intended functions within the scope of license renewal include the following:

- provides rated fire barrier
- provides structural support to NSR components whose failure could prevent accomplishment of SR function(s)
- provides shelter/protection to SR components
- provides structural and/or functional support to SR equipment

In ALRA Table 2.4.A.3-1, the applicant identified the following EYS component types that are within the scope of license renewal and subject to an AMR:

- administration building concrete columns
- administration building concrete floors
- administration building concrete foundation
- administration building concrete walls
- administration building structural beams
- administration building structural columns
- administration building structural fasteners
- concrete encasement of ductlines
- emergency diesel fuel oil tank foundations
- nitrogen tank foundations
- nitrogen tank protective structure
- pipe tunnels .
- SBO equipment foundations
- structural polymer bearing pad

2.4A.1.3.2 Staff Evaluation

The staff reviewed ALRA Section 2.4.A.3 and the UFSAR using the evaluation methodology described in SER Section 2.4. The staff conducted its review in accordance with the guidance described in SRP-LR Section 2.4.

In conducting its review, the staff evaluated the structural component functions described in the ALRA and UFSAR in accordance with the requirements of 10 CFR 54.4(a) to verify that the applicant had not omitted from the scope of license renewal any components with intended functions delineated under 10 CFR 54.4(a). The staff then reviewed those components that the applicant had identified as being within the scope of license renewal to verify that the applicant had not omitted any passive and long-lived components that should be subject to an AMR in accordance with the requirements of 10 CFR 54.21(a)(1).

2.4A.1.3.3 Conclusion

The staff reviewed the ALRA and related structural components to determine whether any SSCs that should be within the scope of license renewal had not been identified by the applicant. No omissions were identified. In addition, the staff performed a review to determine whether any components that should be subject to an AMR had not been identified by the applicant. No omissions were identified. On the basis of its review, the staff concludes that there is reasonable assurance that the applicant had adequately identified the EYS components that are within the scope of license renewal, as required by 10 CFR 54.4(a), and the EYS components that are subject to an AMR, as required by 10 CFR 54.21(a)(1).

2.4A.1.4 NMP1 Fuel Handling System

2.4A.1.4.1 Summary of Technical Information in the Amended Application

In ALRA Section 2.4.A.4, the applicant described the fuel handling system. The fuel handling system involves those components used to move fuel from the time of receipt of new fuel to the storage of spent fuel in the spent fuel storage pool. Components that are evaluated in the fuel handling system include the refueling platform, fuel preparation machines, and spent fuel racks. Although the reactor building crane handles fuel, it is analyzed in the material handling system.

The fuel handling system contains SR components that are relied upon to remain functional during and following DBEs. The failure of NSR SSCs in the fuel handling system could potentially prevent the satisfactory accomplishment of an SR function.

The intended functions within the scope of license renewal include the following:

- provides structural support to NSR components whose failure could prevent accomplishment of SR function(s)

- provides structural and/or functional support to SR equipment

In ALRA Table 2.4.A.4-1, the applicant identified the following fuel handling system component types that are within the scope of license renewal and subject to an AMR:

- fuel preparation machines
- spent fuel rack fasteners
- spent fuel racks

2.4A.1.4.2 Staff Evaluation

The staff reviewed ALRA Section 2.4.A.4 and the UFSAR using the evaluation methodology described in SER Section 2.4. The staff conducted its review in accordance with the guidance described in SRP-LR Section 2.4.

In conducting its review, the staff evaluated the system functions described in the ALRA and UFSAR in accordance with the requirements of 10 CFR 54.4(a) to verify that the applicant had not omitted from the scope of license renewal any components with intended functions delineated under 10 CFR 54.4(a). The staff then reviewed those components that the applicant had identified as being within the scope of license renewal to verify that the applicant had not omitted any passive and long-lived components that should be subject to an AMR in accordance with the requirements of 10 CFR 54.21(a)(1).

2.4A.1.4.3 Conclusion

The staff reviewed the ALRA and related structural components to determine whether any SSCs that should be within the scope of license renewal had not been identified by the applicant. No omissions were identified. In addition, the staff performed a review to determine whether any components that should be subject to an AMR had not been identified by the applicant. No omissions were identified. On the basis of its review, the staff concludes that there is reasonable assurance that the applicant had adequately identified the fuel handling system components that are within the scope of license renewal, as required by 10 CFR 54.4(a), and the fuel handling system components that are subject to an AMR, as required by 10 CFR 54.21(a)(1).

2.4A.1.5 NMP1 Material Handling System

2.4A.1.5.1 Summary of Technical Information in the Amended Application

In ALRA Section 2.4.A.5, the applicant described the material handling system. The material handling system consists of overhead traveling cranes, monorail hoists, platform cranes, jib cranes, and associated mechanical and electrical components. For license renewal purposes, the crane girders and rails are included in the structural steel asset of the structure in which the crane is located.

The material handling system contains SR components that are relied upon to remain functional during and following DBEs. The failure of NSR SSCs in the material handling system could potentially prevent the satisfactory accomplishment of an SR function.

The intended functions within the scope of license renewal include the following:

- provides structural support to NSR components whose failure could prevent accomplishment of SR function(s)
- provides structural and/or functional support to SR equipment

In ALRA Table 2.4.A.5-1, the applicant identified the following material handling system component types that are within the scope of license renewal and subject to an AMR:

- decontamination area monorail hoist
- reactor building crane
- reactor building jib crane
- screen house building crane
- screen house building gate hoists
- turbine building 30 ton capacity crane
- turbine building crane
- turbine building monorail hoist

2.4A.1.5.2 Staff Evaluation

The staff reviewed ALRA Section 2.4.A.5 and the UFSAR using the evaluation methodology described in SER Section 2.4. The staff conducted its review in accordance with the guidance described in SRP-LR Section 2.4.

In conducting its review, the staff evaluated the system functions described in the ALRA and UFSAR in accordance with the requirements of 10 CFR 54.4(a) to verify that the applicant had not omitted from the scope of license renewal any components with intended functions delineated under 10 CFR 54.4(a). The staff then reviewed those components that the applicant had identified as being within the scope of license renewal to verify that the applicant had not omitted any passive and long-lived components that should be subject to an AMR in accordance with the requirements of 10 CFR 54.21(a)(1).

The staff's review of LRA Section 2.4.A.5 identified an area in which additional information was necessary to complete review of the applicant's scoping and screening results. The applicant responded to the staff's RAI as discussed below.

In RAI 2.4.A-5 dated December 9, 2004, the staff stated that LRA Section 2.4.A.5 indicates the only components that require AMR are the screenhouse gate hoists and the 125-ton capacity RB crane. No rail or and crane associated components appear to be included within the scope requiring AMR. Therefore, the staff requested that the applicant clarify the treatment of cranes, rails, and hoists in the scoping and screening and in the AMR. RAI 2.4A.5-5 requested that the applicant submit the following information:

a. A list of all cranes, hoists, rails, and associated components in the scope of license renewal.

b. A list of all cranes, hoists, rails, and associated components requiring aging management review (i.e., passive, long-lived).

c. A list of all cranes, hoists, rails, and associated components requiring aging management and/or TLAA.

In its response by letter dated January 10, 2005, the applicant stated:

(a) The NMP1 125-ton capacity RB overhead crane and the screenhouse gate hoists are the only cranes/hoists that meet 10 CFR 54.4(a) criteria for inclusion within the scope of LR. These components perform safety-related intended functions. LRA Section 2.4.A.5 includes the crane rails and girders as part of the structural steel component type for the building in which the crane is located. Other associated components, annunciators, circuit breakers, switches, motors, relays, resistors, and transformers, are classified as active components and, therefore, are not subject to AMR.

(b) The list of components requiring AMR along with corresponding LRA table locations:

- 125-ton RB Crane - Table 3.5.2.A-5

- Screenhouse Gate Hoists - Table 3.5.2.A-5

- 125-ton RB Crane Girders and Rails - Table 3.5.2.A-2 under the component type "Structural Steel (Carbon and Low Alloy Steel) in Air"

- Screenhouse Gate Hoists Girders and Rails - Table 3.5.2.A-8 under the component type "Structural Steel (Carbon and Low Alloy Steel) in Air"

(c) LRA Section 2.4.A.5 and Table 2.4.A.5-1 list component types within the scope of license renewal and subject to AMR. Because the in-scope NMP1 cranes are designated Service Class A ("Standby or Infrequent Service") by the Crane Manufacturers Association of America none meet the fatigue analysis requirement for a TLAA.

The staff found the above applicant's response acceptable because it provides adequate information. Therefore, the staff's concern described in RAI 2.4A-5 is resolved.

2.4A.1.5.3 Conclusion

The staff reviewed the ALRA, RAI response, and related structural components to determine whether any SSCs that should be within the scope of license renewal had not been identified by the applicant. No omissions were identified. In addition, the staff performed a review to determine whether any components that should be subject to an AMR had not been identified by the applicant. No omissions were identified. On the basis of its review, the staff concludes that there is reasonable assurance that the applicant had adequately identified the material handling system components that are within the scope of license renewal, as required by 10 CFR 54.4(a), and the material handling system components that are subject to an AMR, as required by 10 CFR 54.21(a)(1).

2.4A.1.6 NMP1 Offgas Building

2.4A.1.6.1 Summary of Technical Information in the Amended Application

In ALRA Section 2.4.A.6, the applicant described the offgas building (OGB). The OGB is a seismic Class I structure. The OGB is located adjacent to the TB and the WDB. The OGB substructure is a reinforced concrete structure and is founded on bedrock. The superstructure is structural steel frame with exterior metal walls and masonry block. The interior walls of the substructure are reinforced concrete and concrete block. The OGB contains the piping and equipment associated with the condenser air removal and offgas system.

The OGB contains SR components that are relied upon to remain functional during and following DBEs. The failure of NSR SSCs in the OGB could potentially prevent the satisfactory accomplishment of an SR function. In addition, the OGB performs functions that support fire protection.

The intended functions within the scope of license renewal include the following:

- provides rated fire barrier
- provides structural support to NSR components whose failure could prevent accomplishment of SR function(s)
- provides structural and/or functional support to SR equipment

In ALRA Table 2.4.A.6-1, the applicant identified the following OGB component types that are within the scope of license renewal and subject to an AMR:

- building foundation
- concrete and grout
- concrete columns
- concrete floors
- concrete lean fill
- concrete walls
- doors and framing/hardware
- expansion/grouted anchors
- masonry wall fasteners

- masonry wall framing
- masonry walls
- structural beams
- structural columns
- structural fasteners
- structural steel: platforms, stairways, mezzanines

2.4A.1.6.2 Staff Evaluation

The staff reviewed ALRA Section 2.4.A.6 and the UFSAR using the evaluation methodology described in SER Section 2.4. The staff conducted its review in accordance with the guidance described in SRP-LR Section 2.4.

In conducting its review, the staff evaluated the structural component functions described in the ALRA and UFSAR in accordance with the requirements of 10 CFR 54.4(a) to verify that the applicant had not omitted from the scope of license renewal any components with intended functions delineated under 10 CFR 54.4(a). The staff then reviewed those components that the applicant had identified as being within the scope of license renewal to verify that the applicant had not omitted any passive and long-lived components that should be subject to an AMR in accordance with the requirements of 10 CFR 54.21(a)(1).

2.4A.1.6.3 Conclusion

The staff reviewed the ALRA and related structural components to determine whether any SSCs that should be within the scope of license renewal had not been identified by the applicant. No omissions were identified. In addition, the staff performed a review to determine whether any components that should be subject to an AMR had not been identified by the applicant. No omissions were identified. On the basis of its review, the staff concludes that there is reasonable assurance that the applicant had adequately identified the OGB components that are within the scope of license renewal, as required by 10 CFR 54.4(a), and the OGB components that are subject to an AMR, as required by 10 CFR 54.21(a)(1).

2.4A.1.7 NMP1 Personnel/Equipment Access System

2.4A.1.7.1 Summary of Technical Information in the Amended Application

In ALRA Section 2.4.A.7, the applicant described the personnel/equipment access system. The personnel/equipment access system consists of doors, gates, and the electronic equipment that monitors their positions. The gates and electronic equipment are not in scope for license renewal.

The personnel/equipment access system contains SR components that are relied upon to remain functional during and following DBEs. The failure of NSR SSCs in the personnel/equipment access system could potentially prevent the satisfactory accomplishment of an SR function. In addition, the personnel/equipment access system performs functions that support fire protection.

All doors are addressed in the sections for the structures where the doors are located. There are no other components subject to an AMR for this system.

2.4A.1.7.2 Staff Evaluation

The staff reviewed ALRA Section 2.4.A.7 and the UFSAR using the evaluation methodology described in SER Section 2.4. The staff conducted its review in accordance with the guidance described in SRP-LR Section 2.4.

In conducting its review, the staff evaluated the system functions described in the ALRA and UFSAR in accordance with the requirements of 10 CFR 54.4(a) to verify that the applicant had not omitted from the scope of license renewal any components with intended functions delineated under 10 CFR 54.4(a). The staff then reviewed those components that the applicant had identified as being within the scope of license renewal to verify that the applicant had not omitted any passive and long-lived components that should be subject to an AMR in accordance with the requirements of 10 CFR 54.21(a)(1).

2.4A.1.7.3 Conclusion

The staff reviewed the ALRA and related structural components to determine whether any SSCs that should be within the scope of license renewal had not been identified by the applicant. No omissions were identified. In addition, the staff performed a review to determine whether any components that should be subject to an AMR had not been identified by the applicant. No omissions were identified. On the basis of its review, the staff concludes that there is reasonable assurance that the applicant had adequately identified the personnel/equipment access system components that are within the scope of license renewal, as required by 10 CFR 54.4(a), and the personnel/equipment access system components that are subject to an AMR, as required by 10 CFR 54.21(a)(1).

2.4A.1.8 NMP1 Radwaste Solidification and Storage Building

2.4A.1.8.1 Summary of Technical Information in the Amended Application

In ALRA Section 2.4.A.8, the applicant described the radwaste solidification and storage building (RSSB). The RSSB is a seismic Class I structure located to the east of, and directly adjacent to, the OGB and the WDB. The RSSB is a reinforced concrete structure. The foundation mat is founded on bedrock. During normal operation, maintenance, and loading and unloading operations, the structure provides sufficient environmental isolation.

The RSSB contains SR components that are relied upon to remain functional during and following DBEs. The failure of NSR SSCs in the RSSB could potentially prevent the satisfactory accomplishment of an SR function. In addition, the RSSB performs functions that support fire protection.

The intended functions within the scope of license renewal include the following:

- provides spray shield or curbs for directing flow
- provides rated fire barrier

- provides flood protection barrier
- provides structural support to NSR components whose failure could prevent accomplishment of SR function(s)
- provides shielding against radiation
- provides pressure boundary or essentially leaktight barrier

In ALRA Table 2.4.A.8-1, the applicant identified the following RSSB component types that are within the scope of license renewal and subject to an AMR:

- concrete & grout
- concrete caissons
- concrete curbs
- concrete floors
- concrete slabs
- concrete walls
- doors and framing/hardware
- embedded structural plates
- masonry walls
- penetration sleeves
- pipe tunnels
- radwaste building foundation
- radwaste building sump
- roof hatch
- roof plug lifting pins
- seals and gaskets
- steel liner
- steel shield wall

2.4A.1.8.2 Staff Evaluation

The staff reviewed ALRA Section 2.4.A.8 and the UFSAR using the evaluation methodology described in SER Section 2.4. The staff conducted its review in accordance with the guidance described in SRP-LR Section 2.4.

In conducting its review, the staff evaluated the structural component functions described in the ALRA and UFSAR in accordance with the requirements of 10 CFR 54.4(a) to verify that the applicant had not omitted from the scope of license renewal any components with intended functions delineated under 10 CFR 54.4(a). The staff then reviewed those components that the applicant had identified as being within the scope of license renewal to verify that the applicant had not omitted any passive and long-lived components that should be subject to an AMR in accordance with the requirements of 10 CFR 54.21(a)(1).

2.4A.1.8.3 Conclusion

The staff reviewed the ALRA and related structural components to determine whether any SSCs that should be within the scope of license renewal had not been identified by the applicant. No omissions were identified. In addition, the staff performed a review to determine whether any components that should be subject to an AMR had not been identified by the applicant. No omissions were identified. On the basis of its review, the staff concludes that there is reasonable assurance that the applicant had adequately identified the RSSB components that are within the scope of license renewal, as required by 10 CFR 54.4(a), and the RSSB components that are subject to an AMR, as required by 10 CFR 54.21(a)(1).

2.4A.1.9 NMP1 Screen and Pump House Building

2.4A.1.9.1 Summary of Technical Information in the Amended Application

In ALRA Section 2.4.A.9, the applicant described the screen and pump house (SPH) building. The SPH building is a seismic Class I and Class II structure which is adjacent to the north wall of the RB and TB. The Class II superstructure is framed structural steel supported on a Class I reinforced concrete substructure that is founded on bedrock. The exterior wall is internally-insulated precast concrete panels. The SPH building comprises channels for the flow of very large quantities of raw lake water, gates, stop logs for control of the flow, racks, screens for cleaning the water, and pumps.

The SPH building contains SR components that are relied upon to remain functional during and following DBEs. The failure of NSR SSCs in the SPH building could potentially prevent the satisfactory accomplishment of an SR function. In addition, the SPH building performs functions that support fire protection.

The intended functions within the scope of license renewal include the following:

- provides spray shield or curbs for directing flow

- provides rated fire barrier

- provides structural support to NSR components whose failure could prevent accomplishment of SR function(s)

- provides shelter/protection to SR components

- provides source of cooling water for plant shutdown

- provides structural and/or functional support to SR equipment

In ALRA Table 2.4.A.9-1, the applicant identified the following SPH building component types that are within the scope of license renewal and subject to an AMR:

- building foundation
- concrete and grout
- concrete curbs
- concrete floors
- concrete piers
- concrete slab
- concrete walls
- crane rails and girders
- doors and framing/hardware
- expansion/grouted anchors

- intake structure
- intake structure structural fasteners
- intake structure structural steel
- intake tunnel
- masonry walls
- miscellaneous structural steel
- structural beams
- structural columns
- structural fasteners
- wall shoring

2.4A.1.9.2 Staff Evaluation

The staff reviewed ALRA Section 2.4.A.9 and the UFSAR using the evaluation methodology described in SER Section 2.4. The staff conducted its review in accordance with the guidance described in SRP-LR Section 2.4.

In conducting its review, the staff evaluated the structural component functions described in the ALRA and UFSAR in accordance with the requirements of 10 CFR 54.4(a) to verify that the applicant had not omitted from the scope of license renewal any components with intended functions delineated under 10 CFR 54.4(a). The staff then reviewed those components that the applicant had identified as being within the scope of license renewal to verify that the applicant had not omitted any passive and long-lived components that should be subject to an AMR in accordance with the requirements of 10 CFR 54.21(a)(1).

The staff's review of LRA Section 2.4.A.9 identified an area in which additional information was necessary to complete the review of the applicant's scoping and screening results. The applicant responded to the staff's RAI as discussed below.

In RAI 2.4.A-7 dated December 9, 2004, the staff requested that the applicant confirm for the NMP1 screen and pump house that such items as hatches and plugs, structural steel embedments, reinforced concrete foundation footings, grouted concrete, and water proofing membrane materials are within the scope of license renewal and require AMR. For such items within the scope of license renewal the staff requested that the applicant provide additional information in the format of LRA Table 2.4.A.9-1. For such items not within the scope of license renewal the applicant was asked to provide the basis for omission.

In its response by letter dated January 10, 2005, the applicant stated that the items listed in the RAI are within the scope of license renewal, subject to an AMR, and represented by the component types listed in LRA Table 2.4.A.9-1:

(1) There are no "hatches" in the NMP1 Screen and Pump House.

(2) All doors are included under the component type "Door."

(3) Plugs are concrete and included in the component type "Concrete in Air."

(4) Embedded portions of structural steel embedments are integral with the concrete and included with the component type "Concrete in Air" or "Concrete in Raw Water" depending on the location of the embedment.

(5) Structural steel exposed to atmosphere is included under the component type "Structural Steel (Carbon and Low Alloy Steel) in Air" or "Structural Steel (Carbon and Low Alloy Steel) in Raw Water" depending on the location of the embedment.

(6) Reinforced concrete foundation footings are included in the component type "Concrete in Soil Above the Ground Water Table" (GWT) or "Concrete in Soil Below the GWT" depending on the footing depth.

(7) Grouted concrete is not used at NMP1.

(8) Structural concrete is included in the various concrete component types depending on environment.

(9) Waterproofing membranes are not included because they are applied as coatings. NMPNS does not credit coatings to mitigate aging effects.

The staff found the above applicant's response complete and adequate. Therefore, the staff's concern described in RAI 2.4A.-7 is resolved.

2.4A.1.9.3 Conclusion

The staff reviewed the ALRA, RAI response, and related structural components to determine whether any SSCs that should be within the scope of license renewal had not been identified by the applicant. No omissions were identified. In addition, the staff performed a review to determine whether any components that should be subject to an AMR had not been identified by the applicant. No omissions were identified. On the basis of its review, the staff concludes that there is reasonable assurance that the applicant had adequately identified the SPH building components that are within the scope of license renewal, as required by 10 CFR 54.4(a), and the SPH building components that are subject to an AMR, as required by 10 CFR 54.21(a)(1).

2.4A.1.10 NMP1 Turbine Building

2.4A.1.10.1 Summary of Technical Information in the Amended Application

In ALRA Section 2.4.A.10, the applicant described the TB. The TB is a Class II structure with integrated seismic Class I areas. The reinforced concrete turbine generator foundation pedestal is isolated from the floors of the building to minimize transmission of vibration to the floors. The reinforced concrete TB foundations are supported by concrete column piers founded on bedrock 15 to 25 feet below grade. The TB superstructure consists of an enclosed structural steel frame. The roof is covered with metal decking, insulation, and tar roofing material. Located within the TB are the generating area, the auxiliary equipment area, the feedwater heater area, the auxiliary extension building, and the control room.

The TB contains SR components that are relied upon to remain functional during and following DBEs. The failure of NSR SSCs in the TB could potentially prevent the satisfactory accomplishment of an SR function. In addition, the TB performs functions that support fire protection.

The intended functions within the scope of license renewal include the following:

- provides spray shield or curbs for directing flow

- provides for thermal expansion and/or seismic separation

- provides rated fire barrier

- provides flood protection barrier

- provides missile barrier

- provides structural support to NSR components whose failure could prevent accomplishment of SR function(s)

- provides over-pressure protection

- provides shelter/protection to SR components

- provides shielding against radiation

- provides pressure boundary or essentially leaktight barrier

- provides structural and/or functional support to SR equipment

In ALRA Table 2.4.A.10-1, the applicant identified the following TB component types that are within the scope of license renewal and subject to an AMR:

- auxiliary control room concrete curbs
- auxiliary control room concrete floors
- auxiliary control room concrete walls
- auxiliary control room masonry walls
- beam seats
- bearing plates
- compressible joints and seals
- concrete and grout
- concrete columns
- concrete floors
- concrete slabs
- concrete walls
- control room concrete floors
- control room concrete walls
- control room metal partition wall
- control room structural beams
- control room structural columns
- control room structural fasteners
- control room/auxiliary control room penetration seals
- control room/auxiliary control room penetration sleeves
- diesel generator foundations
- diesel generator room ceiling
- diesel generator room concrete floors
- diesel generator room concrete slabs
- diesel generator room concrete walls
- diesel generator room protection panels

- diesel generator room 102 missile shield
- doors and framing/hardware
- embedded structural plates
- expansion/grouted anchors
- main steam tunnel
- monorail beams
- penetration sleeves
- removable concrete slabs
- removable masonry wall framing
- removable masonry walls
- structural beams
- structural columns
- structural fasteners
- structural steel: platforms, stairways, and mezzanines
- turbine building blowout panels
- turbine building foundation slab
- turbine building manhole cover and frame
- turbine building overhead crane rail and embedded plate
- turbine building overhead crane rail clips and fasteners
- turbine building sump liner fasteners
- turbine building sump liners
- turbine building sump sleeves
- turbine building sumps
- turbine support structure

2.4A.1.10.2 Staff Evaluation

The staff reviewed ALRA Section 2.4.A.10 and the UFSAR using the evaluation methodology described in SER Section 2.4. The staff conducted its review in accordance with the guidance described in SRP-LR Section 2.4.

In conducting its review, the staff evaluated the structural component functions described in the ALRA and UFSAR in accordance with the requirements of 10 CFR 54.4(a) to verify that the applicant had not omitted from the scope of license renewal any components with intended functions delineated under 10 CFR 54.4(a). The staff then reviewed those components that the applicant had identified as being within the scope of license renewal to verify that the applicant had not omitted any passive and long-lived components that should be subject to an AMR in accordance with the requirements of 10 CFR 54.21(a)(1).

The staff's review of LRA Section 2.4.A.10 identified an area in which additional information was necessary to complete the review of the applicant's scoping and screening results. The applicant responded to the staff's RAI as discussed below.

In RAI 2.4.A-6 dated December 9, 2004, the staff stated that in some of the LRA Section 2.4.A tables the staff could not identify the insulation and insulation jacketing included in the license renewal scope nor the specific subsets of insulation and insulation jacketing included in the LRA Section 2.4.A tables. It was also unclear whether insulation and jacketing on the reactor coolant system had been included. Insulation and jacketing are commodities that may perform SR functions as per 10 CFR 50.54(a)(1). Therefore, the staff requested that the applicant:

a. Identify the structures and structural components within the scope of license renewal with insulation and/or insulation jacketing.

b. List all insulation and insulation jacketing materials associated with such structures and structural components that require AMR and the AMR results for each.

c. For insulation and insulation jacketing materials associated with such structures and structural components not requiring aging management, submit the technical basis for this conclusion, including plant-specific operating experience.

d. For insulation and insulation jacketing materials associated with such structures and structural components that require aging management indicate the applicable LRA sections that identify the AMP(s) credited.

In its response by letter dated January 10, 2005, the applicant stated that NMP1 has no structures or structural components within the scope of LR with insulation or insulation jacketing. Therefore, insulation was not included in any LRA Section 2.4.A table. The NMP1 design does not require insulation of structural steel and/or concrete according to CLB documents including safety analyses and plant evaluations. The staff found this plant-specific configuration-based response adequate. Therefore, the staff's concern described in RAI 2.4A-6 is resolved.

2.4A.1.10.3 Conclusion

The staff reviewed the ALRA, RAI response, and related structural components to determine whether any SSCs that should be within the scope of license renewal had not been identified by the applicant. No omissions were identified. In addition, the staff performed a review to determine whether any components that should be subject to an AMR had not been identified by the applicant. No omissions were identified. On the basis of its review, the staff concludes that there is reasonable assurance that the applicant had adequately identified the TB components that are within the scope of license renewal, as required by 10 CFR 54.4(a), and the TB components that are subject to an AMR, as required by 10 CFR 54.21(a)(1).

2.4A.1.11 NMP1 Vent Stack

2.4A.1.11.1 Summary of Technical Information in the Amended Application

In ALRA Section 2.4.A.11, the applicant described the vent stack. The vent stack is a seismic Class I reinforced-concrete chimney, 350 feet high, located 100 feet east of the northeast corner of the RB. The height of the stack and the velocity of discharge provide a high degree of dilution for station effluents. The vent stack's foundation is on a massive reinforced concrete base which extends to bedrock. From this base, it rises through the turbine auxiliary building extension from which it is completely structurally isolated.

The vent stack contains SR components that are relied upon to remain functional during and following DBEs. In addition, the vent stack performs functions that support fire protection.

The intended functions within the scope of license renewal include the following:

- provides for thermal expansion and/or seismic separation
- provides rated fire barrier
- provides path for release of filtered and unfiltered gaseous discharge
- provides structural and/or functional support to SR equipment

In ALRA Table 2.4.A.11-1, the applicant identified the following vent stack component types that are within the scope of license renewal and subject to an AMR:

- compressible joints and seals
- concrete chimney shell
- concrete floors
- vent duct framing
- vent stack foundation

2.4A.1.11.2 Staff Evaluation

The staff reviewed ALRA Section 2.4.A.11 and the UFSAR using the evaluation methodology described in SER Section 2.4. The staff conducted its review in accordance with the guidance described in SRP-LR Section 2.4.

In conducting its review, the staff evaluated the structural component functions described in the ALRA and UFSAR in accordance with the requirements of 10 CFR 54.4(a) to verify that the applicant had not omitted from the scope of license renewal any components with intended functions delineated under 10 CFR 54.4(a). The staff then reviewed those components that the applicant had identified as being within the scope of license renewal to verify that the applicant had not omitted any passive and long-lived components that should be subject to an AMR in accordance with the requirements of 10 CFR 54.21(a)(1).

2.4A.1.11.3 Conclusion

The staff reviewed the ALRA and related structural components to determine whether any SSCs that should be within the scope of license renewal had not been identified by the applicant. No omissions were identified. In addition, the staff performed a review to determine whether any components that should be subject to an AMR had not been identified by the applicant. No omissions were identified. On the basis of its review, the staff concludes that there is reasonable assurance that the applicant had adequately identified the vent stack components that are within the scope of license renewal, as required by 10 CFR 54.4(a), and the vent stack components that are subject to an AMR, as required by 10 CFR 54.21(a)(1).

2.4A.1.12 NMP1 Waste Disposal Building

2.4A.1.12.1 Summary of Technical Information in the Amended Application

In ALRA Section 2.4.A.12, the applicant described the WDB. The WDB and WDB extension are seismic Class I structures located between and adjacent to the RSSB and the turbine auxiliary extension building. The WDB and extension consist of reinforced concrete substructures with steel framed superstructures from grade to their respective roof elevations. The interior walls of the substructure are reinforced concrete. The superstructure walls are also reinforced concrete or concrete masonry units. The reinforced concrete building substructure is founded on bedrock.

The WDB contains SR components that are relied upon to remain functional during and following DBEs. The failure of NSR SSCs in the WDB could potentially prevent the satisfactory accomplishment of an SR function. In addition, the WDB performs functions that support fire protection.

The intended functions within the scope of license renewal include the following:

- provides spray shield or curbs for directing flow

- provides for thermal expansion and/or seismic separation

- provides rated fire barrier

- provides flood protection barrier

- provides structural support to NSR components whose failure could prevent accomplishment of SR function(s)

- provides shelter/protection to SR components

- provides structural and/or functional support to SR equipment

In ALRA Table 2.4.A.12-1, the applicant identified the following WDB component types that are within the scope of license renewal and subject to an AMR:

- building foundation
- compressible joints and seals
- concrete and grout
- concrete curbs
- concrete floors
- concrete sumps
- concrete walls
- doors and framing/hardware
- expansion/grouted anchors
- masonry fasteners
- masonry walls
- miscellaneous structural steel
- steel curbs
- steel sump liner
- steel troughs
- structural beams
- structural columns
- structural fasteners
- structural steel: platforms, stairways, mezzanines

2.4A.1.12.2 Staff Evaluation

The staff reviewed ALRA Section 2.4.A.12 and the UFSAR using the evaluation methodology described in SER Section 2.4. The staff conducted its review in accordance with the guidance described in SRP-LR Section 2.4.

In conducting its review, the staff evaluated the structural component functions described in the ALRA and UFSAR in accordance with the requirements of 10 CFR 54.4(a) to verify that the applicant had not omitted from the scope of license renewal any components with intended functions delineated under 10 CFR 54.4(a). The staff then reviewed those components that the applicant had identified as being within the scope of license renewal to verify that the applicant had not omitted any passive and long-lived components that should be subject to an AMR in accordance with the requirements of 10 CFR 54.21(a)(1).

2.4A.1.12.3 Conclusion

The staff reviewed the ALRA and related structural components to determine whether any SSCs that should be within the scope of license renewal had not been identified by the applicant. No omissions were identified. In addition, the staff performed a review to determine whether any components that should be subject to an AMR had not been identified by the applicant. No omissions were identified. On the basis of its review, the staff concludes that there is reasonable assurance that the applicant had adequately identified the WDB components that are within the scope of license renewal, as required by 10 CFR 54.4(a), and the WDB components that are subject to an AMR, as required by 10 CFR 54.21(a)(1).

2.4B NMP2 Structures

In ALRA Section 2.4.B, the applicant identified the NMP2 structures that are subject to an AMR for license renewal.

The applicant described, in the ALRA, the following sections for supporting systems and structures for the NMP2 structures:

- 2.4.B.1   NMP2 primary containment structure
- 2.4.B.2   NMP2 reactor building
- 2.4.B.3   NMP2 auxiliary service building
- 2.4.B.4   NMP2 control room building
- 2.4.B.5   NMP2 diesel generator building
- 2.4.B.6   NMP2 essential yard structures
- 2.4.B.7   NMP2 fuel handling system
- 2.4.B.8   NMP2 main stack
- 2.4.B.9   NMP2 material handling system
- 2.4.B.10   NMP2 motor operated doors system
- 2.4.B.11   NMP2 radwaste building
- 2.4.B.12   NMP2 screenwell building
- 2.4.B.13   NMP2 standby gas treatment building
- 2.4.B.14   NMP2 turbine building

The staff's review findings regarding the above ALRA Sections 2.4.B.1 through 2.4.B.14 are presented in SER Sections 2.4B.1 through 2.4B.14, respectively.

2.4B.1 NMP2 Primary Containment Structure

2.4B.1.1 Summary of Technical Information in the Amended Application

In ALRA Section 2.4.B.1, the applicant described the primary containment structure (PCS). The PCS is a seismic Category I structure consisting of a drywell chamber, located above a suppression chamber, and a drywell floor, which separates the drywell chamber from the suppression chamber. It also includes the structural portions of primary containment penetrations. The PCS is supported on a 10-ft thick reinforced concrete mat, which also supports the reactor building (RB). A series of 24-in diameter downcomer vent pipes penetrates the drywell floor. The drywell is a steellined reinforced concrete vessel in the shape of a frustum of two cones, closed by a dome with a torispherical head. The PCS contains a Mark II pressure suppression system. The pressure suppression chamber is a cylindrical stainless steel clad steel-lined reinforced concrete vessel located below the drywell. The PCS houses the RPV, the reactor recirculation system, and other branch connections of the RCPB.

The PCS contains SR components that are relied upon to remain functional during and following DBEs. The failure of NSR SSCs in the PCS could potentially prevent the satisfactory accomplishment of an SR function.

The intended functions within the scope of license renewal include the following:

- provides spray shield or curbs for directing flow

- provides shielding against high energy line breaks

- provides missile barrier

- provides structural support to NSR components whose failure could prevent accomplishment of SR function(s)

- provides pipe whip restraint

- provides pressure retaining boundary

- provides shelter/protection to SR components

- provides shielding against radiation

- provides pressure boundary or essentially leaktight barrier

- provides structural and/or functional support to SR equipment

In ALRA Table 2.4.B.1-1, the applicant identified the following PCS component types that are within the scope of license renewal and subject to an AMR:

- beam seats
- bearing plates
- biological shield wall
- biological shield wall door radiation shields

- biological shield wall doors
- concrete & grout
- concrete slabs
- containment penetrations (electrical)
- containment penetrations (instrument)
- containment penetrations (mechanical)
- control rod drive removal hatch
- downcomers
- drywell
- drywell coating
- drywell emergency escape lock
- drywell equipment hatch
- drywell floor
- drywell floor concrete insulation
- drywell floor supplementary steel
- drywell head
- drywell head closure pins
- drywell head fasteners
- drywell head stainless steel elements
- drywell liner
- drywell personnel airlock
- embedded structural plates
- equipment hatch ring beam
- impingement and jet shielding
- inner refueling seal
- insulation support framing
- lubrite plate
- monorail beams
- pipe whip restraint fasteners
- pipe whip restraints
- precast concrete beams
- radiation shields
- reactor pedestal
- reactor pedestal anchor bolts
- reactor stabilizers
- refueling bulkhead
- sealing compounds
- seals and gaskets
- star truss
- structural beams
- structural columns
- structural fasteners
- structural girders
- structural insulation liner
- structural plates
- structural steel: platforms, stairways, mezzanines, removable curbs
- suppression chamber seal
- suppression pool
- suppression pool access hatches

- suppression pool liner
- water level indicator shields

2.4B.1.2 Staff Evaluation

The staff reviewed ALRA Section 2.4.B.1 and the USAR using the evaluation methodology described in SER Section 2.4. The staff conducted its review in accordance with the guidance described in SRP-LR Section 2.4.

In conducting its review, the staff evaluated the structural component functions described in the ALRA and USAR in accordance with the requirements of 10 CFR 54.4(a) to verify that the applicant had not omitted from the scope of license renewal any components with intended functions delineated under 10 CFR 54.4(a). The staff then reviewed those components that the applicant had identified as being within the scope of license renewal to verify that the applicant had not omitted any passive and long-lived components that should be subject to an AMR in accordance with the requirements of 10 CFR 54.21(a)(1).

The staff's review of the original LRA Section 2.4.B.1 identified areas in which additional information was necessary to complete review of the applicant's scoping and screening results. The applicant responded to the staff's RAIs as discussed below.

In RAI 2.4.B-1 dated December 9, 2004, the staff stated that the NMP2 PCS encloses the reactor vessel and such structures as the biological shield wall, concrete pedestal, and the barrier floor between the drywell and the suppression chamber. The original LRA Table 2.4.B.1-1 does not include these structures in the scope of license renewal though they perform SR functions per 10 CFR 50.54(a)(1). Therefore, the staff requested, if they were not included by oversight, that the applicant describe its scoping and AMR or indicate their location in the LRA if somewhere else. The staff further requested that, if omitted from the scope of license renewal, the applicant provide the basis for omitting these items.

The applicant responded by letter dated January 10, 2005, and stated that the components listed in the staff RAI are within the scope of license renewal and are subject to AMR, they are included in the original LRA Table 2.4.B.1-1. Since the structures in question are comprised of multiple materials, they have been captured in the LRA Table 2.4.B.1-1 as a function of their materials of construction intead of their functional names. This information has been incorporated in the ALRA as discussed below.

In its ALRA dated July 14, 2005, the staff found the above response acceptable because the applicant correctly identified these structures and components within the scope of license renewal and included them in ALRA Table 2.4.B.1-1. Therefore, the staff's concern described in RAI 2.4.B-1 is resolved.

In RAI 2.4.B-2 dated December 9, 2004, the staff noted that the Group 2 structures defined in GALL Report Chapter III include the BWR RB with steel superstructure (enclosure building) and should be within the scope of license renewal. LRA Table 2.4.B.1-1 was unclear whether the entire enclosure building (including the concrete structure, metal panels) is within the scope of license renewal. The staff requested that the applicant clarify the extent to which the enclosure building is within the scope of license renewal and the locations where its components are included in AMR in Table 3.5.2.B-2. The applicant's response by letter dated January 10, 2005, has been incorporated in the ALRA as discussed below.

In its ALRA dated July 14, 2005, the applicant included the essential components of the enclosure building in ALRA Table 2.4.B.1-1. Therefore, the staff's concern described in RAI 2.4.B-2 is resolved.

2.4B.1.3 Conclusion

The staff reviewed the ALRA, RAI responses, and related structural components to determine whether any SSCs that should be within the scope of license renewal had not been identified by the applicant. No omissions were identified. In addition, the staff performed a review to determine whether any components that should be subject to an AMR had not been identified by the applicant. No omissions were identified. On the basis of its review, the staff concludes that there is reasonable assurance that the applicant had adequately identified the PCS components that are within the scope of license renewal, as required by 10 CFR 54.4(a), and the PCS components that are subject to an AMR, as required by 10 CFR 54.21(a)(1).

2.4B.2 NMP2 Reactor Building

2.4B.2.1 Summary of Technical Information in the Amended Application

In ALRA Section 2.4.B.2, the applicant described the RB. The RB is a seismic Category I structure that encloses the PCS. The RB wall is a reinforced concrete cylinder with varying wall thickness, extending from the top of the mat to the polar crane level. The wall from the crane rail elevation to the roof is steel framing with insulated metal siding. The metal siding panels have sealed joints to minimize air leakage. The RB, including the auxiliary bays, is founded on a rock-bearing, reinforced concrete mat. The mat acts to support the RB, auxiliary bays, and the primary containment. The auxiliary bays are rigidly attached to the RB and considered part of the secondary containment structure. The RB houses the refueling and reactor servicing equipment, new and spent fuel storage facilities, and other reactor auxiliary or service equipment, including the RCIC system, RWCU system, standby liquid control system, CRD system equipment, core standby cooling systems, RHR systems, and electrical equipment components. Included within the RB for the purposes of license renewal are the secondary containment, the north and south auxiliary bays, and the main steam tunnel east of the turbine building. Civil/structural components from the fuel nuclear transfer system and the vents – turbine and RB system are also evaluated as part of the RB.

The RB contains SR components that are relied upon to remain functional during and following DBEs. The failure of NSR SSCs in the RB could potentially prevent the satisfactory accomplishment of an SR function. In addition, the RB performs functions that support fire protection.

The intended functions within the scope of license renewal include the following:

- provides spray shield or curbs for directing flow
- provides for thermal expansion and/or seismic separation

- provides rated fire barrier

- provides flood protection barrier

- provides shielding against high energy line breaks

- provides missile barrier

- provides structural support to NSR components whose failure could prevent accomplishment of SR function(s)

- provides pipe whip restraint

- provides over-pressure protection

- provides shelter/protection to SR components

- provides shielding against radiation

- provides pressure boundary or essentially leaktight barrier

- provides structural and/or functional support to SR equipment

In ALRA Table 2.4.B.2-1, the applicant identified the following RB component types that are within the scope of license renewal and subject to an AMR:

- auxiliary bay foundations
- auxiliary bay sumps
- beam pockets
- bearing plates
- cask pit cavity
- cask pit cavity liner
- cask washdown pit
- cask washdown pit liner
- compressible joints and seals
- concrete and grout
- concrete columns
- concrete curbs
- concrete floors
- concrete lean fill
- concrete slabs
- concrete walls
- doors and framing/hardware
- embedded structural plates
- expansion/grouted anchors
- fuel pool canal
- fuel pool canal liner
- fuel pool gates
- fuel transfer shielding bridge
- main steam tunnel
- miscellaneous structural steel framing
- monorail beams
- overpressurization vent panel fasteners
- overpressurization vent panels

- penetration seal clamps
- penetration seals
- penetration sleeves
- pipe whip restraint fasteners
- pipe whip restraints
- porous concrete pipe
- radiation shields
- rail track and support beams
- reactor building foundation mat
- reactor building metal siding
- reactor building metal siding fasteners
- reactor building polar crane rail and embedded plate
- reactor building polar crane rail clips and fasteners
- reactor building sumps
- reactor head cavity pit
- reactor head cavity pit liner
- reactor head cavity plug liners
- reactor head cavity plugs
- reactor internal storage pool
- reactor internal storage pool liner
- refueling bridge crane rail and embedded plate
- refueling bridge crane rail clips and fasteners
- refueling canal/dryer-separator canal plug liners
- refueling canal/dryer-separator canal plugs
- removable concrete slabs
- removable masonry wall anchors
- removable masonry wall framing
- sealing compounds
- seals and gaskets
- spent fuel pool girders
- spent fuel pool structural steel framing
- spent fuel storage pool
- spent fuel storage pool liner
- spent fuel storage pool structural fasteners
- structural beams
- structural columns
- structural fasteners
- structural plates
- structural steel: platforms, stairways, mezzanines
- sump liner fasteners (reactor building and auxiliary bay)
- sump liners (reactor building and auxiliary bay)

2.4B.2.2 Staff Evaluation

The staff reviewed ALRA Section 2.4.B.2 and the USAR using the evaluation methodology described in SER Section 2.4. The staff conducted its review in accordance with the guidance described in SRP-LR Section 2.4.

In conducting its review, the staff evaluated the structural component functions described in the ALRA and USAR in accordance with the requirements of 10 CFR 54.4(a) to verify that the applicant had not omitted from the scope of license renewal any components with intended functions delineated under 10 CFR 54.4(a). The staff then reviewed those components that the applicant had identified as being within the scope of license renewal to verify that the applicant had not omitted any passive and long-lived components that should be subject to an AMR in accordance with the requirements of 10 CFR 54.21(a)(1).

The staff's review of LRA Section 2.4.B.2 identified areas in which additional information was necessary to complete evaluation of the applicant's scoping and screening results. The applicant responded to the staff's RAIs as discussed below.

In RAI 2.4.B-3 dated December 9, 2004, the staff stated that it was not clear from LRA Table 2.4.B.2-1 whether the entire enclosure building (including the steel framing, metal siding, sealer materials, the overhead crane, and its railing) is within the scope of license renewal. Therefore, the staff requested that the applicant clarify the extent to which items of the enclosure building above the operating floor are within the scope of license renewal and describe in the format of LRA Table 2.4.B.2-1 scoping and AMR for the applicable components.

In its response by letter dated January 10, 2005, the applicant stated that the RB is a concrete structure up to the refueling floor elevation and above this elevation a steel-framed structure with metal wall panels. The steel framing members above the refueling floor are included in LRA Table 2.4.B.2-1 in the component type "Structural Steel (Carbon and Low Alloy Steel) in Air." The metal panels are included in LRA Table 2.4.B.2-1 in the component type "Metal Siding in Air." Sealer materials are included in the component type "Polymer in Air" in LRA Table 2.4.B.2-1. The RB overhead crane is included in LRA Table 2.4.B.9-1 as the component type "Polar Crane." The crane rails are included in LRA Table 2.4.B.2-1 in the component type "Structural Steel (Carbon and Low Alloy Steel) in Air." The applicant stated that these components and component types are within the scope of LR and subject to AMR. The applicant also pointed it out that the last sentence on LRA page 2.4-28 states: "The entire RB is made up of components that require an AMR," indicating that all components that comprise the RB are within the scope of license renewal and subject to AMR.

The staff found the applicant's response adequate and acceptable and its concern described in RAI 2.4B.2-3 is resolved.

In RAI 2.4.B-6 dated December 9, 2004, the staff stated that it could not identify from some of the tables of LRA Section 2.4.B the insulation and insulation jacketing included within the scope of license renewal nor the specific subsets of insulation and insulation jacketing included in the Section 2.4.B tables. It was also unclear whether insulation and jacketing on the reactor coolant system had been included. Therefore, the staff requested that the applicant:

- Identify the structures and structural components within the scope of license renewal with insulation and/or insulation jacketing.

- List all insulation and insulation jacketing materials associated with such structures and structural components that require aging management review and the results of aging management review for each.

- For insulation and insulation jacketing materials associated with such structures and structural components that do not require aging management submit the technical basis for this omission including plant-specific operating experience.

- For insulation and insulation jacketing materials associated with such structures and structural components that require aging management indicate the LRA sections that identify the AMPs credited to manage aging.

The applicant responded that NMP2 has no structures or structural components within the scope of license renewal with insulation and/or insulation jacketing. Therefore, the applicant did not include insulation in any LRA Section 2.4.B table. The NMP2 design does not require insulation of structural steel or concrete according to current licensing basis documents including safety analyses and plant evaluations.

Since NMP2 has no structures or structural components within the scope of license renewal with insulation and/or insulation jacketing, the staff found the applicant's response acceptable. Therefore, the staff's concern described in RAI 2.4.B-6 is resolved.

2.4B.2.3 Conclusion

The staff reviewed the ALRA, RAI responses, and related structural components to determine whether any SSCs that should be within the scope of license renewal had not been identified by the applicant. No omissions were identified. In addition, the staff performed a review to determine whether any components that should be subject to an AMR had not been identified by the applicant. No omissions were identified. On the basis of its review, the staff concludes that there is reasonable assurance that the applicant had adequately identified the RB components that are within the scope of license renewal, as required by 10 CFR 54.4(a), and the RB components that are subject to an AMR, as required by 10 CFR 54.21(a)(1).

2.4B.3 NMP2 Auxiliary Service Building

2.4B.3.1 Summary of Technical Information in the Amended Application

In ALRA Section 2.4.B.3, the applicant described the auxiliary service building (ASB). The ASB is a reinforced concrete and steel-framed structure. The ASB is surrounded by the RB, TB, and CRB. The ASB below elevation 261 ft is classified as seismic Category I. The basement floor is a reinforced concrete slab poured over electrical tunnels. The floor at elevation 261 ft is a concrete slab on steel deck supported by structural steel. The ASB contains the HVAC room, instrument calibration facility, and decontamination and shower facilities for personnel.

The ASB contains SR components that are relied upon to remain functional during and following DBEs. The failure of NSR SSCs in the ASB could potentially prevent the satisfactory accomplishment of an SR function. In addition, the ASB performs functions that support fire protection.

The intended functions within the scope of license renewal include the following:

- provides spray shield or curbs for directing flow

- provides for thermal expansion and/or seismic separation

- provides rated fire barrier

- provides flood protection barrier

- provides missile barrier

- provides structural support to NSR components whose failure could prevent accomplishment of SR function(s)

- provides shelter/protection to SR components

- provides structural and/or functional support to SR equipment

In ALRA Table 2.4.B.3-1, the applicant identified the following ASB component types that are within the scope of license renewal and subject to an AMR:

- ASB foundation
- compressible joints and seals
- concrete & grout
- concrete curbs
- concrete floors
- concrete slabs
- concrete walls
- doors and framing/hardware
- penetration seals
- penetration sleeves
- removable concrete slabs
- structural beams
- structural columns
- structural fasteners
- structural steel: platforms, stairways, mezzanines

2.4B.3.2 Staff Evaluation

The staff reviewed ALRA Section 2.4.B.3 and the USAR using the evaluation methodology described in SER Section 2.4. The staff conducted its review in accordance with the guidance described in SRP-LR Section 2.4.

In conducting its review, the staff evaluated the structural component functions described in the ALRA and USAR in accordance with the requirements of 10 CFR 54.4(a) to verify that the applicant had not omitted from the scope of license renewal any components with intended functions delineated under 10 CFR 54.4(a). The staff then reviewed those components that the applicant had identified as being within the scope of license renewal to verify that the applicant had not omitted any passive and long-lived components that should be subject to an AMR in accordance with the requirements of 10 CFR 54.21(a)(1).

2.4B.3.3 Conclusion

The staff reviewed the ALRA and related structural components to determine whether any SSCs that should be within the scope of license renewal had not been identified by the applicant. No omissions were identified. In addition, the staff performed a review to determine whether any components that should be subject to an AMR had not been identified by the applicant. No omissions were identified. On the basis of its review, the staff concludes that there is reasonable assurance that the applicant had adequately identified the ASB components that are within the scope of license renewal, as required by 10 CFR 54.4(a), and the ASB components that are subject to an AMR, as required by 10 CFR 54.21(a)(1).

2.4B.4 NMP2 Control Room Building

2.4B.4.1 Summary of Technical Information in the Amended Application

In ALRA Section 2.4.B.4, the applicant described the control room building (CRB). The CRB is a seismic Category I structure. It is a five-story reinforced concrete and steel structure. The exterior walls and roof are constructed of reinforced concrete. The interior floors are concrete decking supported by steel framing. The building is founded on bedrock and is supported by a reinforced concrete mat. The upper four floors are reinforced concrete slabs on steel deck supported by structural steel. Underground concrete tunnels connect the CRB to the RB. The CRB contains the control room, SR switchgear, batteries, and associated equipment.

The CRB contains SR components that are relied upon to remain functional during and following DBEs. The failure of NSR SSCs in the CRB could potentially prevent the satisfactory accomplishment of an SR function. In addition, the CRB performs functions that support fire protection.

The intended functions within the scope of license renewal include the following:

- provides spray shield or curbs for directing flow
- provides for thermal expansion and/or seismic separation
- provides rated fire barrier
- provides flood protection barrier
- provides missile barrier
- provides structural support to NSR components whose failure could prevent accomplishment of SR function(s)
- provides shelter/protection to SR components
- provides shielding against radiation
- provides pressure boundary or essentially leaktight barrier
- provides structural and/or functional support to SR equipment

In ALRA Table 2.4.B.4-1, the applicant identified the following CRB component types that are within the scope of license renewal and subject to an AMR:

- compressible joints and seals
- concrete and grout
- concrete columns
- concrete curbs
- concrete floors
- concrete lean fill
- concrete slabs
- concrete walls
- control room building foundation
- control room building sump
- doors and framing/hardware
- embedded structural plates
- expansion/grouted anchors
- miscellaneous structural steel
- penetration seal clamp
- penetration seals
- penetration sleeves
- removable concrete slabs
- sealing compounds
- structural beams
- structural columns
- structural fasteners
- structural steel: platforms, stairways, mezzanines
- supplemental structural steel
- suspended seismic support framing
- tornado-proof steel duct
- ventilation duct framing

2.4B.4.2 Staff Evaluation

The staff reviewed ALRA Section 2.4.B.4 and the USAR using the evaluation methodology described in SER Section 2.4. The staff conducted its review in accordance with the guidance described in SRP-LR Section 2.4.

In conducting its review, the staff evaluated the structural component functions described in the ALRA and USAR in accordance with the requirements of 10 CFR 54.4(a) to verify that the applicant had not omitted from the scope of license renewal any components with intended functions delineated under 10 CFR 54.4(a). The staff then reviewed those components that the applicant had identified as being within the scope of license renewal to verify that the applicant had not omitted any passive and long-lived components that should be subject to an AMR in accordance with the requirements of 10 CFR 54.21(a)(1).

2.4B.4.3 Conclusion

The staff reviewed the ALRA and related structural components to determine whether any SSCs that should be within the scope of license renewal had not been identified by the applicant. No omissions were identified. In addition, the staff performed a review to determine whether any components that should be subject to an AMR had not been identified by the applicant. No omissions were identified. On the basis of its review, the staff concludes that there is reasonable assurance that the applicant had adequately identified the CRB components that are within the scope of license renewal, as required by 10 CFR 54.4(a), and the CRB components that are subject to an AMR, as required by 10 CFR 54.21(a)(1).

2.4B.5 NMP2 Diesel Generator Building

2.4B.5.1 Summary of Technical Information in the Amended Application

In ALRA Section 2.4.B.5, the applicant described the diesel generator building (DGB). The DGB is a seismic Category I reinforced concrete structure enclosing the three diesel generators (DGs) and their associated equipment. The DGs are supported on reinforced concrete pedestals. The building is divided into three rooms separated by fire walls, each housing one DG. Fuel oil storage tanks are located below the building, with their fuel oil pumps housed in the individual DG rooms. The DGB is founded on bedrock and supported by reinforced concrete wall footings.

The DGB contains SR components that are relied upon to remain functional during and following DBEs. The failure of NSR SSCs in the DGB could potentially prevent the satisfactory accomplishment of an SR function. In addition, the DGB performs functions that support fire protection.

The intended functions within the scope of license renewal include the following:

- provides spray shield or curbs for directing flow

- provides for thermal expansion and/or seismic separation

- provides rated fire barrier

- provides flood protection barrier

- provides missile barrier

- provides structural support to NSR components whose failure could prevent accomplishment of SR function(s)

- provides shelter/protection to SR components

- provides structural and/or functional support to SR equipment

In ALRA Table 2.4.B.5-1, the applicant identified the following DGB component types that are within the scope of license renewal and subject to an AMR:

- compressible joints and seals
- concrete & grout
- concrete curbs
- concrete floors
- concrete slabs
- concrete walls
- concrete lean fill
- crane rails/girders

- diesel generator building foundation
- diesel generator pedestals
- diesel generator tank foundations and encasements
- doors and framing/hardware
- expansion/grouted anchors
- manhole cover and frame
- missile logs
- penetration seal clamp
- penetration seals
- penetration sleeves
- oil sump
- sealing compounds
- structural beams
- structural columns
- structural fasteners
- structural steel: platforms, stairways, mezzanines

2.4B.5.2 Staff Evaluation

The staff reviewed ALRA Section 2.4.B.5 and the USAR using the evaluation methodology described in SER Section 2.4. The staff conducted its review in accordance with the guidance described in SRP-LR Section 2.4.

In conducting its review, the staff evaluated the structural component functions described in the ALRA and USAR in accordance with the requirements of 10 CFR 54.4(a) to verify that the applicant had not omitted from the scope of license renewal any components with intended functions delineated under 10 CFR 54.4(a). The staff then reviewed those components that the applicant had identified as being within the scope of license renewal to verify that the applicant had not omitted any passive and long-lived components that should be subject to an AMR in accordance with the requirements of 10 CFR 54.21(a)(1).

2.4B.5.3 Conclusion

The staff reviewed the ALRA and related structural components to determine whether any SSCs that should be within the scope of license renewal had not been identified by the applicant. No omissions were identified. In addition, the staff performed a review to determine whether any components that should be subject to an AMR had not been identified by the applicant. No omissions were identified. On the basis of its review, the staff concludes that there is reasonable assurance that the applicant had adequately identified the DGB components that are within the scope of license renewal, as required by 10 CFR 54.4(a), and the DGB components that are subject to an AMR, as required by 10 CFR 54.21(a)(1).

2.4B.6 NMP2 Essential Yard Structures

2.4B.6.1 Summary of Technical Information in the Amended Application

In ALRA Section 2.4.B.6, the applicant described the EYS. The EYS include, but are not limited to, electrical, piping, and vent tunnels; manholes; underground duct banks; and earth berms and ditches used for flood control. Seismic Category I electrical tunnels and piping tunnels contain Category I systems and are constructed of reinforced concrete. Included in the essential yard structures are all Class 1E duct banks and manholes. Earthen berms are located around the perimeter of the site to provide flood protection to the site. A stone-faced dike was constructed along the shoreline. The dike prevents flooding of the plant from high lake water levels and the effects of the probable maximum windstorm. The EYS also include the structures that support the equipment and high voltage lines in the 115KV switchyard and Scriba substation for SBO. The SBO components are evaluated in the switchyard system.

The EYS contain SR components that are relied upon to remain functional during and following DBEs. The failure of NSR SSCs in the EYS could potentially prevent the satisfactory accomplishment of an SR function. In addition, the EYS perform functions that support fire protection and SBO.

The intended functions within the scope of license renewal include the following:

- provides spray shield or curbs for directing flow

- provides for thermal expansion and/or seismic separation

- provides rated fire barrier

- provides flood protection barrier

- provides path for release of filtered and unfiltered gaseous discharge

- provides missile barrier

- provides structural support to NSR components whose failure could prevent accomplishment of SR function(s)

- provides shelter/protection to SR components

- provides pressure boundary or essentially leaktight barrier

- provides structural and/or functional support to SR equipment

In ALRA Table 2.4.B.6-1, the applicant identified the following EYS component types that are within the scope of license renewal and subject to an AMR:

- bus duct enclosure
- Class 1E manhole sumps
- compressible joints and seals
- concrete and grout
- concrete encasement of ductlines
- concrete lean fill
- doors and framing/hardware
- earthen berm
- electrical and radwaste tunnels removable concrete
- slabs
- electrical and radwaste tunnels steel beams
- electrical and radwaste tunnels steel columns
- electrical and radwaste tunnels

2-294

- embedded structural plates
- expansion/grouted anchors
- manhole covers and frames
- manholes
- penetration seal clamp
- penetration seals
- penetration sleeves
- pipe tunnel structural framing
- pipe tunnels
- pipe tunnel sumps
- revetment ditch
- sealing compounds
- service water tunnel
- service water tunnel removable concrete slabs
- service water valve pit
- service water valve pit removable concrete slabs
- service water valve pit sealants
- stone-faced dike
- structural fasteners
- structural steel: platforms, stairways, and mezzanines
- transformer area walls
- transformer curbs
- transformer foundation pads
- vent tunnel
- vent tunnel fill concrete
- 115 KV steel transmission towers
- 115 KV steel transmission tower foundations
- 115 KV wooden transmission towers

2.4B.6.2 Staff Evaluation

The staff reviewed ALRA Section 2.4.B.6 and the USAR using the evaluation methodology described in SER Section 2.4. The staff conducted its review in accordance with the guidance described in SRP-LR Section 2.4.

In conducting its review, the staff evaluated the structural component functions described in the ALRA and USAR in accordance with the requirements of 10 CFR 54.4(a) to verify that the applicant had not omitted from the scope of license renewal any components with intended functions delineated under 10 CFR 54.4(a). The staff then reviewed those components that the applicant had identified as being within the scope of license renewal to verify that the applicant had not omitted any passive and long-lived components that should be subject to an AMR in accordance with the requirements of 10 CFR 54.21(a)(1).

The staff's review of LRA Section 2.4.B.6 identified an area in which additional information was necessary to complete the evaluation of the applicant's scoping and screening results. The applicant responded to the staff's RAI as discussed below.

In RAI 2.4.B-4 dated December 9, 2004, the staff stated that LRA Section 2.4.B.6 states that the NMP2 EYS include electrical equipment, piping, vent tunnels, manholes, underground duct banks, and earth berms and ditches used for flood control. The EYS also are said to include structures that support the equipment and high voltage lines in the switchyard and Scriba substation for SBO. LRA Table 2.4.B.6-1 does not indicate that some associated structural steel supports or embedments are in the scope of license renewal though they perform SR functions per 10 CFR 50.54(a)(1). Therefore, the staff requested that, if they were within the scope of license renewal, the applicant provide a description of its scoping and include the structural steel items in the format of LRA Table 2.4.B.6-1. The staff also requested that the applicant provide the basis for omission from the scope of license renewal.

In its response by letter dated January 10, 2005, the applicant stated that the equipment and high voltage line supports in the switchyard and Scriba substation are within the scope of license renewal under 10 CFR 54.4(a)(3) for SBO. In LRA Section 2.4.B.6 and Table 2.4.B.6-1, structures that support the equipment and high voltage lines in the switchyard and the Scriba substation are included in the component types "Structural Steel (Carbon and Low Alloy Steel) in Air," "Treated Wood in Air," "Treated Wood in Soil Above the GWT," and "Treated Wood in Soil Below the GWT." Embedments are included as part of the "Concrete in Air" component type with the exposed portions of anchor bolts included in the component type "Fasteners (Carbon and Low Alloy Steel) in Air." Hilti bolts are included in the "Expansion/Grouted Anchors (Carbon and Low Alloy Steel) in Air" component type.

The staff found the applicant's explanation adequate and acceptable. Therefore, the staff's concern described in RAI 2.4.B-4 is resolved.

2.4B.6.3 Conclusion

The staff reviewed the ALRA, RAI response, and related structural components to determine whether any SSCs that should be within the scope of license renewal had not been identified by the applicant. No omissions were identified. In addition, the staff performed a review to determine whether any components that should be subject to an AMR had not been identified by the applicant. No omissions were identified. On the basis of its review, the staff concludes that there is reasonable assurance that the applicant had adequately identified the EYS components that are within the scope of license renewal, as required by 10 CFR 54.4(a), and the EYS components that are subject to an AMR, as required by 10 CFR 54.21(a)(1).

2.4B.7 NMP2 Fuel Handling System

2.4B.7.1 Summary of Technical Information in the Amended Application

In ALRA Section 2.4.B.7 the applicant described the fuel handling system. The fuel handling system involves those components used to move fuel from the time of receipt of new fuel to the storage of spent fuel in the spent fuel storage pool. Components that are evaluated in the fuel handling system include the channel handling boom, the fuel preparation machines, the fuel transfer shielding bridge, the refueling crane platform and equipment, the new fuel storage vault, lifting and handling equipment, and spent fuel pool storage racks. Although the RB polar crane handles fuel, it is analyzed in the material handling system. Civil/structural components from the fuel nuclear refueling, the fuel nuclear storage, and the materials handling fuel storage area subsystems are also evaluated as part of the fuel handling system.

The fuel handling system contains SR components that are relied upon to remain functional during and following DBEs. The failure of NSR SSCs in the fuel handling system could potentially prevent the satisfactory accomplishment of an SR function.

The intended functions within the scope of license renewal include the following:

- provides structural support to NSR components whose failure could prevent accomplishment of SR function(s)

- provides structural and/or functional support to SR equipment

In ALRA Table 2.4.B.7-1, the applicant identified the following fuel handling system component types that are within the scope of license renewal and subject to an AMR:

- channel handling boom
- control blade storage frame
- fuel preparation machines
- fuel storage racks
- head strongback carousel
- in-vessel storage rack
- new fuel storage rack
- new fuel storage vault cover
- recirculation pump motor lifting lugs
- refueling crane and platform equipment
- steam dryer primary lifting beam

2.4B.7.2 Staff Evaluation

The staff reviewed ALRA Section 2.4.B.7 and the USAR using the evaluation methodology described in SER Section 2.4. The staff conducted its review in accordance with the guidance described in SRP-LR Section 2.4.

In conducting its review, the staff evaluated the system functions described in the ALRA and USAR in accordance with the requirements of 10 CFR 54.4(a) to verify that the applicant had not omitted from the scope of license renewal any components with intended functions delineated under 10 CFR 54.4(a). The staff then reviewed those components that the applicant had identified as being within the scope of license renewal to verify that the applicant had not omitted any passive and long-lived components that should be subject to an AMR in accordance with the requirements of 10 CFR 54.21(a)(1).

2.4B.7.3 Conclusion

The staff reviewed the ALRA and related structural components to determine whether any SSCs that should be within the scope of license renewal had not been identified by the applicant. No omissions were identified. In addition, the staff performed a review to determine whether any components that should be subject to an AMR had not been identified by the applicant. No omissions were identified. On the basis of its review, the staff concludes that there is reasonable assurance that the applicant had adequately identified the fuel handling system components that are within the scope of license renewal, as required by 10 CFR 54.4(a), and the fuel handling system components that are subject to an AMR, as required by 10 CFR 54.21(a)(1).

2.4B.8 NMP2 Main Stack

2.4B.8.1 Summary of Technical Information in the Amended Application

In ALRA Section 2.4.B.8, the applicant described the main stack. The main stack is a seismic Category I reinforced-concrete chimney, approximately 430-ft high, located on the northeast side of the power station. The main stack is designed and constructed to provide elevated release of offgas, standby gas treatment, turbine building ventilation, and other systems. The main stack foundation is a on a reinforced concrete base, which extends to bedrock.

The main stack contains SR components that are relied upon to remain functional during and following DBEs. The failure of NSR SSCs in the main stack could potentially prevent the satisfactory accomplishment of an SR function.

The intended functions within the scope of license renewal include the following:

- provides spray shield or curbs for directing flow

- provides for thermal expansion and/or seismic separation

- provides flood protection barrier

- provides path for release of filtered and unfiltered gaseous discharge

- provides structural support to NSR components whose failure could prevent accomplishment of SR function(s)

- provides shelter/protection to SR components

- provides structural and/or functional support to SR equipment

In ALRA Table 2.4.B.8-1, the applicant identified the following main stack component types that are within the scope of license renewal and subject to an AMR:

- compressible joints and seals
- concrete chimney shell
- concrete curbs
- concrete floors
- concrete slabs
- concrete lean fill
- embedded steel
- expansion/grouted anchors
- main stack foundation
- penetration seal clamp
- penetration seals

- penetration sleeves
- structural fasteners
- structural steel: platforms, stairways, and mezzanines

2.4B.8.2 Staff Evaluation

The staff reviewed ALRA Section 2.4.B.8 and the USAR using the evaluation methodology described in SER Section 2.4. The staff conducted its review in accordance with the guidance described in SRP-LR Section 2.4.

In conducting its review, the staff evaluated the structural component functions described in the ALRA and USAR in accordance with the requirements of 10 CFR 54.4(a) to verify that the applicant had not omitted from the scope of license renewal any components with intended functions delineated under 10 CFR 54.4(a). The staff then reviewed those components that the applicant had identified as being within the scope of license renewal to verify that the applicant had not omitted any passive and long-lived components that should be subject to an AMR in accordance with the requirements of 10 CFR 54.21(a)(1).

2.4B.8.3 Conclusion

The staff reviewed the ALRA and related structural components to determine whether any SSCs that should be within the scope of license renewal had not been identified by the applicant. No omissions were identified. In addition, the staff performed a review to determine whether any components that should be subject to an AMR had not been identified by the applicant. No omissions were identified. On the basis of its review, the staff concludes that there is reasonable assurance that the applicant had adequately identified the main stack components that are within the scope of license renewal, as required by 10 CFR 54.4(a), and the main stack components that are subject to an AMR, as required by 10 CFR 54.21(a)(1).

2.4B.9 NMP2 Material Handling System

2.4B.9.1 Summary of Technical Information in the Amended Application

In ALRA Section 2.4.B.9, the applicant described the material handling system. The material handling system consists of overhead traveling cranes, monorail hoists, platform cranes, jib cranes, and associated mechanical and electrical components. For license renewal purposes, the crane girders and rails are included in the structural steel asset of the structure in which the crane is located.

The material handling system contains SR components that are relied upon to remain functional during and following DBEs. The failure of NSR SSCs in the material handling system could potentially prevent the satisfactory accomplishment of an SR function.

The intended functions within the scope of license renewal include the following:

- provides structural support to NSR components whose failure could prevent accomplishment of SR function(s)

- provides structural and/or functional support to SR equipment

In ALRA Table 2.4.B.9-1, the applicant identified the following material handling system component types that are within the scope of license renewal and subject to an AMR:

- control building equipment hoist (el. 306)
- emergency diesel generator cranes
- main steam isolation valve crane
- main steam isolation valve hoist
- main turbine area traveling crane
- reactor building polar crane
- recirculation motor handling cranes
- safety relief valve hoists
- screenwell area traveling crane
- stop log area crane

2.4B.9.2 Staff Evaluation

The staff reviewed ALRA Section 2.4.B.9 and the USAR using the evaluation methodology described in SER Section 2.4. The staff conducted its review in accordance with the guidance described in SRP-LR Section 2.4.

In conducting its review, the staff evaluated the system functions described in the ALRA and USAR in accordance with the requirements of 10 CFR 54.4(a) to verify that the applicant had not omitted from the scope of license renewal any components with intended functions delineated under 10 CFR 54.4(a). The staff then reviewed those components that the applicant had identified as being within the scope of license renewal to verify that the applicant had not omitted any passive and long-lived components that should be subject to an AMR in accordance with the requirements of 10 CFR 54.21(a)(1).

2.4B.9.3 Conclusion

The staff reviewed the ALRA and related structural components to determine whether any SSCs that should be within the scope of license renewal had not been identified by the applicant. No omissions were identified. In addition, the staff performed a review to determine whether any components that should be subject to an AMR had not been identified by the applicant. No omissions were identified. On the basis of its review, the staff concludes that there is reasonable assurance that the applicant had adequately identified the material handling system components that are within the scope of license renewal, as required by 10 CFR 54.4(a), and the material handling system components that are subject to an AMR, as required by 10 CFR 54.21(a)(1).

2.4B.10 NMP2 Motor Operated Doors System

2.4B.10.1 Summary of Technical Information in the Amended Application

In ALRA Section 2.4.B.10, the applicant described the motor operated doors system. The motor operated doors system consists of various motor operated doors and the associated electronic equipment that monitors their positions.

The motor operated doors system contains SR components that are relied upon to remain functional during and following DBEs. In addition, the motor operated doors system performs functions that support fire protection.

All doors have been identified and are addressed in appropriate tables describing structures in which the doors are physically located. The remaining electrical components are active components. There are no other components subject to an AMR for this system.

2.4B.10.2 Staff Evaluation

The staff reviewed ALRA Section 2.4.B.10 and the USAR using the evaluation methodology described in SER Section 2.4. The staff conducted its review in accordance with the guidance described in SRP-LR Section 2.4.

In conducting its review, the staff evaluated the system functions described in the ALRA and USAR in accordance with the requirements of 10 CFR 54.4(a) to verify that the applicant had not omitted from the scope of license renewal any components with intended functions delineated under 10 CFR 54.4(a). The staff then reviewed those components that the applicant had identified as being within the scope of license renewal to verify that the applicant had not omitted any passive and long-lived components that should be subject to an AMR in accordance with the requirements of 10 CFR 54.21(a)(1).

2.4B.10.3 Conclusion

The staff reviewed the ALRA and related structural components to determine whether any SSCs that should be within the scope of license renewal had not been identified by the applicant. No omissions were identified. In addition, the staff performed a review to determine whether any components that should be subject to an AMR had not been identified by the applicant. No omissions were identified. On the basis of its review, the staff concludes that there is reasonable assurance that the applicant had adequately identified the motor operated doors system components that are within the scope of license renewal, as required by 10 CFR 54.4(a), and the motor operated doors system components that are subject to an AMR, as required by 10 CFR 54.21(a)(1).

2.4B.11 NMP2 Radwaste Building

2.4B.11.1 Summary of Technical Information in the Amended Application

In ALRA Section 2.4.B.11, the applicant described the radwaste building (RWB). The RWB is a seismic Category I structure and contains the radioactive waste system. It is a five-story, concrete and steel building. The exterior walls are reinforced concrete. A rolling steel door is provided in the north wall for truck access into the building. The basement floor is a concrete mat on bedrock. The upper four floors are concrete supported by steel deck and beams. The roof consists of steel framing with steel deck, insulation, and four-ply, built-up roofing. The decontamination area is located south of the RWB, and is an extension of the TB and the RWB.

The RWB contains SR components that are relied upon to remain functional during and following DBEs. The failure of NSR SSCs in the RWB could potentially prevent the satisfactory accomplishment of an SR function. In addition, the RWB performs functions that support fire protection.

The intended functions within the scope of license renewal include the following:

- provides spray shield or curbs for directing flow

- provides for thermal expansion and/or seismic separation

- provides rated fire barrier

- provides flood protection barrier

- provides structural support to NSR components whose failure could prevent accomplishment of SR function(s)

- provides shelter/protection to SR components

- provides shielding against radiation

- provides pressure boundary or essentially leaktight barrier

- provides structural and/or functional support to SR equipment

In ALRA Table 2.4.B.11-1, the applicant identified the following RWB component types that are within the scope of license renewal and subject to an AMR:

- concrete and grout
- concrete floors
- concrete lean fill
- concrete slabs
- concrete walls
- decontamination area compressible joints and seals
- decontamination area concrete floors
- decontamination area concrete lean fill
- decontamination area concrete slabs
- decontamination area concrete walls
- decontamination area foundation
- doors and framing/hardware
- embedded structural plates
- expansion/grouted anchors
- hotline trough
- penetration seal clamp
- penetration seals
- penetration sleeves
- radwaste building foundation
- radwaste building sump
- radwaste building sump flange plate
- steel liner
- structural beams
- structural columns
- structural fasteners

2.4B.11.2 Staff Evaluation

The staff reviewed ALRA Section 2.4.B.11 and the USAR using the evaluation methodology described in SER Section 2.4. The staff conducted its review in accordance with the guidance described in SRP-LR Section 2.4.

In conducting its review, the staff evaluated the structural component functions described in the ALRA and USAR in accordance with the requirements of 10 CFR 54.4(a) to verify that the applicant had not omitted from the scope of license renewal any components with intended functions delineated under 10 CFR 54.4(a). The staff then reviewed those components that the applicant had identified as being within the scope of license renewal to verify that the applicant had not omitted any passive and long-lived components that should be subject to an AMR in accordance with the requirements of 10 CFR 54.21(a)(1).

2.4B.11.3 Conclusion

The staff reviewed the ALRA and related structural components to determine whether any SSCs that should be within the scope of license renewal had not been identified by the applicant. No omissions were identified. In addition, the staff performed a review to determine whether any components that should be subject to an AMR had not been identified by the applicant. No omissions were identified. On the basis of its review, the staff concludes that there is reasonable assurance that the applicant had adequately identified the RWB components that are within the scope of license renewal, as required by 10 CFR 54.4(a), and the RWB components that are subject to an AMR, as required by 10 CFR 54.21(a)(1).

2.4B.12 NMP2 Screenwell Building

2.4B.12.1 Summary of Technical Information in the Amended Application

In ALRA Section 2.4.B.12, the applicant described the screenwell building (SWB). The SWB consists of a concrete substructure and a steel frame superstructure. The substructure, below grade elevation 261'-0", including the service water pump room, is designated seismic Category I, whereas the steel frame superstructure, including the circulating water pump and water treatment area, is designed as a non-Category I area. The SWB includes the service water pump rooms, the diesel and electric fire pump rooms, the water treatment area, the circulating water pump area, and other associated equipment. Stop logs, traveling screens, trash rakes, etc., are set in the concrete walls, as required to divert the flow of water. These components are built-up structures of steel and concrete guided and supported by the reinforced concrete walls and floors. For license renewal purposes, the SWB also includes the intake structures and the intake/discharge tunnels.

The SWB contains SR components that are relied upon to remain functional during and following DBEs. The failure of NSR SSCs in the SWB could potentially prevent the satisfactory accomplishment of an SR function. In addition, the SWB performs functions that support fire protection, ATWS, and SBO.

The intended functions within the scope of license renewal include the following:

2-303

- provides spray shield or curbs for directing flow

- provides for thermal expansion and/or seismic separation

- provides filtration

- provides rated fire barrier

- provides flood protection barrier

- provides missile barrier

- provides structural support to NSR components whose failure could prevent accomplishment of SR function(s)

- provides shelter/protection to SR components

- provides source of cooling water for plant shutdown

- provides structural and/or functional support to SR equipment

In ALRA Table 2.4.B.12-1, the applicant identified the following SWB component types that are within the scope of license renewal and subject to an AMR:

- compressible joints and seals
- concrete and grout
- concrete curbs
- concrete floors
- concrete slabs
- concrete walls
- crane rails/girders
- doors and framing/hardware
- embedded structural plates
- expansion/grouted anchors
- hot line tunnel
- intake shaft access door and framing
- intake shaft concrete lean fill
- intake shafts
- intake structure anchor bolts
- intake structure bar racks
- intake structure concrete and grout
- intake structure fasteners
- intake structure hatch cover and manhole
- intake structure structural steel and embedments
- intake structure Tremie concrete
- intake structures
- intake tunnel compressible material
- intake tunnel concrete lean fill
- intake tunnels
- masonry walls
- penetration seal clamp
- penetration seals
- penetration sleeves

- removable concrete slabs
- removable steel nose piece
- screenwell building sumps
- screenwell building foundation
- service water pump bay sumps
- service water tunnel
- service water valve missile protection
- stop log fasteners
- stop log seals
- stop logs and guides
- structural beams
- structural columns
- structural fasteners
- structural foundation piles
- structural steel: platforms, stairways, and mezzanines
- trash racks and guides

2.4B.12.2 Staff Evaluation

The staff reviewed ALRA Section 2.4.B.12 and the USAR using the evaluation methodology described in SER Section 2.4. The staff conducted its review in accordance with the guidance described in SRP-LR Section 2.4.

In conducting its review, the staff evaluated the structural component functions described in the ALRA and USAR in accordance with the requirements of 10 CFR 54.4(a) to verify that the applicant had not omitted from the scope of license renewal any components with intended functions delineated under 10 CFR 54.4(a). The staff then reviewed those components that the applicant had identified as being within the scope of license renewal to verify that the applicant had not omitted any passive and long-lived components that should be subject to an AMR in accordance with the requirements of 10 CFR 54.21(a)(1).

The staff's review of LRA Section 2.4.B.12 identified areas in which additional information was necessary to complete the evaluation of the applicant's scoping and screening results. The applicant responded to the staff's RAIs as discussed below.

In RAI 2.4.B-5 dated December 9, 2004, the staff stated that LRA Table 2.4.B.11-1 lists structural steel foundation piles (carbon and low alloy steel) in undisturbed soil as one of the component types in the NMP2 screenwell building requiring AMR. Because these piles are inaccessible, the staff requested that the applicant discuss results of the AMR of the piles and indicate where in the LRA the aging management of the piles is addressed.

In its response by letter dated January 10, 2005, the applicant stated that the structural steel foundation piles in undisturbed soil are subject to AMR; however, the component type "Structural Steel Foundation Piles (Carbon and Low Alloy Steel) in Undisturbed Soil" has no aging effects and, as stated in Section 4.3.1.1 of EPRI TR-103842, undisturbed soils are so deficient in oxygen at levels a few feet below the surface or the water table that steel piles are not appreciably affected by corrosion regardless of the soil type or properties. The NMPNS site ground water and soil are both non-aggressive in nature as defined by SRP-LR. The applicant indicated that previous LRAs (e.g., Fort Calhoun) that the staff had reviewed have also not identified any aging effects requiring management for carbon steel foundation piles.

The staff found the above response adequate. Therefore, the concern described in RAI 2.4.B-5 is resolved.

In RAI 2.4.B-7 dated December 9, 2004, the staff requested that the applicant confirm that such Screenwell Building items as hatches and plugs, structural steel embedments, reinforced concrete foundation footings, grouted concrete, and water proofing membrane materials are within the scope of license renewal and require AMR and, if within the scope of license renewal, that the applicant provide additional information in the format of LRA Table 2.4.B.11-1. If not within the scope of license renewal, the applicant was to provide the basis for their omission.

In its response by letter dated January 10, 2005, the applicant responded that the items listed in the RAI are within the scope of license renewal and subject to AMR with the exception of the waterproofing membranes, which are not included because they are applied as coatings. NMPNS does not credit coatings to mitigate aging effects. The LRA lists in Table 2.4.B.11-1 component types that represent these items:

- There are no "hatches" in the NMP2 Screenwell Building. All doors are included in the component type "Door."

- Plugs are concrete and included in the component type "Concrete in Air."

- Embedded portions of structural steel embedments are integral with the concrete and included in the component type "Concrete in Air" or "Concrete in Raw Water" depending on location. Structural steel exposed to atmosphere is included in the component type "Structural Steel (Carbon and Low Alloy Steel) in Air" or "Structural Steel (Carbon and Low Alloy Steel) in Raw Water" depending on location.

- Reinforced concrete foundation footings are included in the component type "Concrete in Soil Above the GWT" or "Concrete in Soil Below the GWT" depending on footing depth.

- Grouted concrete is not used at NMP2. Structural concrete is included in the various concrete component types depending on its environment.

The staff found the applicant's response complete and adequate. Therefore, the concern described in RAI 2.4.B-7 is resolved.

2.4B.12.3 Conclusion

The staff reviewed the ALRA, RAI responses, and related structural components to determine whether any SSCs that should be within the scope of license renewal had not been identified by the applicant. No omissions were identified. In addition, the staff performed a review to determine whether any components that should be subject to an AMR had not been identified by the applicant. No omissions were identified. On the basis of its review, the staff concludes that there is reasonable assurance that the applicant had adequately identified the SWB components that are within the scope of license renewal, as required by 10 CFR 54.4(a), and the SWB components that are subject to an AMR, as required by 10 CFR 54.21(a)(1).

2.4B.13 NMP2 Standby Gas Treatment Building

2.4B.13.1 *Summary of Technical Information in the Amended Application*

In ALRA Section 2.4.B.13, the applicant described the standby gas treatment building (SGTB). The SGTB and railroad access area contain the standby gas treatment (SGT) filters and associated equipment and allow access for spent fuel shipping. This structure is classified a seismic Category I structure up to elevation 286 ft. The portion of the building above elevation 286 ft is classified as nonseismic. The SGTB is a two-story, reinforced concrete and steel-framed structure. The structure shares a common wall with the railroad access lock adjacent to the RB. The reinforced concrete floor slab is provided at the grade level of elevation 261 ft. A railroad access lock approximately 25 x 90 ft is provided adjacent to the RB. This building is a reinforced concrete and steel-framed structure and shares a common wall with the SGTB.

The SGTB contains SR components that are relied upon to remain functional during and following DBEs. The failure of NSR SSCs in the SGTB could potentially prevent the satisfactory accomplishment of an SR function. In addition, the SGTB performs functions that support fire protection.

The intended functions within the scope of license renewal include the following:

- provides for thermal expansion and/or seismic separation

- provides rated fire barrier

- provides flood protection barrier

- provides shielding against high energy line breaks

- provides missile barrier

- provides structural support to NSR components whose failure could prevent accomplishment of SR function(s)

- provides pressure retaining boundary

- provides shelter/protection to SR components

- provides pressure boundary or essentially leaktight barrier

- provides structural and/or functional support to SR equipment

In ALRA Table 2.4.B.13-1, the applicant identified the following SGTB component types that are within the scope of license renewal and subject to an AMR:

- compressible joints and seals
- concrete and grout
- concrete floors
- concrete lean fill
- concrete slabs
- concrete walls
- doors and framing/hardware

- embedded rail girders
- embedded structural plates
- penetration seals
- penetration sleeves
- SGTB foundation
- structural beams
- structural columns
- structural fasteners
- structural steel: platforms, stairways, and mezzanines

2.4B.13.2 Staff Evaluation

The staff reviewed ALRA Section 2.4.B.13 and the USAR using the evaluation methodology described in SER Section 2.4. The staff conducted its review in accordance with the guidance described in SRP-LR Section 2.4.

In conducting its review, the staff evaluated the structural component functions described in the ALRA and USAR in accordance with the requirements of 10 CFR 54.4(a) to verify that the applicant had not omitted from the scope of license renewal any components with intended functions delineated under 10 CFR 54.4(a). The staff then reviewed those components that the applicant had identified as being within the scope of license renewal to verify that the applicant had not omitted any passive and long-lived components that should be subject to an AMR in accordance with the requirements of 10 CFR 54.21(a)(1).

2.4B.13.3 Conclusion

The staff reviewed the ALRA and related structural components to determine whether any SSCs that should be within the scope of license renewal had not been identified by the applicant. No omissions were identified. In addition, the staff performed a review to determine whether any components that should be subject to an AMR had not been identified by the applicant. No omissions were identified. On the basis of its review, the staff concludes that there is reasonable assurance that the applicant had adequately identified the SGTB components that are within the scope of license renewal, as required by 10 CFR 54.4(a), and the SGTB components that are subject to an AMR, as required by 10 CFR 54.21(a)(1).

2.4B.14 NMP2 Turbine Building

2.4B.14.1 Summary of Technical Information in the Amended Application

In ALRA Section 2.4.B.14, the applicant described the TB. The TB complex includes the TB, heater bays, main steam tunnel, and condensate demineralizer regenerative and offgas area. A portion of the TB, main steam tunnel area, and offgas area are analyzed to seismic conditions, whereas the remaining portions are designed as nonseismic. The complex houses the turbine generator, condenser, moisture separator, etc., in the TB areas, heaters and related pumps and accessories in heater bay areas, and offgas system equipment and tanks in offgas areas. The main steam tunnel connects the TB with the RB. The TB complex is constructed partially on spread footings and partially on a mat foundation. This building complex is constructed of reinforced concrete floors and walls up to the operating floor level. The TB's operating floor is concrete supported by steel deck and beams. The structure above the operating floor level is constructed of a structural steel framing system braced by vertical and horizontal bracing systems up to roof level, enclosed by metal siding. A steel roof deck with roofing is provided at the top of the structure.

The TB contains SR components that are relied upon to remain functional during and following DBEs. The failure of NSR SSCs in the TB could potentially prevent the satisfactory accomplishment of an SR function. In addition, the TB performs functions that support fire protection.

The intended functions within the scope of license renewal include the following:

- provides spray shield or curbs for directing flow

- provides for thermal expansion and/or seismic separation

- provides rated fire barrier

- provides flood protection barrier

- provides shielding against high energy line breaks

- provides structural support to NSR components whose failure could prevent accomplishment of SR function(s)

- provides pipe whip restraint

- provides shelter/protection to SR components

- provides shielding against radiation

- provides structural and/or functional support to SR equipment

In ALRA Table 2.4.B.14-1, the applicant identified the following TB component types that are within the scope of license renewal and subject to an AMR:

- compressible joints and seals
- concrete and grout
- concrete columns
- concrete curbs
- concrete floors
- concrete slabs
- concrete walls
- concrete lean fill
- crane rails/girders
- doors and framing/hardware
- embedded structural plates
- expansion/grouted anchors
- main steam tunnel
- masonry walls
- penetration seal clamp
- penetration seals
- penetration sleeves

- pipe whip restraints
- pipe whip restraint fasteners
- removable concrete slabs
- removable masonry wall framing
- removable masonry walls
- structural beams
- structural columns
- structural fasteners
- structural steel: platforms, stairways, and mezzanines
- TB foundation
- TB sumps
- turbine support mat
- turbine support structure

2.4B.14.2 Staff Evaluation

The staff reviewed ALRA Section 2.4.B.14 and the USAR using the evaluation methodology described in SER Section 2.4. The staff conducted its review in accordance with the guidance described in SRP-LR Section 2.4.

In conducting its review, the staff evaluated the structural component functions described in the ALRA and USAR in accordance with the requirements of 10 CFR 54.4(a) to verify that the applicant had not omitted from the scope of license renewal any components with intended functions delineated under 10 CFR 54.4(a). The staff then reviewed those components that the applicant had identified as being within the scope of license renewal to verify that the applicant had not omitted any passive and long-lived components that should be subject to an AMR in accordance with the requirements of 10 CFR 54.21(a)(1).

2.4B.14.3 Conclusion

The staff reviewed the ALRA and related structural components to determine whether any SSCs that should be within the scope of license renewal had not been identified by the applicant. No omissions were identified. In addition, the staff performed a review to determine whether any components that should be subject to an AMR had not been identified by the applicant. No omissions were identified. On the basis of its review, the staff concludes that there is reasonable assurance that the applicant had adequately identified the TB components that are within the scope of license renewal, as required by 10 CFR 54.4(a), and the TB components that are subject to an AMR, as required by 10 CFR 54.21(a)(1).

2.4C NMPNS Structural Commodities

In ALRA Section 2.4.C, the applicant identified the NMPNS structural commodities that are subject to an AMR for license renewal.

The applicant described the supporting structures and components of the structural commodities in the following sections of the ALRA:

- 2.4.C.1   component supports
- 2.4.C.2   fire stops and seals

The staff's review findings regarding ALRA Sections 2.4.C.1 and 2.4.C.2 are presented in SER Sections 2.4A.2.1 and 2.4A.2.2, respectively.

2.4C.1 Component Supports

2.4C.1.1 Summary of Technical Information in the Amended Application

In ALRA Section 2.4.C.1, the applicant described the component supports commodity. Component supports are connections between a system component and a plant structural member such as a concrete wall or floor or structural steel beam or column. Supports for both the distributive portions of systems and equipment like pumps and pressure vessels are included as parts of this commodity group. Supported components include vessels, piping, passive pump components, and heat exchangers. Supports for electrical cables, cable trays, cable tray missile shields, conduits, HVAC ducting, motor control center cabinets, electrical enclosures, fans, filters, and heaters are also included in this commodity. Seismic restraints, which may or may not provide support during normal operation, are also considered parts of this commodity.

The component supports commodity contains SR components that are relied upon to remain functional during and following DBEs. The failure of NSR SSCs in the component supports commodity could potentially prevent the satisfactory accomplishment of an SR function.

The intended functions within the scope of license renewal include the following:

- provides missile barrier

- provides structural support to NSR components whose failure could prevent accomplishment of SR function(s)

- provides structural and/or functional support to SR equipment

In ALRA Table 2.4.C.1-1, the applicant identified the following component supports commodity component types that are within the scope of license renewal and subject to an AMR:

- ASME Class 1, 2, 3 and MC hangers and supports
- cable trays and supports
- cable tray missile shields (NMP1 only)
- conduit

- electrical panels, racks, cabinets, and other enclosures
- equipment supports and foundations
- instrumentation racks, frames, panels, enclosures
- lubrite plates
- non-ASME class hangers and supports
- tube track
- vibration isolating elements

2.4C.1.2 Staff Evaluation

The staff reviewed ALRA Section 2.4.C.1, the UFSAR, and the USAR using the evaluation methodology described in SER Section 2.4. The staff conducted its review in accordance with the guidance described in SRP-LR Section 2.4.

In conducting its review, the staff evaluated the system functions described in the ALRA, UFSAR, and USAR in accordance with the requirements of 10 CFR 54.4(a) to verify that the applicant had not omitted from the scope of license renewal any components with intended functions delineated under 10 CFR 54.4(a). The staff then reviewed those components that the applicant had identified as being within the scope of license renewal to verify that the applicant had not omitted any passive and long-lived components that should be subject to an AMR in accordance with the requirements of 10 CFR 54.21(a)(1).

The staff's review of LRA Section 2.4.C.1 identified an area in which additional information was necessary to complete review of the applicant's scoping and screening results. The applicant responded to the staff's RAI as discussed below.

In RAI 2.4.C.1-1 dated October 11, 2005, the staff stated that in ALRA Table 2.4.C-1 the applicant included ASME Classes 1, 2, 3, and MC hangers and supports. The staff assumed that the drywell and torus external supports were included within the Class MC supports. Therefore, the staff requested that the applicant confirm the staff's assumption.

In its response by letter dated October 28, 2005, the applicant clarified:

(1) The drywell supports are included in ALRA Section 2.4.C.1, "Component Supports," under the Component Type "ASME Class 1, 2, 3, and MC Hangers and Supports" in Table 2.4.C.1-1.

(2) The torus supports are included in ALRA Section 2.4.A.2, "NMP1 Reactor Building." They are listed in Table 2.4.A.2-1 under the Component Type "Torus Support Columns."

The applicant explained that these two component types encompass all drywell and torus supports.

The applicant further stated that the aging management of the drywell supports is addressed under the Component Type "Structural Steel (Carbon and Low Alloy Steel) in Air" in ALRA Table 3.5.2.C-1. These supports are managed by the ASME Section XI Inservice Inspection (Subsection IWF) Program consistent with GALL Report Item III.B1.3.1-a.

The applicant finally stated that the aging management of the Torus Support Columns is addressed in ALRA Table 3.5.2.A-2 under the Component Type "Torus Support Columns." These supports are managed by the ASME Section XI Inservice Inspection (Subsection IWF) Program consistent with GALL Report Item III.B1.3.1-a.

The staff found the clarification provided by the applicant consistent with the staff's assumption and with GALL item III.B1.3.1-a. Therefore, the staff's concern described in RAI 2.4.C.1-1 is resolved.

2.4C.1.3 Conclusion

The staff reviewed the ALRA, RAI response, and related structural components to determine whether any SSCs that should be within the scope of license renewal had not been identified by the applicant. No omissions were identified. In addition, the staff performed a review to determine whether any components that should be subject to an AMR had not been identified by the applicant. No omissions were identified. On the basis of its review, the staff concludes that there is reasonable assurance that the applicant had adequately identified the component supports commodity components that are within the scope of license renewal, as required by 10 CFR 54.4(a), and the component supports commodity components that are subject to an AMR, as required by 10 CFR 54.21(a)(1).

2.4C.2 Fire Stops and Seals

2.4C.2.1 Summary of Technical Information in the Amended Application

In ALRA Section 2.4.C.2 the applicant described the fire stops and seals commodity. The fire stops and seals commodity addresses penetration fire stop/seal and structural fire seal materials. The following items are not included under this commodity: (1) process piping, electrical cables, or conduits running through the fire penetration (included under the associated mechanical or electrical systems), (2) cast in place penetration sleeves and any flanges or welds (evaluated as part of the structural steel asset associated with the structure), (3) embedded portions of cast-in-place sleeves (included under the concrete asset for the structure), and (4) fire barrier walls, which are included under the structure.

The fire stops and seals commodity contains SR components that are relied upon to remain functional during and following DBEs. In addition, the fire stops and seals commodity performs functions that support fire protection.

The intended function within the scope of license renewal is to provide rated fire barrier.

In ALRA Table 2.4.C.2-1, the applicant identified the following fire stops and seals commodity component types that are within the scope of license renewal and subject to an AMR:

- aluminum spacers
- stainless steel clamps
- fire stop materials
- fire wrap materials
- penetration extensions

2.4C.2.2 Staff Evaluation

The staff reviewed ALRA Section 2.4.C.2, the UFSAR, and the USAR using the evaluation methodology described in SER Section 2.4. The staff conducted its review in accordance with the guidance described in SRP-LR Section 2.4.

In conducting its review, the staff evaluated the structural component functions described in the ALRA, UFSAR, and USAR in accordance with the requirements of 10 CFR 54.4(a) to verify that the applicant had not omitted from the scope of license renewal any components with intended functions delineated under 10 CFR 54.4(a). The staff then reviewed those components that the applicant had identified as being within the scope of license renewal to verify that the applicant had not omitted any passive and long-lived components that should be subject to an AMR in accordance with the requirements of 10 CFR 54.21(a)(1).

2.4C.2.3 Conclusion

The staff reviewed the ALRA and related structural components to determine whether any SSCs that should be within the scope of license renewal had not been identified by the applicant. No omissions were identified. In addition, the staff performed a review to determine whether any components that should be subject to an AMR had not been identified by the applicant. No omissions were identified. On the basis of its review, the staff concludes that there is reasonable assurance that the applicant had adequately identified the fire stops and seals commodity components that are within the scope of license renewal, as required by 10 CFR 54.4(a), and the fire stops and seals commodity components that are subject to an AMR, as required by 10 CFR 54.21(a)(1).

2.5 Scoping and Screening Results: Electrical and Instrumentation and Controls Systems

This section documents the staff's review of the applicant's scoping and screening results for electrical systems and instrumentation and controls (I&C) systems. Specifically, ALRA Section 2.5 discusses the following electrical and I&C systems that are within the scope of license renewal:

- NMP1 24V DC electrical distribution system
- NMP1 125V DC electrical distribution system
- NMP1 120V AC electrical distribution system
- NMP1 600V AC electrical distribution system
- NMP1 4.16KV AC electrical distribution system
- NMP1 115KV AC electrical distribution system
- NMP1 anticipated transients without scram system
- NMP1 communications system
- NMP1 plant lighting system
- NMP1 plant process computer system
- NMP1 reactor protection system
- NMP1 remote shutdown system
- NMP1 Neutron monitoring system
- NMP2 13.8KV AC electrical distribution system

- NMP2 4.16KV AC electrical distribution system
- NMP2 battery-24V-station system
- NMP2 common electrical system
- NMP2 communications paging system
- NMP2 communications telephone system
- NMP2 emergency DC distribution system
- NMP2 emergency uninterruptible power supplies system
- NMP2 feedwater control system
- NMP2 heat tracing system
- NMP2 information handling annunciator system
- NMP2 motor control center emergency system
- NMP2 normal AC high voltage distribution system
- NMP2 normal DC distribution system
- NMP2 normal UPS system
- NMP2 process computer system
- NMP2 reactor protection motor generator system
- NMP2 reactor protection system
- NMP2 redundant reactivity control system
- NMP2 remote shutdown system
- NMP2 reserve station service transformers system
- NMP2 standby and emergency AC distribution system
- NMP2 standby diesel generator protection (breaker) system
- NMP2 startup transient analysis system
- NMP2 station control bus nonvital AC supply system
- NMP2 station control bus vital AC supply system
- NMP2 station lighting system
- NMP2 switchyard system
- NMP2 synchronizing - diesel generator system
- NMP2 unit substation emergency AC controls and heater supply
- NMP2 unit substation emergency system
- NMP2 unit substation system
- NMP2 uninterruptible power supplies distribution system
- NMP2 standby diesel generator protection (generator) system

In accordance with the requirements of 10 CFR 54.21(a)(1), the applicant identified and listed passive, long-lived SCs that are within the scope of license renewal and subject to an AMR. To verify that the applicant properly implemented its methodology, the staff focused its review on the implementation results. This approach allowed the staff to confirm that there were no omissions of electrical and I&C system components that meet the scoping criteria and are subject to an AMR.

Staff Evaluation Methodology. The staff's evaluation of the information provided in the ALRA was performed in the same manner for all electrical and I&C systems and commodities. The objective of the review was to determine if the components and supporting structures for a specific electrical and I&C system or commodity, that appeared to meet the scoping criteria specified in 10 CFR Part 54, were identified by the applicant as within the scope of license renewal, in accordance with 10 CFR 54.4. Similarly, the staff evaluated the applicant's screening results to verify that all long-lived, passive components were subject to an AMR in accordance with 10 CFR 54.21(a)(1).

Scoping. To perform its evaluation, the staff reviewed the applicable ALRA section and associated component drawings, focusing its review on components that had not been identified as within the scope of license renewal. The staff reviewed relevant licensing basis documents, including the UFSAR and USAR, for each electrical and I&C system component to determine if the applicant had omitted components with intended functions delineated under 10 CFR 54.4(a) from the scope of license renewal. The staff also reviewed the licensing basis documents to determine if all intended functions delineated under 10 CFR 54.4(a) were specified in the ALRA. If omissions were identified, the staff requested additional information to resolve the discrepancies.

Screening. Once the staff completed its review of the scoping results, the staff evaluated the applicant's screening results. For those systems and components with intended functions, the staff sought to determine: (1) if the functions are performed with moving parts or a change in configuration or properties, or (2) if they are subject to replacement based on a qualified life or specified time period, as described in 10 CFR 54.21(a)(1). For those that did not meet either of these criteria, the staff sought to confirm that these electrical and I&C systems and components were subject to an AMR as required by 10 CFR 54.21(a)(1). If discrepancies were identified, the staff requested additional information to resolve them.

2.5.1 NMPNS Electrical Commodities

In LRA Section 2.5.C, the applicant described the components and systems included in the commodity group:

- 2.5.C.1   Cables and Connectors
- 2.5.C.2   Non-Segregated/switchyard Bus
- 2.5.C.3   Containment Electrical Penetrations
- 2.5.C.4   Switchyard Components

The commodity group is within the scope of license renewal under 10 CFR 54.4(a)(1) because it provides electrical power to safety Class 1, 2, and 3 components. Some SSCs in the system are considered within the scope of license renewal because their failure could affect the capability of safety-related SSCs per 10 CFR 54.4(a)(2). Others are within the scope of license renewal because they support fire protection, anticipated transient without scram, and station blackout per 10 CFR 54.4(a)(3). An intended function within the scope of license renewal is to electrically connect specified sections of an electrical circuit to deliver voltage, current, or signal. Additional intended functions are to isolate electrically and provide structural support to transmission conductors, phase buses, and switchyard buses.

In ALRA Tables 2.5.C.1-1, 2.5.C.2-1, 2.5.C.3-1, and 2.5.C.4-1, the applicant identified the following commodity group component types within the scope of license renewal and subject to an AMR:

- conductor insulation for electrical cables and connector

- conductor insulation for electrical cables used in circuits sensitive to reduction in conductor insulation resistance (IR)

- fuse holders (not part of a larger assembly)

- insulators
- non-segregated busses
- switchyard busses
- electrical penetrations
- high-voltage insulators
- transmission conductors
- transmission conductor connectors

In ALRA Section 2.5.C the applicant identified the NMPNS electrical commodities subject to an AMR for license renewal.

After applying the scoping and screening methodology the applicant categorized the components requiring AMR into passive commodity groups. In ALRA Section 2.5.C, the applicant identified the SCs of the electrical and I&C systems subject to an AMR for license renewal. The staff's review findings regarding ALRA Sections 2.5.C.1 through 2.5.C.4 are presented in SER Sections 2.5.1.1 through 2.5.1.4, respectively, for both NMP1 and NMP2.

2.5.1.1 Cables and Connectors

2.5.1.1.1 Summary of Technical Information in the Amended Application

In ALRA Section 2.5.C.1, the applicant described the cables and connectors commodity. The components addressed in this commodity are electrical cables, connectors, splices, terminal blocks, and fuse blocks. Cables are identified on a plant-wide basis, and are not identified as being associated with a particular system.

The intended function, within the scope of license renewal, is to provide continuity to deliver electrical signals or power (includes insulation).

In ALRA Table 2.5.C.1-1, the applicant identified the following cables and connectors commodity component types that are within the scope of license renewal and subject to an AMR:

- conductor insulation for electrical cables and connectors
- conductor insulation for electrical cables used in circuits that are to reduction in conductor insulation resistance (IR)
- fuse holders (not part of a larger assembly)

2.5.1.1.2 Staff Evaluation

The staff reviewed ALRA Section 2.5.C.1, UFSAR, and USAR using the evaluation methodology described in SER Section 2.5. The staff conducted its review in accordance with the guidance described in SRP-LR Section 2.5.

In conducting its review, the staff evaluated the commodity functions described in the ALRA, UFSAR, and USAR in accordance with the requirements of 10 CFR 54.4(a) to verify that the applicant had not omitted from the scope of license renewal any components with intended functions delineated under 10 CFR 54.4(a). The staff then reviewed those components that the applicant had identified as being within the scope of license renewal to verify that the applicant had not omitted any passive and long-lived components that should be subject to an AMR in accordance with the requirements of 10 CFR 54.21(a)(1).

The applicant evaluated the cables and connectors plant-wide as commodities across system boundaries. In ALRA Section 2.5.C.1 the applicant stated that the components addressed in this commodity are electrical cables, connectors, splices, terminal blocks, and fuse blocks. Cables are identified plant-wide and are not identified as associated with particular systems. Cables and their associated connectors provide electrical continuity to specified sections of an electrical circuit to deliver voltage, current, and signals to various equipment and components throughout the plant to enable them to perform their intended functions.

The staff found that the applicant correctly identified the cables and connectors as components that perform their functions without moving parts or change in configuration or properties (passive and long-lived) and, therefore, subject to AMR.

2.5.1.1.3 Conclusion

The staff reviewed the ALRA, UFSAR, and USAR to determine whether any SSCs that should be within the scope of license renewal had not been identified by the applicant. No omissions were identified. In addition, the staff performed a review to determine whether any components that should be subject to an AMR had not been identified by the applicant. No omissions were identified. On the basis of its review, the staff concludes that there is reasonable assurance that the applicant had adequately identified the cables and connectors commodity components that are within the scope of license renewal, as required by 10 CFR 54.4(a), and the cables and connectors commodity components that are subject to an AMR, as required by 10 CFR 54.21(a)(1).

2.5.1.2 Non-Segregated/Switchyard Bus

2.5.1.2.1 Summary of Technical Information in the Amended Application

In ALRA Section 2.5.C.2, the applicant described the non-segregated/switchyard bus commodity. The components evaluated in this commodity encompass the electrical switchyard and non-segregated busses, as well as their associated insulators. Electrical busses perform the function of providing electrical continuity to specified sections of an electrical circuit voltage and current to various equipment and components throughout the plant to enable them to perform their intended functions.

The intended functions within the scope of license renewal include the following:

- provides continuity to deliver electrical signals or power (includes insulation)
- insulates and supports an electrical conductor

In ALRA Table 2.5.C.2-1, the applicant identified the following non-segregated/switchyard bus commodity component types that are within the scope of license renewal and subject to an AMR:

- insulators
- non-segregated bus
- switchyard bus

2.5.1.2.2 Staff Evaluation

The staff reviewed ALRA Section 2.5.C.2, UFSAR, and USAR using the evaluation methodology described in SER Section 2.5. The staff conducted its review in accordance with the guidance described in SRP-LR Section 2.5.

In conducting its review, the staff evaluated the commodity functions described in the ALRA, UFSAR, and USAR in accordance with the requirements of 10 CFR 54.4(a) to verify that the applicant had not omitted from the scope of license renewal any components with intended functions delineated under 10 CFR 54.4(a). The staff then reviewed those components that the applicant had identified as being within the scope of license renewal to verify that the applicant had not omitted any passive and long-lived components that should be subject to an AMR in accordance with the requirements of 10 CFR 54.21(a)(1).

The non-segregated/switchyard bus identified by the applicant requiring an AMR includes insulators, a non-segregated bus, and a switchyard bus. Electrical busses provide electrical continuity to an electrical circuit and voltage and current to equipment and components throughout the plant to enable them to perform their intended functions. The intended function of the insulators is electrical insulation and NSR functional support through separation of busses and conductors from other components and structures.

The staff reviewed these component categories and found them subject to 10 CFR 54.4(a)(1) and 10 CFR 54.4(b) requirements. The staff reviewed the information in the UFSAR and USAR and found that the applicant had identified correctly the non-segregated/switchyard bus commodity component types that perform intended functions without moving parts or change in configuration or properties (passive and long lived) and are, therefore, subject to an AMR.

2.5.1.2.3 Conclusion

The staff reviewed the ALRA, UFSAR, and USAR to determine whether any SSCs that should be within the scope of license renewal had not been identified by the applicant. No omissions were identified. In addition, the staff performed a review to determine whether any components that should be subject to an AMR had not been identified by the applicant. No omissions were identified. On the basis of its review, the staff concludes that there is reasonable assurance that the applicant had adequately identified the non-segregated/switchyard bus commodity components that are within the scope of license renewal, as required by 10 CFR 54.4(a), and the non-segregated/switchyard bus commodity components that are subject to an AMR, as required by 10 CFR 54.21(a)(1).

2.5.1.3 Containment Electrical Penetrations

2.5.1.3.1 Summary of Technical Information in the Amended Application

In ALRA Section 2.5.C.3, the applicant described the containment electrical penetrations commodity. The components evaluated in this commodity encompass the non-EQ electrical penetrations that form part of the containment pressure boundary. An electrical penetration provides an electrical connection between two sections of the electrical/I&C circuit. The pigtail at each end of the penetration is connected to the field cable in various ways and is included in this evaluation. The connector or connection method is included in the cables and connectors commodity group. The structural steel portion of the primary containment electrical penetrations is evaluated in the NMP1 primary containment structure and the NMP2 primary containment structure.

The intended functions within the scope of license renewal include the following:

- provides continuity to deliver electrical signals or power (includes insulation)
- provides pressure retaining boundary

In ALRA Table 2.5.C.3-1, the applicant identified that the electrical penetrations component type of the containment electrical penetrations commodity is within the scope of license renewal and subject to an AMR.

2.5.1.3.2 Staff Evaluation

The staff reviewed ALRA Section 2.5.C.3, UFSAR, and USAR using the evaluation methodology described in SER Section 2.5. The staff conducted its review in accordance with the guidance described in SRP-LR Section 2.5.

In conducting its review, the staff evaluated the commodity functions described in the ALRA, UFSAR, and USAR in accordance with the requirements of 10 CFR 54.4(a) to verify that the applicant had not omitted from the scope of license renewal any components with intended functions delineated under 10 CFR 54.4(a). The staff then reviewed those components that the applicant had identified as being within the scope of license renewal to verify that the applicant had not omitted any passive and long-lived components that should be subject to an AMR in accordance with the requirements of 10 CFR 54.21(a)(1).

The containment electrical penetrations identified by the applicant are non-EQ and form part of the containment pressure boundary. They also provide electrical continuity to an electrical circuit to deliver voltage, current, and signals across the containment boundary (either continuously or intermittently) to equipment and components throughout the plant to enable them to perform their intended functions. An electrical penetration is a device used to provide an electrical connection between two sections of the electrical/I&C circuit, inside and outside of the containment. This evaluation includes the pigtail located at each end of the penetration. The pigtail connects the end of the penetration to the field cable in various ways. The connector or connection method is addressed in the cables and connectors commodity group section.

The staff reviewed these component categories and found them subject to 10 CFR 54.4(a)(1) and 10 CFR 54.4(b) requirements. The staff reviewed the information in the UFSAR and USAR and found that the applicant had identified correctly the containment electrical penetrations that perform their intended functions without moving parts or change in configuration or properties (passive and long lived) and are, therefore, subject to an AMR.

2.5.1.3.3 Conclusion

The staff reviewed the ALRA, UFSAR, and USAR to determine whether any SSCs that should be within the scope of license renewal had not been identified by the applicant. No omissions were identified. In addition, the staff performed a review to determine whether any components that should be subject to an AMR had not been identified by the applicant. No omissions were identified. On the basis of its review, the staff concludes that there is reasonable assurance that the applicant had adequately identified the containment electrical penetrations commodity components that are within the scope of license renewal, as required by 10 CFR 54.4(a), and the containment electrical penetrations commodity components that are subject to an AMR, as required by 10 CFR 54.21(a)(1).

2.5.1.4 Switchyard Components

2.5.1.4.1 Summary of Technical Information in the Amended Application

In ALRA Section 2.5.C.4, the applicant described the switchyard components commodity. The switchyard components commodity was developed to address the addition of the 115KV switchyards for SBO recovery to the scope of license renewal. Cables, connectors, and busbars are evaluated in their respective commodity groups.

The intended functions within the scope of license renewal include the following:

*   provides continuity to deliver electrical signals or power (includes insulation)
*   insulates and supports an electrical conductor

In ALRA Table 2.5.C.4-1, the applicant identified the following switchyard components commodity component types that are within the scope of license renewal and subject to an AMR:

*   high voltage insulators
*   transmission conductors
*   transmission conductor connectors

2.5.1.4.2 Staff Evaluation

The staff reviewed ALRA Section 2.5.C.4, UFSAR, and USAR using the evaluation methodology described in SER Section 2.5. The staff conducted its review in accordance with the guidance described in SRP-LR Section 2.5.

In conducting its review, the staff evaluated the commodity functions described in the ALRA, UFSAR, and USAR in accordance with the requirements of 10 CFR 54.4(a) to verify that the applicant had not omitted from the scope of license renewal any components with intended functions delineated under 10 CFR 54.4(a). The staff then reviewed those components that the applicant had identified as being within the scope of license renewal to verify that the applicant had not omitted any passive and long-lived components that should be subject to an AMR in accordance with the requirements of 10 CFR 54.21(a)(1).

As identified by the applicant the switchyard component commodity was developed to address the addition of the 115kV switchyard for SBO recovery to the scope of license renewal. The components subject to AMR within the yard are the transmission conductors, insulators, and associated connectors. Cables, connectors, and bus bars are evaluated in their respective commodity group sections. Switchyard transmission conductors and associated connectors provide electrical connections to electrical circuits to deliver voltage, current, and signals to equipment and components throughout the switchyard to enable them to perform their intended functions. The intended function of the high-voltage insulators is electrical insulation and NSR function support through separation of the busses and conductors from other components and structures.

The staff reviewed these component categories and found them subject to 10 CFR 54.4(a)(1) and 10 CFR 54.4(b) requirements. The staff reviewed the information in the UFSAR and USAR and found that the applicant had identified correctly the switchyard components that perform their intended functions without moving parts or change in configuration or properties (passive and long lived) and are, therefore, subject to an AMR.

2.5.1.4.3 Conclusion

The staff reviewed the ALRA, UFSAR, and USAR to determine whether any SSCs that should be within the scope of license renewal had not been identified by the applicant. No omissions were identified. In addition, the staff performed a review to determine whether any components that should be subject to an AMR had not been identified by the applicant. No omissions were identified. On the basis of its review, the staff concludes that there is reasonable assurance that the applicant had adequately identified the switchyard components commodity components that are within the scope of license renewal, as required by 10 CFR 54.4(a), and the switchyard components commodity components that are subject to an AMR, as required by 10 CFR 54.21(a)(1).

2.6 Conclusion for Scoping and Screening

The staff reviewed the information in ALRA Section 2, "Scoping and Screening Methodology for Identifying Structures and Components Subject to Aging Management Review, and Implementation Results." The staff determined that the applicant's scoping and screening, including its supplement 10 CFR 54.4(a)(2) review, which brought additional NSR piping segments and associated components within the scope of license renewal, was consistent with the requirements of 10 CFR 54.21(a)(1) and the staff's position on the treatment of SR and NSR SSCs within the scope of license renewal and the structures and components requiring an AMR is consistent with the requirements of 10 CFR 54.4 and 10 CFR 54.21(a)(1).

On the basis of its review, the staff concludes that there is reasonable assurance that the applicant had adequately identified those systems and components that are within the scope of license renewal, as required by 10 CFR 54.4(a), and those systems and components that are subject to an AMR, as required by 10 CFR 54.21(a)(1).

the staff concludes that there is reasonable assurance that the activities authorized by the renewed licenses will continue to be conducted in accordance with the CLB, and any changes made to the CLB, in order to comply with 10 CFR 54.29(a), are in accordance with the Atomic Energy Act of 1954 and the NRC's regulations.

NRC FORM 335 (9-2004) NRCMD 3.7	U.S. NUCLEAR REGULATORY COMMISSION **BIBLIOGRAPHIC DATA SHEET** *(See instructions on the reverse)*	1. REPORT NUMBER (Assigned by NRC, Add Vol., Supp., Rev., and Addendum Numbers, if any.) NUREG-1900 Vol. 1

2. TITLE AND SUBTITLE Safety Evaluation Report Related to the License Renewal of Nine Mile Point Nuclear Station, Units 1 and 2	3. DATE REPORT PUBLISHED

	MONTH	YEAR
	September	2006

4. FIN OR GRANT NUMBER

5. AUTHOR(S) Tommy Le	6. TYPE OF REPORT

7. PERIOD COVERED *(Inclusive Dates)*

8. PERFORMING ORGANIZATION - NAME AND ADDRESS *(If NRC, provide Division, Office or Region, U.S. Nuclear Regulatory Commission, and mailing address; if contractor, provide name and mailing address.)*

Division of License Renewal
Office of Nuclear Reactor Regulation
U.S. Nuclear Regulatory Commission
Washington, DC 20555-0001

9. SPONSORING ORGANIZATION - NAME AND ADDRESS *(If NRC, type "Same as above"; if contractor, provide NRC Division, Office or Region, U.S. Nuclear Regulatory Commission, and mailing address.)*

Same as above

10. SUPPLEMENTARY NOTES

11. ABSTRACT *(200 words or less)*

This safety evaluation report (SER) documents the technical review of the Nine Mile Point Nuclear Station Units, 1 and 2 (NMPNS), license renewal application (LRA) by the staff of the U.S. Nuclear Regulatory Commission (NRC) (the staff). By letter dated May 26, 2004, Constellation Energy Group, LLC submitted the LRA for NMPNS in accordance with Title10, Part 54, of the Code of Federal Regulations (10CFRPart54). Due to concerns with the adequacy of support for and documentation of the license renewal activities in the initial submission, the applicant submitted an amended LRA (ALRA) on July14, 2005. Constellation Energy Group, LLC is requesting renewal of the operating licenses for NMPNS (Facility Operating License NumbersDPR-63 and NPF-69, respectively), for a period of 20 years beyond the current expiration dates of midnight August 22, 2009, for Unit 1 (NMP1) and midnight October 31, 2026, for Unit 2 (NMP2).

This SER presents the status of the staff's review of information submitted to the staff through April 21, 2006, the cutoff date for consideration in this SER. On March 3, 2006, the staff issued a draft SER which identified two open items that had to be resolved before the staff makes a final determination on the application. The two open items have now been resolved and SER Section1.5 summarizes these items and their resolutions. SER Section 6 provides the staff's final conclusion on the review of the NMPNS License Renewal Application dated May 26, 2004, as amended July 14, 2005, and all its subsequent supplemental letters as listed in SER Appendix B.

12. KEY WORDS/DESCRIPTORS *(List words or phrases that will assist researchers in locating the report.)*

license renewal, nuclear power

13. AVAILABILITY STATEMENT
unlimited

14. SECURITY CLASSIFICATION
(This Page)
unclassified
(This Report)
unclassified

15. NUMBER OF PAGES

16. PRICE